The VLSI Designer's Library

The VLSI Systems Series
LYNN CONWAY, *Editor*

The VLSI Designer's Library	John Newkirk and Robert Mathews, 1983
Lecture Notes in cMOS VLSI Design	Neil Weste and Kamram Eshragian, 1983
Principles of cMOS VLSI Design: A Systems Perspective	Neil Weste and Kamran Eshraghian, 1984

The VLSI Designer's Library

nMOS, $\lambda = 2.5\mu$, with Butting Contacts

JOHN A. NEWKIRK
ROBERT MATHEWS
Stanford University
Stanford, California

1983

ADDISON-WESLEY PUBLISHING COMPANY, INC.
Advanced Book Program
Reading, Massachusetts

LONDON · AMSTERDAM · DON MILLS, ONTARIO · SYDNEY · TOKYO

Library of Congress Cataloging in Publication Data

Newkirk, John A.
The VLSI designer's library.

(The VLSI systems series)
Includes index.
1. Microelectronics--Diagrams. I. Mathews, Robert G. II. Title. III. Title: V.L.S.I. designer's library. IV Series.
TK7874.N44 1983 621.381′73′0223 83-14967
ISBN 0-201-05444-2

Manufactured in the United States of America.

ABCDEFGHIJ-KP-89876543

Contents

Preface

Since the fall of 1979, we have taught graduate-level nMOS LSI design classes at Stanford based on the Mead and Conway text, *Introduction to VLSI Systems* (Addison-Wesley, Reading, MA, 1980). In one quarter, students must learn the basis of design, select a project, and carry it through to a fabrication-ready layout. The class is intensive, demanding 100-150 hours for the project alone. A library of tested, working designs is critical in helping students complete projects and in decreasing the chances that the projects will fail catastrophically. Accordingly, this book is an important companion to Mead and Conway.

This cell library collects together designs we have found useful in class and research projects over the past three years. The documentation is intended to provide the minimum amount of information required by our average designer: a brief description of a cell's function, its physical dimensions, and its interface specifications. The designs all follow in the spirit of the Mead and Conway text, using λ design rules and 2-phase clocking; we assume that the reader is familiar with the text.

The library serves 3 purposes. First, it does serve as a source of tested, working cells and subsystems to be used as is. Second, as a catalogue of working designs, it provides the reader examples of design and layout styles to ponder and layouts of working cells to modify for his own purposes. Third, it provides a model of documentation and a standard for checking and testing designs.

This collection is not a cell library in the sense of, say, a library of standard cells. In particular, it is not complete, and we do not claim that library cells alone will suffice for most custom designs. We have attempted to obtain a spectrum of designs, but serious gaps remain, for example, a family of compatible data-path components. Moreover, most of the designs are small, and the documentation required for system-level designs will doubtless be different.

If you find mistakes, please inform us. If you have documented, tested designs to contribute, send them to us. The library is far from finished, and we expect it to grow and change over time.

Rob Mathews
John Newkirk
Electrical Engineering Department
Stanford University
Stanford, CA 94305
Spring, 1983

Acknowledgments

This library is a collection of designs from many sources. It therefore represents two kinds of hard work: creating the original cells and subsystems in the first place, and documenting and validating them. Dick Lyon, then working in Lynn Conway's group at PARC, seeded the library by providing a large number of working, correct designs in the fall of 1979. Since then, many others have provided designs, and we have tried to credit them.

However, checking and documenting a design is usually several times the work that creating it is. We owe a big debt to John Redford on this score—he spearheaded the effort of picking up the pieces and assembling them into a first working document. Charlie Burns carried through the first and second revisions of the library; without his help, we would never have gotten the library to press. Many others have made major contributions, including C.Y. Chu, Alan Eustace, H.V. Jagadish, Scott Macomber, Dave Noice, Tim Saxe, Lyle Smith, and Wayne Wolf.

Finally, much of the prerequisite tools and research work was supported by the Department of Defense Advanced Research Projects Agency under contract MDA-903-79-C-0680. Col. Duane Adams and Paul Losleben provided much-needed encouragement and guidance.

Permuted Index

Conventions and Comments

Organization of this Book
Each chapter of the *Designer's Library* documents a family of related components. The chapter's introduction explains the family, including how to combine cells to form subsystems. The chapter's body consists of manual pages describing each component. These manual pages are organized alphabetically by component name; the chapters, alphabetically by family name. Thus, AddOut(ALU) is in the ALU chapter between Adder and AddRestore.

Manual Pages
The manual page for each cell describes the cell's function, origin, and physical characteristics. A layout associates names and locations with cell terminals. Where appropriate, additional diagrams, such as schematics, also appear. Here we explain the notations used in describing the cells.

Layouts
Layouts are shown as color plots, with layers represented as in Mead and Conway:

Layer	*Color*
Diffusion	Green
Poly	Red
Metal	Blue
Implant	Yellow
Cut	Black
Glass cut	Black

For a black-and-white representation, we recommend the stipple patterns shown in Fig. 1 (due to Dick Lyon).

Properties—Interface
This section refers to cell input and output pins. Each pin has associated with it a name and a variety of characteristics. Signals are classified as one of the types:

Inputs	*Outputs*
4:1 ratio	gate
8:1 ratio	superbuffer
switch control	switch-logic
switched	precharged
analog	

Gate 4:1 inputs and switch control inputs share the property that they require restored signal levels; they may therefore be driven by gates or superbuffers. Gate 8:1 inputs and switched inputs will tolerate degraded signal levels.

Switched inputs have the property that they present an unpredictable load that can cause charge-sharing problems; they may not in general be driven from precharged outputs, which are high-impedance on one clock phase. Feeding any unbuffered storage node into a switched input is also an invitation to disaster.

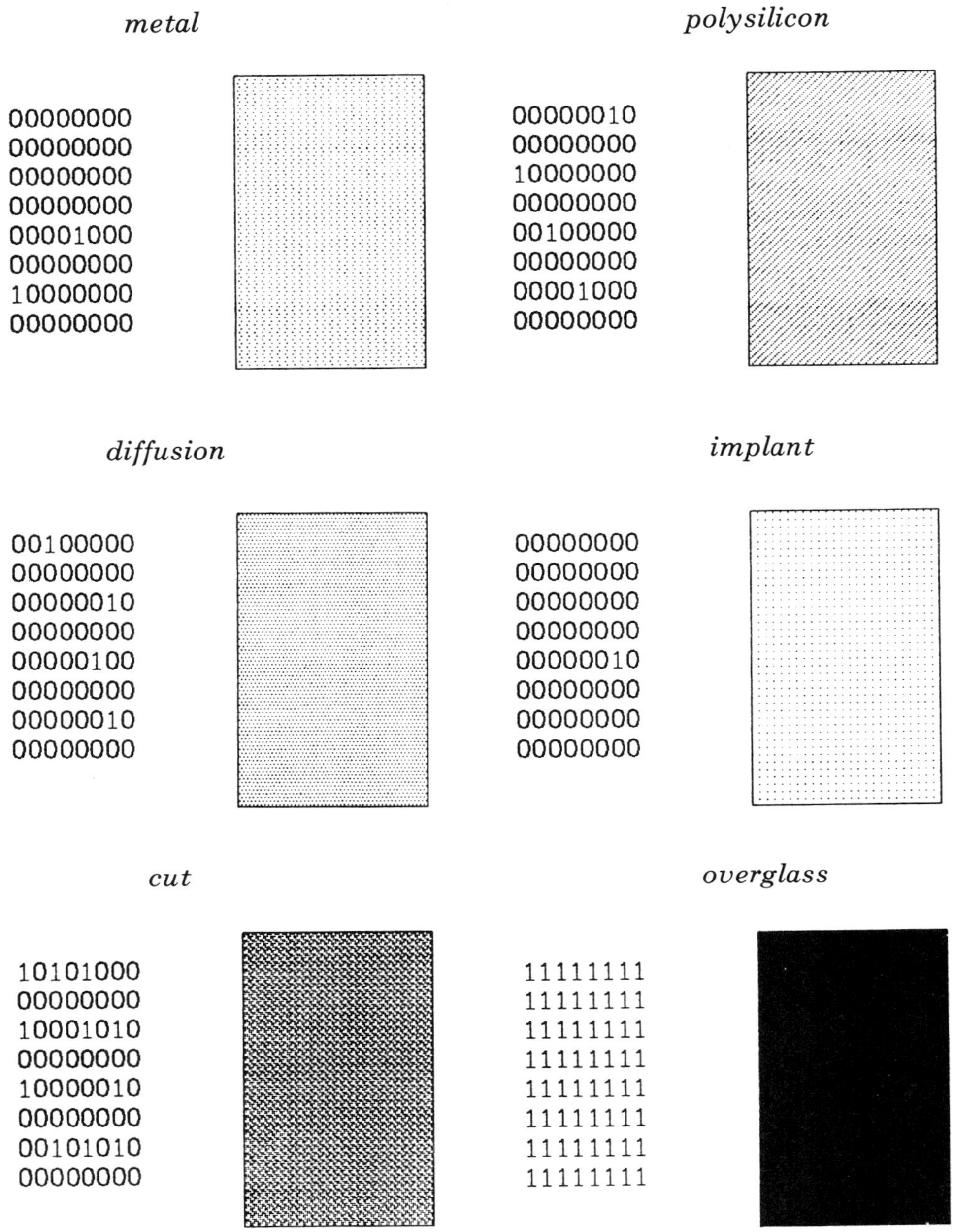

Figure 1. Recommended stipple patterns.

Two notations are used in describing electrical characteristics. Ratios are given as a Z_{pu}/Z_{pd}, *where the individual numbers are channel lengths or widths in units of 2λ.* Capacitance is given in units of □ C_g where 1 □ C_g is the capacitance of a minimum-sized gate region, $4\lambda^2$ in area. For relative capacitances, we have used the data on page 51 of Mead and Conway.

Properties—DC current
Currents given for gate logic are calculated under the approximations that the pulldown's resistance is negligible and that the pullup is in saturation. Thus,

$$I_{sat} = k' \frac{W}{L}(V_{gs} - V_{th})^2 \approx .2 \frac{W}{L}\text{mA}, \; V_{ds} \geq (V_{gs} - V_{th}),$$

where we assume:

$$\begin{aligned} k' &\approx 10\mu A/V^2, \\ V_{gs} &\approx 0V, \\ V_{th} &\approx -4V, \text{ and} \\ V_{ds} &\approx 5V. \end{aligned}$$

W and L are the width and length of the active area of the pullup. The same I_{sat} applies for both 4:1 and 8:1 ratio inverters, since we are assuming the pulldown resistance is negligible.

There are occasional comments about current-carrying capability of metal lines. We are assuming a conservative metal-migration limit of $1\text{mA}/\mu^2$, where the metal thickness is 1μ for $\lambda = 2.5\mu$. Thus, a 4λ metal line can carry 10mA.

CIF ID
The CIF ID is just the CIF Symbol number.

Tesselation
This section makes comments on stacking and abutting cells. For example, "abuts horizontally" means the cells can be laid side by side.

Status
We have defined categories to try to standardize the state of each design:

1. Checked. DRC, static checks, and switch simulation are okay. This is the minimum required.
2. Tested. It was extensively tested, and it works!
3. Iterated. It was tested, found wanting, and redesigned.
4. Generally Used. It was tested and found working in at least three designs.
5. Revised. A working design has been minorly modified and checked but not yet tested. The circuit has not changed. The nature of the revision will be noted.

Design Rules
The library was designed and checked for a $\lambda = 2.5\mu$, depletion-load nMOS process with butting contacts. Most of the cells, including the pads, are known to function properly on commonly available $\lambda = 2\mu$ processes. The pads become increasingly difficult to bond for smaller λ, of course.

The major obstacle to scaling is our use of butting contacts. Small-linewidth processes define precisely self-aligned transistors; the result is that metal spikes from the cover of a butting contact can make a connection to the substrate under the edge of the transistor in the contact. Butting contacts have posed us no serious problems to date. In a future version of this library, buried contacts will replace them.

We have tried to observe design rules at least as conservative as those presented in Mead and Conway. Among the additional restrictions are:

1. The implant spacing has been increased from 1.5λ to 2λ nearly everywhere; the exception is that we only require the implant to surround the channel of pullups by 2λ (and not the active area defined by the edges of the associated butting contact).
2. Layer separations are enforced for adjacent features even if they are electrically connected (to prevent masking problems).
3. Contacts over active area are not allowed (because of scaling considerations), except for butting contacts over depletion-mode pullups.
4. Metal signal paths are not allowed to cover storage nodes (because of capacitive coupling).

We liberalized one rule so that butting contacts, which are effectively institutionalized design-rule violations, do not cause DRC headaches:

5. Poly/diffusion spacing is 0λ for electrically connected features. Thus, a 2λ diffusion wire may connect to the side of a butting contact touching the polysilicon in the contact; the following are all acceptable:

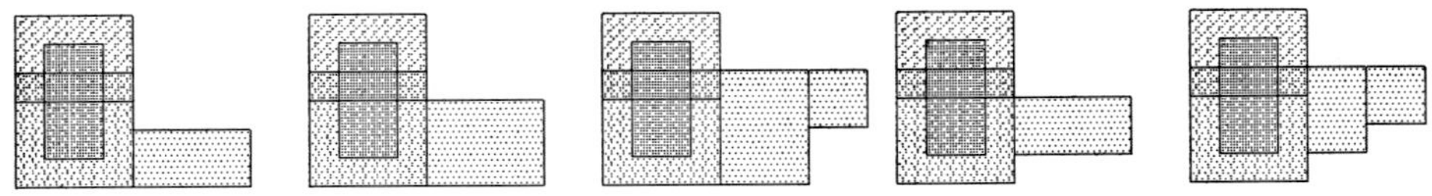

All cells have been design-rule checked with a high-quality checker (see D. Noice, J. Newkirk, R. Mathews, "A Polygon Package for Analyzing Integrated Circuit Designs," *VLSI Design*, vol. 2(3), third quarter, 1981, pp. 33–36) and found to be correct. However, the additional rules 3) and 4) were not checked.

All designs are located in the first quadrant, with the lower-left corner of the minimal bounding box at the origin. All designs use only right-angle geometry, with all features falling on a λ grid. (Wire centers may fall on a ½λ grid.)

CIF Conventions

Layouts are described in CIF, Caltech Intermediate Form (see Mead and Conway, pp. 115–127). The CIF descriptions appear in the appendix. They are also available in machine-readable form from Addison-Wesley.

The CIF subset we use is as follows:

1. All geometry is specified as boxes without rotations.
2. Symbol definitions for layouts all begin with a DS scaled 125/1. The only other scaling is for printed-circuit boards; it is 1270/2, because 25.4μ is 1 mil.
3. Symbols are never deleted or redefined.
4. Calls appear only in the design frames. All other symbols are self-contained.
5. Comments may be present, but noise characters, such as lower-case letters, are absent.
6. The machine-readable version has each statement on its own line.

As a consequence of (1), (4), (5), and (6), most of the CIF may be easily parsed or converted to other formats. (2) means that the CIF may be easily scaled down to λ = 2μ by replacing "125/1" with "100/1" everywhere in the text; of course, the pads may need rework.

1 / Arithmetic/ Logical Unit

NAME

intro—introduction to arithmetic-logical unit and adder

SYNOPSIS

This chapter describes cells that can be used to construct a bit-sliced ALU. *Adder* is a stripped-down version that sacrifices some flexibility for a decrease in size. Both are based on the ALU for the OM2 chip presented in Mead and Conway.

DESCRIPTION

The ALU's constituent parts are described in *ALU*, *ALUBufPair*, and *ALURestore*: *ALU* is the bit-slice, *ALUBufPair* drives the function block control lines, and *ALURestore* buffers the carry chain. The Adder components are: *Adder_M*, a one bit full adder/subtractor with pre-charged carry-out; *Adder*, a one bit full adder/subtractor without pre-charged carry-out; *AddRestore*, which buffers the carry chain; *AddBufPair*, which drives the function block control lines; *AddOut*, an output buffer; and *AddClockOut*, a clocked output buffer.

OVERVIEW

A typical commercial ALU is controlled by 3 or 4 function select lines; the OM2 bit-slice is not so simple. It requires a significant number of control inputs, and the result of any given input combination is often not obvious. Because of this generality and flexibility, the designer has to carefully program the ALU for each application.

This is the structure of the ALU slice:

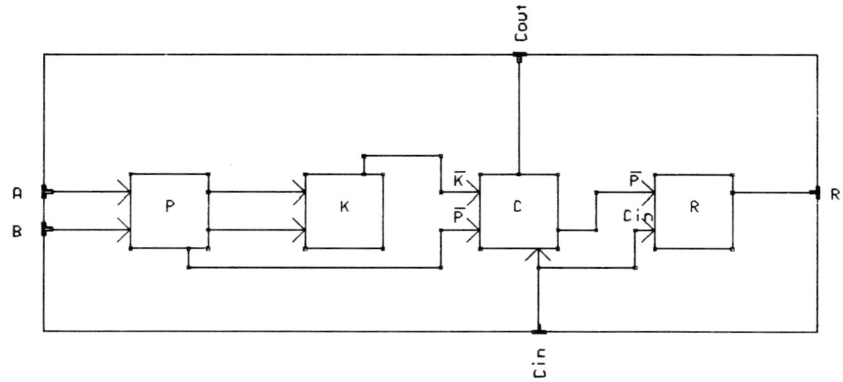

A and B are buffered by a pair of inverters (which also provide $\overline{A}$ and $\overline{B}$); these are combined into two new functions, carry-propagate P and carry-kill K by the **P** and **K** function blocks. These two functions are then combined with the carry-in, **Cin**, to generate the carry-out, **Cout**. Finally, P and **Cin** are combined to generate the output R from the **R** function block.

To look at it from the opposite perspective, the output is not a direct function of A and B. Instead, it is calculated from the carry-in and the carry-propagate. Carry-propagate along with carry-kill also determine the carry-out. Thus, the output is a function of the carry-in to the first stage and the three function blocks (each of which requires 4 control lines).

It is important to realize that P and K are not strict complements of each other. Consider the schematic of the carry chain block:

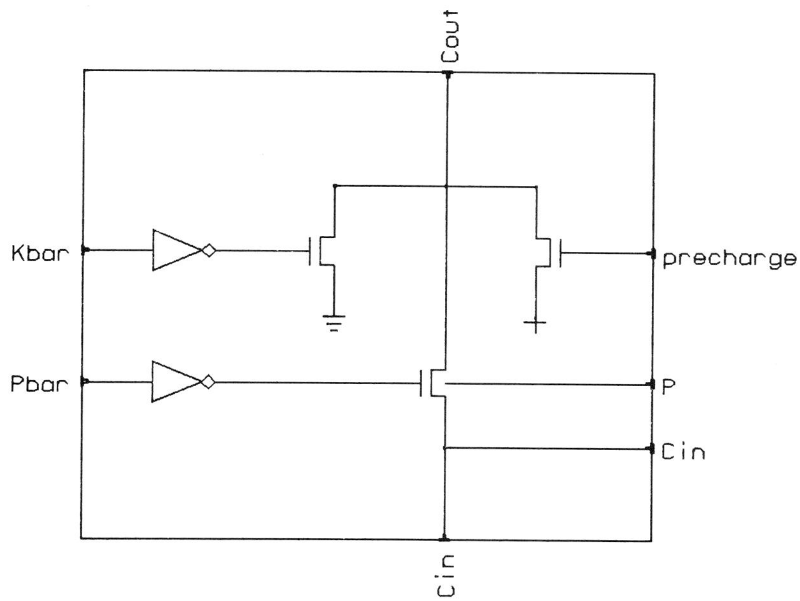

The carry-out is determined during $\phi 2$. During the previous $\phi 1$, the carry-out line is charged to 1 by the precharge transistor. (Note that the next stage's carry-in gate prevents the charge from flowing farther up the carry chain.) After the carry-out is charged, P and K are used to selectively drain the line. Since a line can be discharged faster than it can be charged, precharging and selective discharging is quicker than selectively charging the line through a load transistor.

Now consider the effects of P and K:

a) If K = 1 and P = 0, the carry-out line is drained and a 0 is passed to the next stage.

b) If P = 1 and K = 0, the node is exposed to the state of the previous stage. If that stage is also 1, then the carry-out will be 1; if the previous stage is passing a 0, then its carry-out line will be grounded and the charge in the present stage will also be drained through that ground.

c) If P = 0 and K = 0, there is no way for the carry-out line to be drained; thus a 1 is unconditionally passed to the next stage.

d) Finally, if P = 1 and K = 1, the carry-out line is both drained and connected to the previous stage. This control setting makes no sense if the carry chain is ever buffered.

An example will help explain how to program the function blocks. First, tabulate R and Cout as a function of A, B, and Cin, then derive P and K.

For the function A + B:

A	*B*	*Cin*	*P*	*K*	*R*	*Cout*
	inputs				*outputs*	
0	0	0	0	1	0	0
0	0	1	0	1	1	0
0	1	0	1	0	1	0
0	1	1	1	0	0	1
1	0	0	1	0	1	0
1	0	1	1	0	0	1
1	1	0	0	0	0	1
1	1	1	0	0	1	1

Now find expressions for $\overline{P}$, $\overline{K}$, and R (the *P* and *K* function blocks generate $\overline{P}$ and $\overline{K}$, not P and K). We see that:

$$\overline{P} = \overline{A \text{ xor } B}$$
$$\overline{K} = \overline{A \text{ or } B}$$
$$R = P \text{ xor } Cin$$

We can also see that the carry-in to the first stage should be 0. From here, inspect the programming of the function blocks to determine which combination of input lines will generate the desired functions.

Since the carry chain is nominally a series of pass transistors, the chain must be periodically buffered to minimize propagation delay. The level restoring cell is *ALURestore*, and should be used after every three cells for low delay.

Precharging and Control-line Drivers

The ALU and adder slices use precharged function blocks and carry chains for speed. A high level on the carry chain is maintained only by charge storage, not an active pullup; if the P or K lines glitch and drain this charge off, it will not come back. Accordingly, proper clocking is critical for these cells, although there are several legal variants.

Intended use. During $\phi 1$, the function blocks and carry chain are precharged, the control lines are held high to permit the precharging, and the operands and control values are latched. During $\phi 2$, the data lines must remain constant. Only those control lines with low control values are brought low, selectively discharging some function blocks and hence sections of the carry chain. High-valued control lines must remain high, never glitching throughout $\phi 2$. *ALUBufPair* is a driver pair that enforces the control-line convention, and the latches on the front of the slice enforce the data-line convention, ensuring proper operation.

Constant-value control lines. If, for the functions you need, a control should always be high, you may simply tie it high. If it should be low, you may tie it to $\phi 2$.

Simplest control. At the expense of some speed and of fighting the precharging, control lines can be driven by signals that settle during $\phi 1$ and remain stable throughout $\phi 2$ (notice that VDD and GND qualify). During $\phi 1$, the precharge transistor for a function block may fight one of the control lines, producing indeterminate output and control-line values. However, the proper output, a low value, will always be computed during $\phi 2$. The improper $\phi 1$ output may also cause fighting in the carry chain, but only in cases where the carry will be killed during $\phi 2$ anyway!

This form of control requires the least logic, especially when all control lines should be constant. On the other hand, it is conceptually complex, so be sure to understand it before using it.

Cascading. The ALU and adder slices may be stacked up to form parallel units with any desired word width. They may not, however, be cascaded without incurring a delay of at least one clock cycle, since the inputs must be stable through $\phi 2$ for the precharging to work correctly. Adder units built out of *StaticAdders* may be cascaded, since the logic used there is all static.

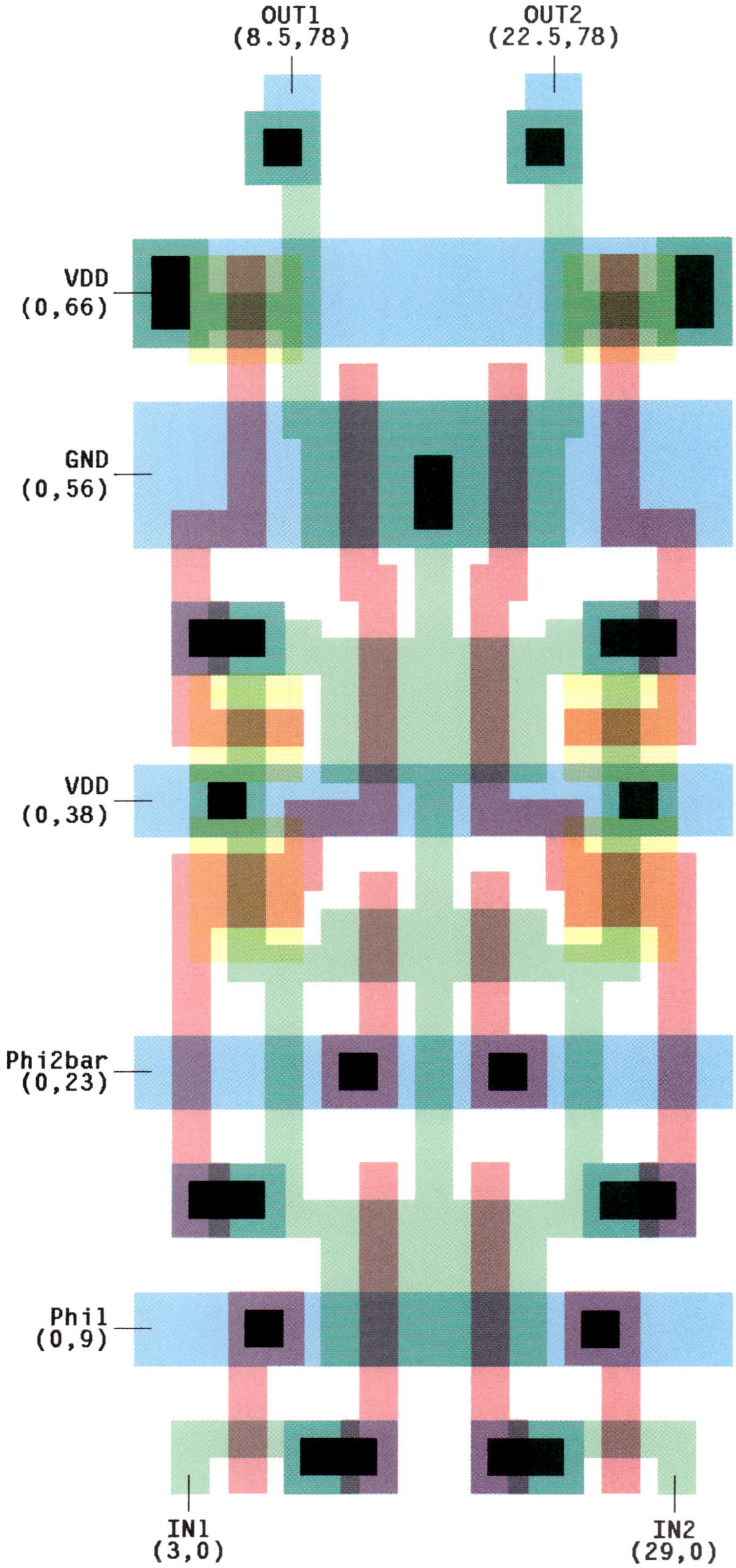
OUT1
(8.5,78)
OUT2
(22.5,78)
VDD
(0,66)
GND
(0,56)
VDD
(0,38)
Phi2bar
(0,23)
Phi1
(0,9)
IN1
(3,0)
IN2
(29,0)

NAME

AddBufPair—Adder driver

SYNOPSIS

AddBufPair is the function block driver for the Adder.

PROPERTIES

Size

33 x 78 λ

Interface

IN1	input,	switched	5 □ C_g	
IN2	input,	switched	5 □ C_g	
$\phi 1$	input,	switch control	5 □ C_g	
$\overline{\phi 2}$	input,	4:1 ratio	7.3 □ C_g	
OUT1	output,	super-buffered	3.75 □ C_g	[1 / ¼] □ *R*
OUT2	output,	super-buffered	3.75 □ C_g	[1 / ¼] □ *R*

DC Current

0.6 mA.

DESCRIPTION

This cell is exactly the same circuit as *ALUBufPair*. It was designed to pitch match with the *Adder* and *StaticAdder*. Notice that it is asymmetric: if one cell is flipped and then abutted with another so that the gap between the central two lines is the larger of the two possibilities, then the pair will drive alternate lines from one side of the adder. See *ALUBufPair*.

TESSELATION

Abuts horizontally, without overlap. Pitch matches *Adder* and *StaticAdder*. On the right side of the adder cell, it overlaps by 2λ; on the left side there is no overlap.

ORIGIN

Designed at CalTech (Dave Johannsen); implemented at Stanford (John Redford).

STATUS

Revised.

SEE ALSO

ALUBufPair, Adder, AddRestore, StaticAdder.

LAYOUT SCALE

10 λ per inch.

CIF ID

311

LOGIC DIAGRAM

See ALUBufPair(ALU).

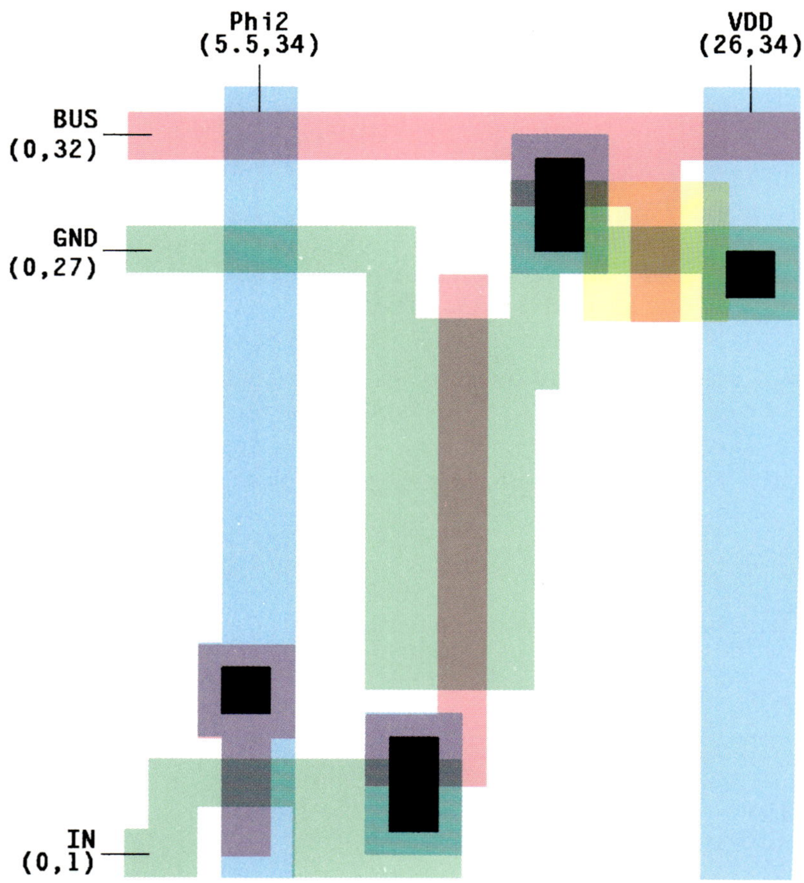
Phi2
(5.5,34)
VDD
(26,34)
BUS
(0,32)
GND
(0,27)
IN
(0,1)

NAME
AddClkOut—clocked adder buffer

SYNOPSIS
A clocked output buffer for the *Adder* cell.

PROPERTIES
Size

28 x 34λ

Interface

IN	input,	switched	9 □ C_g
ϕB	input,	switch control	3 □ C_g
BUS	output,	gate	6 □ C_g [1 / 1/8] □ *R*

DC Current

0.1 mA

DESCRIPTION
The output of the *Adder* cell is only valid during ϕB, if ϕA is the precharging clock phase. This cell provides a holding register if you need to use the adder output on ϕA.

TESSELATION
Abuts vertically, without overlap.

LOGIC DIAGRAM

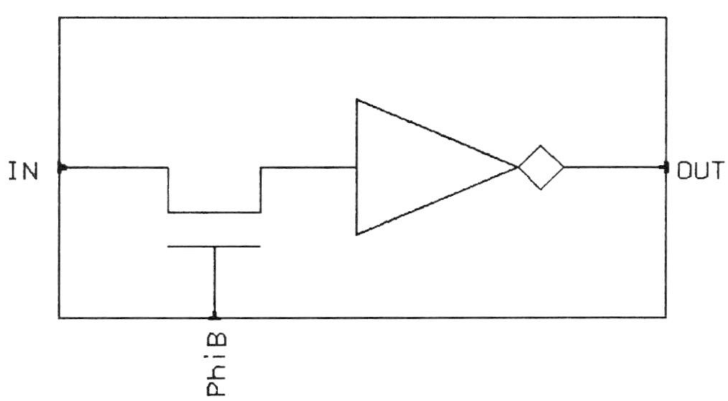

ORIGIN
Stanford University (John Redford).

STATUS
Generally used.

SEE ALSO
Adder, AddOut.

LAYOUT SCALE
8 λ per inch.

CIF ID
173

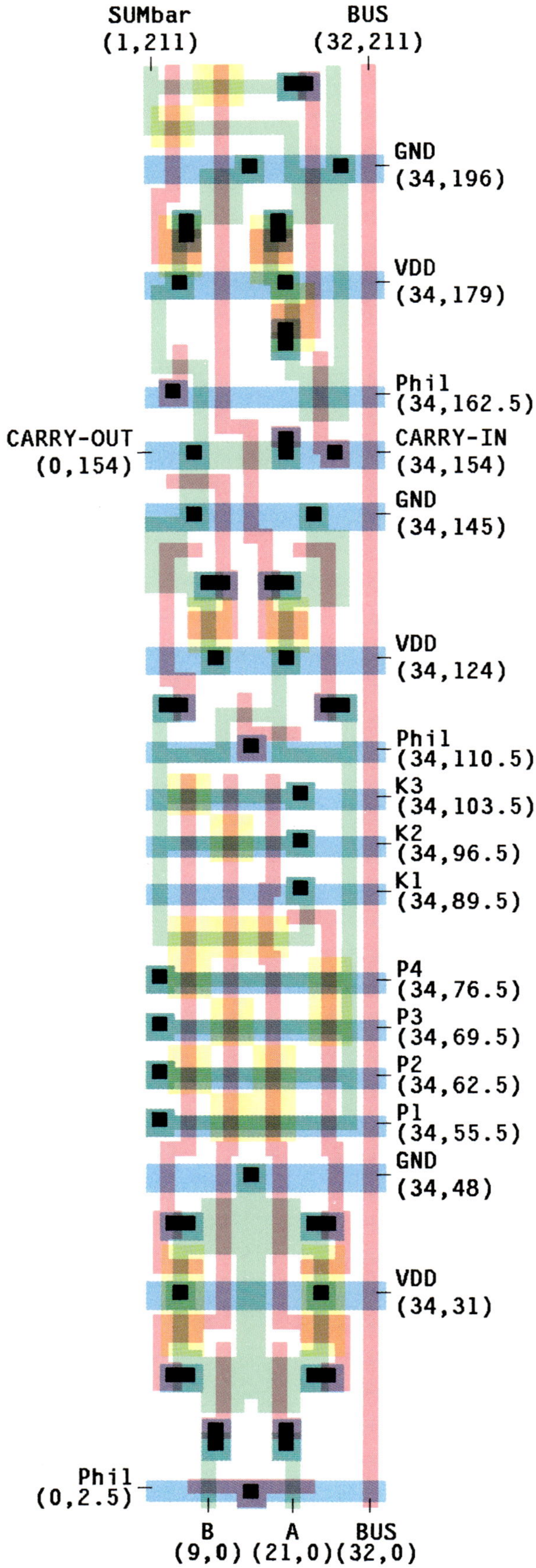
SUMbar
(1,211)
BUS
(32,211)
GND
(34,196)
VDD
(34,179)
Phil
(34,162.5)
CARRY-OUT
(0,154)
CARRY-IN
(34,154)
GND
(34,145)
VDD
(34,124)
Phil
(34,110.5)
K3
(34,103.5)
K2
(34,96.5)
K1
(34,89.5)
P4
(34,76.5)
P3
(34,69.5)
P2
(34,62.5)
P1
(34,55.5)
GND
(34,48)
VDD
(34,31)
Phil
(0,2.5)
B
(9,0)
A
(21,0)
BUS
(32,0)

NAME

Adder—adder/subtractor

SYNOPSIS

An adder/subtractor based on the OM2 ALU.

PROPERTIES

Size

34 x 211λ

Interface

A	input,	switched	$5\ \square\ C_g$
B	input,	switched	$5\ \square\ C_g$

P1 is typical of the P and K control lines.

P1	input,	switched	$2.75\ \square\ C_g$
Carry-In	input,	switched	$7\ \square\ C_g$
Carry-Out	output,	precharged	$4.5\ \square\ C_g$
$\overline{SUM}$	output,	switch-logic	$1.5\ \square\ C_g$

DC Current

0.3 mA

DESCRIPTION

This cell trades some of the flexibility of the OM2 ALU for a 40% decrease in area. The R function block has been replaced by an XOR circuit, one bus has been eliminated, and the K block has been narrowed. However, the cell can still do addition, subtraction, and logic operations, with the following control line settings:

OP ($\overline{SUM}$)	P	K	LSB Cin
Zero	0	0	0
One	15	0	0
$\overline{A}$	3	0	0
$\overline{B}$	5	0	0
A	12	0	0
B	10	0	0
A *and* B	1	0	0
A *or* B	7	0	0
A *xor* B	6	0	0
$\overline{\text{A and B}}$	14	0	0
$\overline{\text{A } or \text{ B}}$	8	0	0
$\overline{\text{A } xor \text{ B}}$	9	0	0
$\overline{B => A}$	4	0	0
$\overline{A => B}$	2	0	0
A => B	13	0	0
$\overline{A+1}$	12	1	1
$\overline{A-B}$	6	5	1
$\overline{-A}$	3	6	1
$\overline{A+B}$	9	3	0

Note that this adder can not compute B − A. The numbers are decimal representations of the binary numbers that are fed to the function blocks, where P1 and K1 are the least significant bits. All the data words are assumed to be in two's complement form. The *ALUBufPair cell*, or one with the same clocking, should be used to drive the control lines. The output $\overline{\text{SUM}}$ is inverted so that the buffers *AddOut* and *AddClkOut* can be attached to it.

TESSELATION

Abuts vertically, without overlap.

LOGIC DIAGRAM

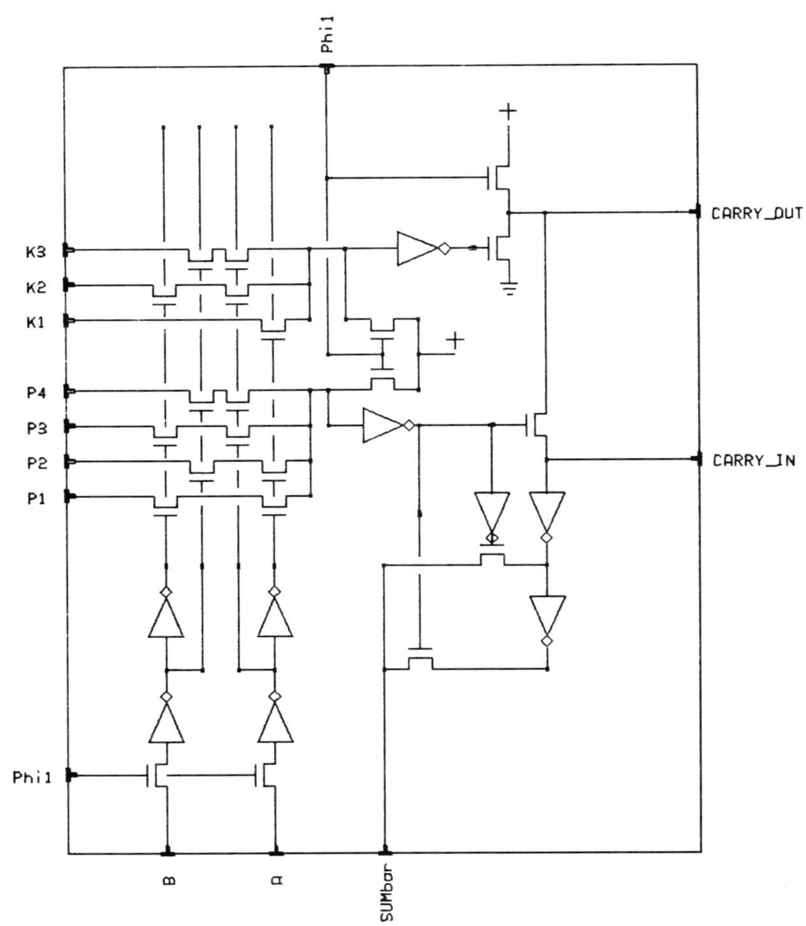

ORIGIN

Stanford University (John Redford).

STATUS

Generally used.

SEE ALSO

ALU, AddRestore, AddOut, AddClkOut, ALUBufPair.

LAYOUT SCALE

25 λ per inch.

CIF ID

170

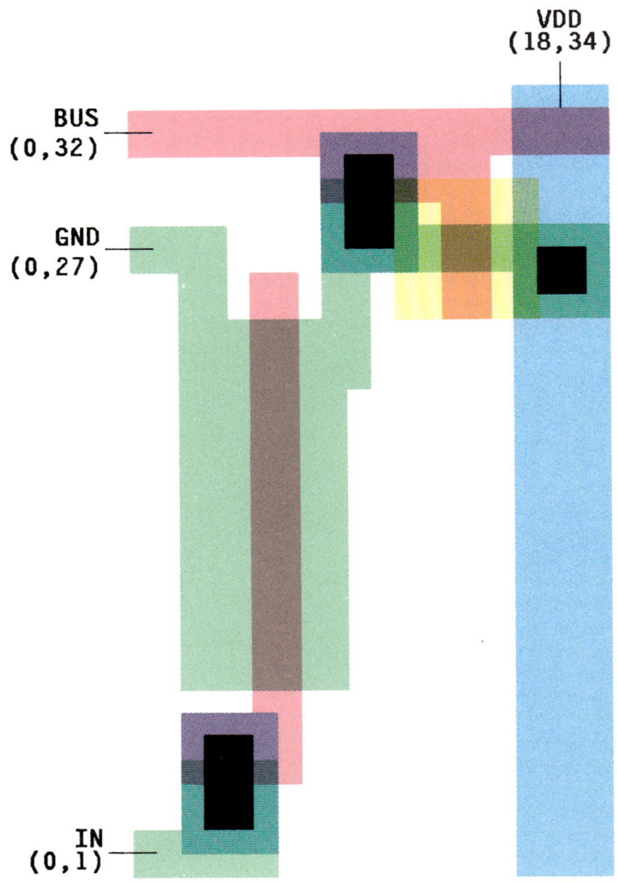
VDD
(18,34)
BUS
(0,32)
GND
(0,27)
IN
(0,1)

NAME

AddOut—adder output buffer

SYNOPSIS

An output buffer for the *Adder* cell.

PROPERTIES

Size

20 x 34λ

Interface

IN	input,	8:1 ratio	10 □ C_g	
BUS	output,	gate	6 □ C_g	[1 / 1/8] □ *R*

DC Current

0.1 mA

DESCRIPTION

AddOut buffers the output of the *Adder* cell and connects it to the bus line that runs next to it.

TESSELATION

Abuts vertically, without overlap.

LOGIC DIAGRAM

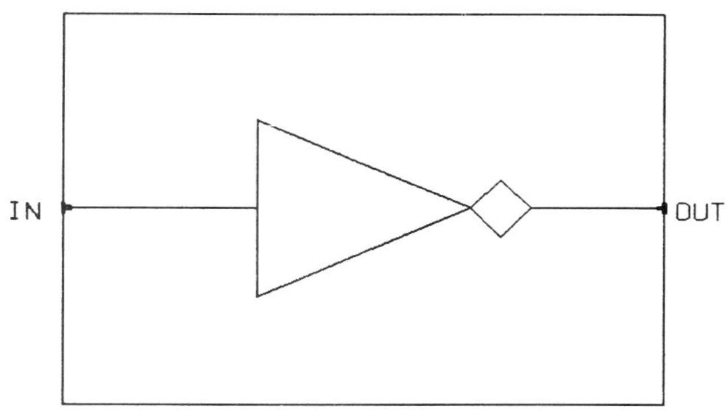

ORIGIN

Stanford University (John Redford, modified by C. Y. Chu).

STATUS

Revised to eliminate design-rule errors.

SEE ALSO

Adder

LAYOUT SCALE

8 λ per inch.

CIF ID

172

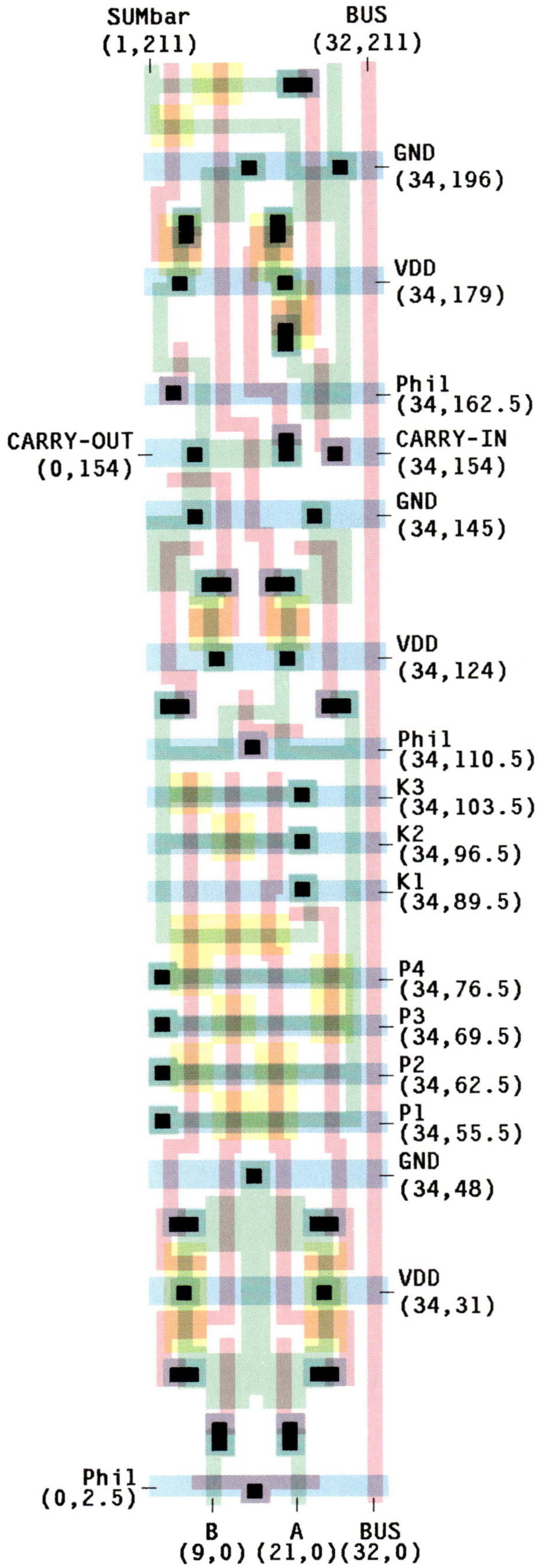
SUMbar
(1,211)
BUS
(32,211)
GND
(34,196)
VDD
(34,179)
Phil
(34,162.5)
CARRY-OUT
(0,154)
CARRY-IN
(34,154)
GND
(34,145)
VDD
(34,124)
Phil
(34,110.5)
K3
(34,103.5)
K2
(34,96.5)
K1
(34,89.5)
P4
(34,76.5)
P3
(34,69.5)
P2
(34,62.5)
P1
(34,55.5)
GND
(34,48)
VDD
(34,31)
Phil
(0,2.5)
B
(9,0)
A
(21,0)
BUS
(32,0)

NAME

AddRestore—carry-buffered adder/subtractor

SYNOPSIS

An adder/subtractor based on the OM2 ALU with buffering for the carry chain.

PROPERTIES

Size

34 x 211λ

Interface

A	input,	switched	5 □ C_g
B	input,	switched	5 □ C_g

P1 is typical of the P and K control lines.

P1	input,	switched	2.75 □ C_g
Carry-In	input,	switched	5 □ C_g
$\overline{\textit{Carry-Out}}$	output,	precharged	4.5 □ C_g
$\overline{\textit{SUM}}$	output,	switch-logic	1.5 □ C_g

DC Current

0.3 mA

DESCRIPTION

Same as *Adder*, except that the carry-in signal is restored by sending it through a pair of inverters before going to the propagate and kill transistors. It should be used after every three *Adder* cells.

TESSELATION

Abuts vertically, without overlap.

ORIGIN

Stanford University (John Redford).

STATUS

Generally used.

SEE ALSO

Adder, AddOut, AddClkOut.

LAYOUT SCALE

25 λ per inch.

CIF ID

171

LOGIC DIAGRAM

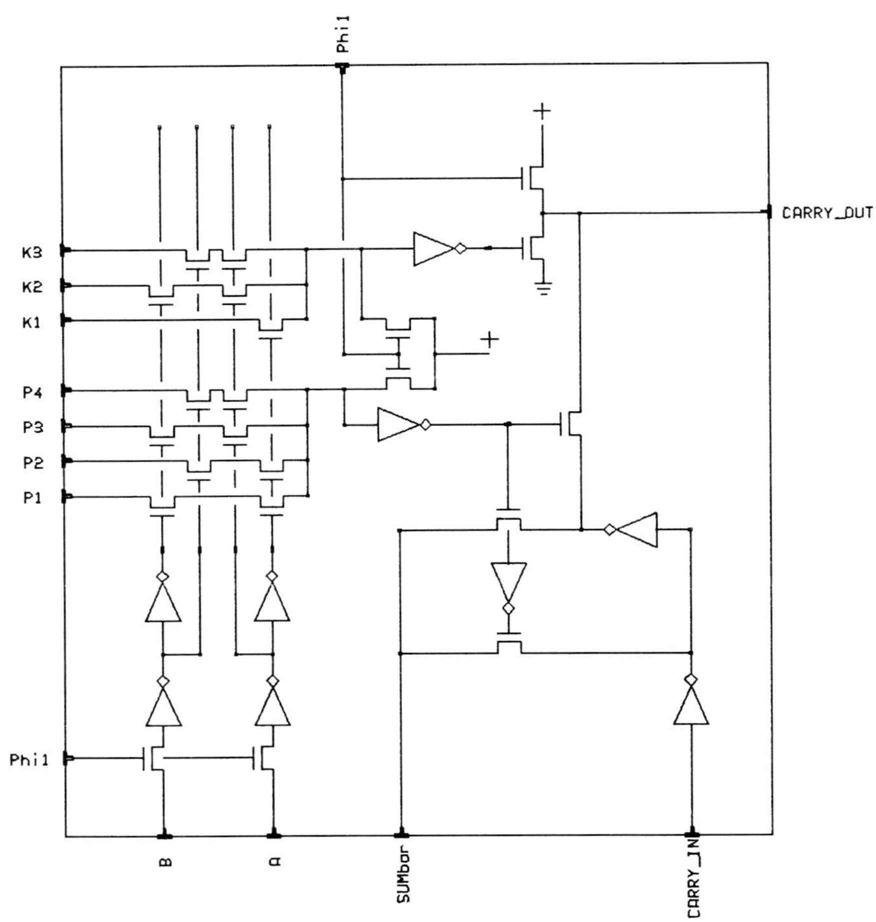

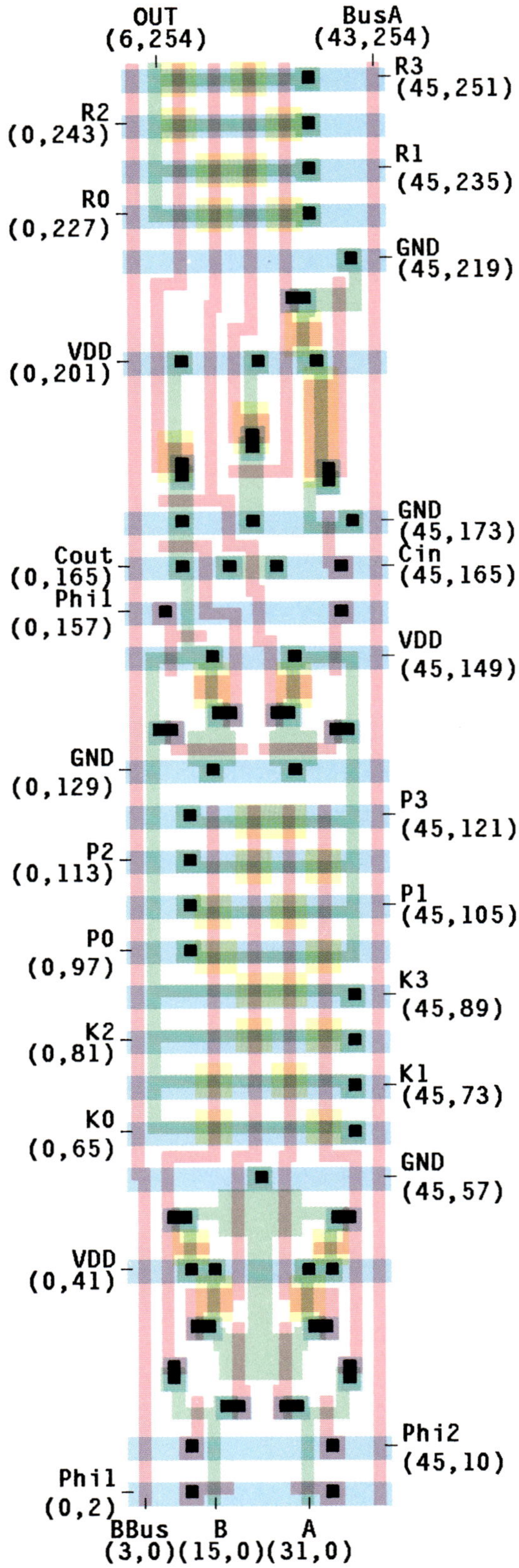
OUT
(6,254)
BusA
(43,254)
R3
(45,251)
R2
(0,243)
R1
(45,235)
R0
(0,227)
GND
(45,219)
VDD
(0,201)
GND
(45,173)
Cout
(0,165)
Cin
(45,165)
Phi1
(0,157)
VDD
(45,149)
GND
(0,129)
P3
(45,121)
P2
(0,113)
P1
(45,105)
P0
(0,97)
K3
(45,89)
K2
(0,81)
K1
(45,73)
K0
(0,65)
GND
(45,57)
VDD
(0,41)
Phi2
(45,10)
Phi1
(0,2)
BBus
(3,0)
B
(15,0)
A
(31,0)

NAME

ALU—ALU bit-slice

SYNOPSIS

ALU is one bit of the OM2 Data Path Chip ALU.

PROPERTIES

Size

45 x 254 λ

Interface

A	input,	switched	3 □ C_g
B	input,	switched	3 □ C_g

K0 is typical of the function block lines:

K0	input,	switched	4.25 □ C_g

CIN	input,	switched	3.75 □ C_g
COUT	output,	precharged	4.5 □ C_g
OUT	output,	switch-logic	4.5 □ C_g

DC Current

0.4 mA.

DESCRIPTION

This is one bit of the OM2 Data Path ALU. It employs a very general structure to generate arbitrary functions of its two inputs and a carry-in. The carry is propagated using a Manchester-type carry chain. Extensive use is made of precharging.

Note that the output comes directly out of the R function block. It should be buffered before it is used anywhere.

ALU Operations

Here is a table of ALU operations that have been checked by simulation:

OP	K	P	R	LSB Cin	Operation
A+B	14	9	6	—	Add with carry
A−B	11	6	6	1	$A+\overline{B}+1$
B−A	13	6	6	1	$\overline{A}+B+1$
−A	5	10	6	1	$\overline{A}+1$
−B	3	12	6	1	$\overline{B}+1$
A+Cin	10	5	6	—	Increment A if LSB Cin=1
B+Cin	12	3	6	—	Increment B if LSB Cin=1
A-Cin	5	10	9	—	Decrement A if LSB Cin=1
B-Cin	3	12	9	—	Decrement B if LSB Cin=1
A *and* B	0	8	3	0	Bit-wise AND
A *or* B	0	1	4	0	Bit-wise OR
A *xor* B	0	6	3	0	Bit-wise XOR
$\overline{A}$	0	5	3	0	Not A
$\overline{B}$	0	3	3	0	Not B
$\overline{A\ or\ B}$	0	1	1	0	Bit-wise NOR
$\overline{A\ xor\ B}$	0	9	1	0	XNOR
B=>A	0	4	12	0	B implies A
A=>B	0	2	12	0	A implies B

$\overline{A \& B}$	0	8	12	0	Bit-wise NAND
Zero	0	0	0	0	Zero function
One	0	0	4	0	One function
$\overline{A=>B}$	0	2	1	0	$A\overline{B}$
$\overline{B=>A}$	0	4	1	0	$\overline{A}B$
A	0	10	3	0	Just A
B	0	12	3	0	Just B

The numbers are decimal representations of the four-bit binary numbers that are fed to the function blocks, where P0, K0, and R0 are the least significant bits. All the data words are assumed to be in two's complement form. Logical operations allow 0 or 1 for LSB carry-in; however, a 0 input ensures a 0 output in carry out. Other operations such as shifts and mixed logic functions are possible.

TESSELATION

Abuts horizontally, without overlap.

LOGIC DIAGRAM

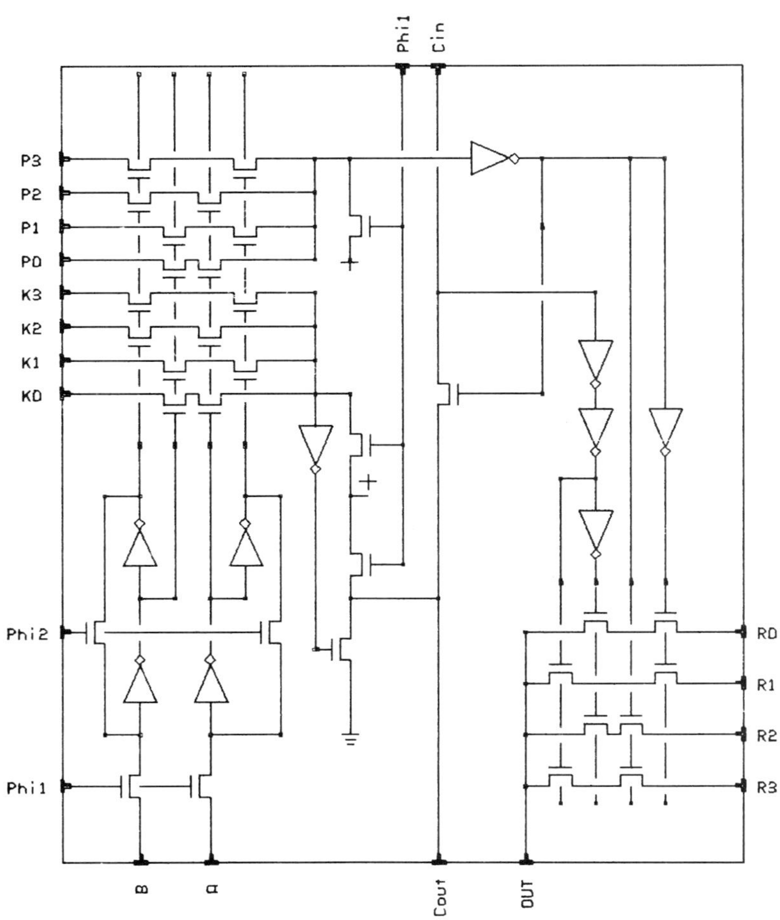

ORIGIN

Designed at CalTech (Dave Johannsen); implemented at Stanford (John Redford).

STATUS

Generally used.

SEE ALSO

ALUBufPair, ALURestore.

LAYOUT SCALE

32 λ per inch.

CIF ID

302

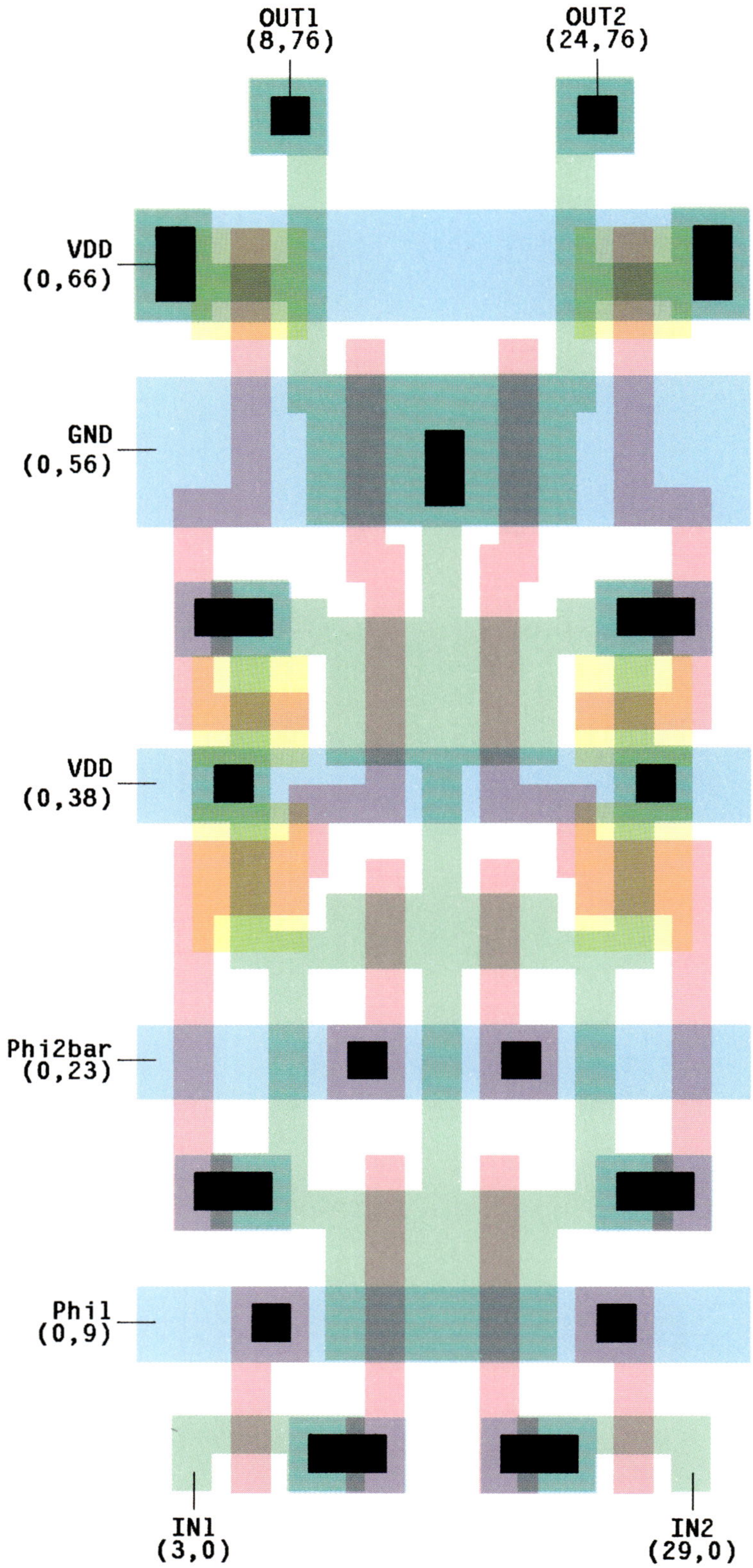
OUT1
(8,76)
OUT2
(24,76)
VDD
(0,66)
GND
(0,56)
VDD
(0,38)
Phi2bar
(0,23)
Phi1
(0,9)
IN1
(3,0)
IN2
(29,0)

NAME

ALUBufPair—ALU driver

SYNOPSIS

ALUBufPair is the function block driver for the ALU.

PROPERTIES

Size

32 x 76 λ

Interface

IN1	input,	switched	5 □ C_g	
IN2	input,	switched	5 □ C_g	
$\phi 1$	input,	switch control	5 □ C_g	
$\overline{\phi 2}$	input,	4:1 ratio	7.3 □ C_g	
OUT1	output,	super-buffered	3.75 □ C_g	[1 / ¼] □ *R*
OUT2	output,	super-buffered	3.75 □ C_g	[1 / ¼] □ *R*

DC Current

0.6 mA.

DESCRIPTION

This is a pair of superbuffer drivers for the ALU function block control lines. The drivers pull the lines high during the precharge clock phase ($\phi 1$), so they will only need to be pulled down during the active phase. Their inputs are loaded during $\phi 1$. The function block lines are on a pitch of 8λ, but this cell's outputs come out at 16λ intervals, so they must be put on alternate sides of the ALU if all the lines are to be driven. They come in pairs so that no overlapping will be needed. If you wish to design your own drivers, be sure to make the clocking similar to that here, or the ALU's precharging scheme will cause problems.

LOGIC DIAGRAM

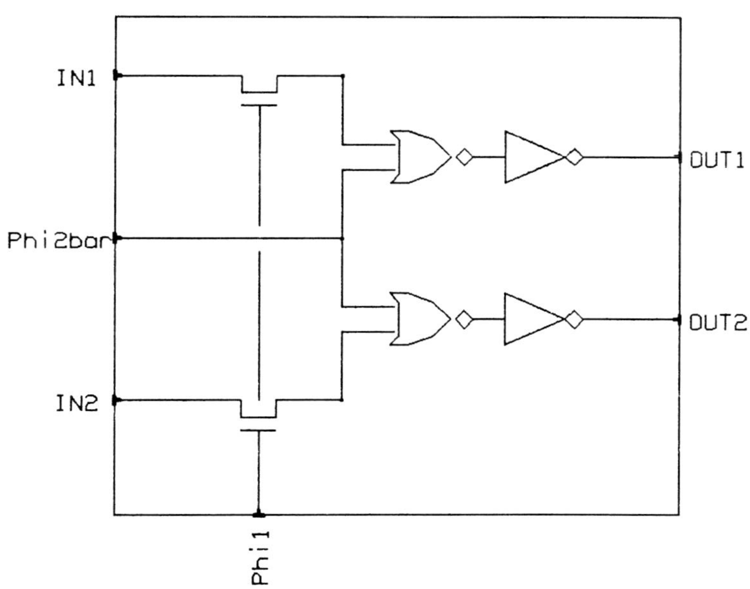

TESSELATION

Abuts horizontally, without overlap. Pitch matches *ALU*.

ORIGIN

Designed at CalTech (Dave Johannsen); implemented at Stanford (John Redford).

STATUS

Generally used.

SEE ALSO

ALU, ALURestore.

LAYOUT SCALE

10 λ per inch.

CIF ID

310

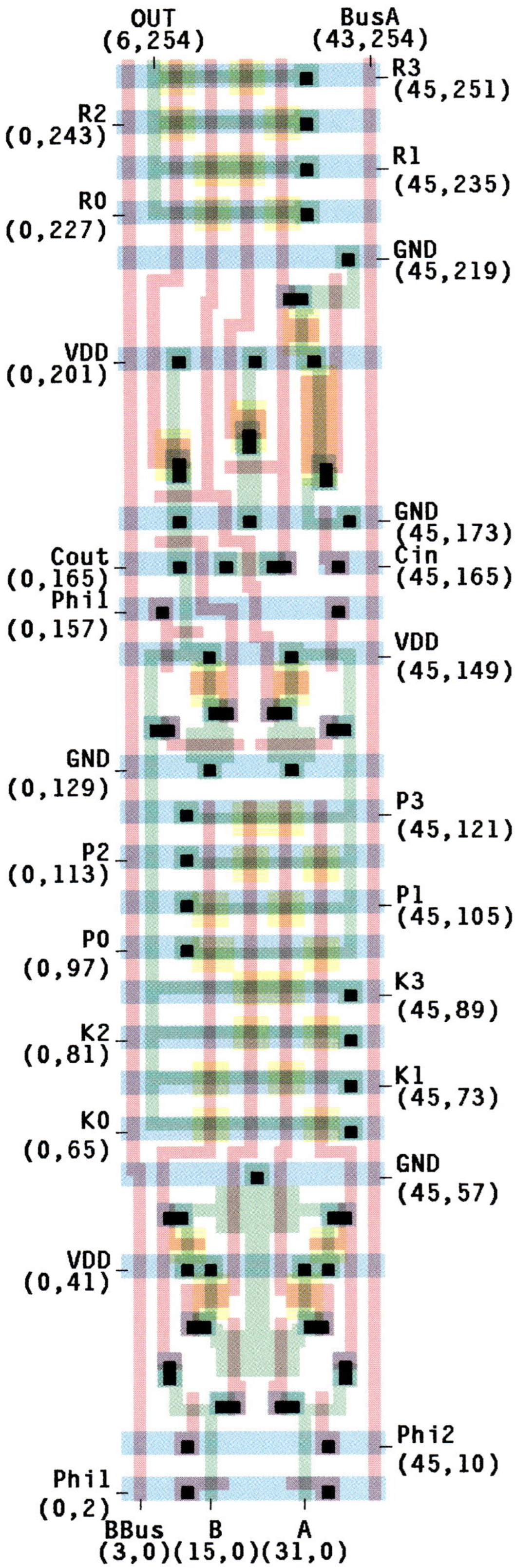
OUT
(6,254)
BusA
(43,254)
R3
(45,251)
R2
(0,243)
R1
(45,235)
R0
(0,227)
GND
(45,219)
VDD
(0,201)
GND
(45,173)
Cout
(0,165)
Cin
(45,165)
Phi1
(0,157)
VDD
(45,149)
GND
(0,129)
P3
(45,121)
P2
(0,113)
P1
(45,105)
P0
(0,97)
K3
(45,89)
K2
(0,81)
K1
(45,73)
K0
(0,65)
GND
(45,57)
VDD
(0,41)
Phi2
(45,10)
Phi1
(0,2)
BBus
(3,0)
B
(15,0)
A
(31,0)

NAME

ALURestore—carry-chain buffer

SYNOPSIS

ALURestore is inserted periodically to buffer the pass transistor carry chain.

PROPERTIES

Size

45 x 254 λ

Interface

A	input,	switched	3 □ C_g
B	input,	switched	3 □ C_g

K0 is typical of the function block lines:

K0	input,	switched	4.25 □ C_g

CIN	input,	switched	2.25 □ C_g
COUT	output,	precharged	4.5 □ C_g
OUT	output,	switch-logic	4.5 □ C_g

DC Current

0.4 mA.

DESCRIPTION

This is the same as *ALU* except that CIN is routed through a pair of inverters before it goes to the propagate and kill transistors.

TESSELATION

Abuts horizontally, without overlap.

ORIGIN

Designed at CalTech (Dave Johannsen); implemented at Stanford (John Redford).

STATUS

Generally used.

SEE ALSO

ALU, ALUBufPair.

LAYOUT SCALE

32 λ per inch.

CIF ID

303

LOGIC DIAGRAM

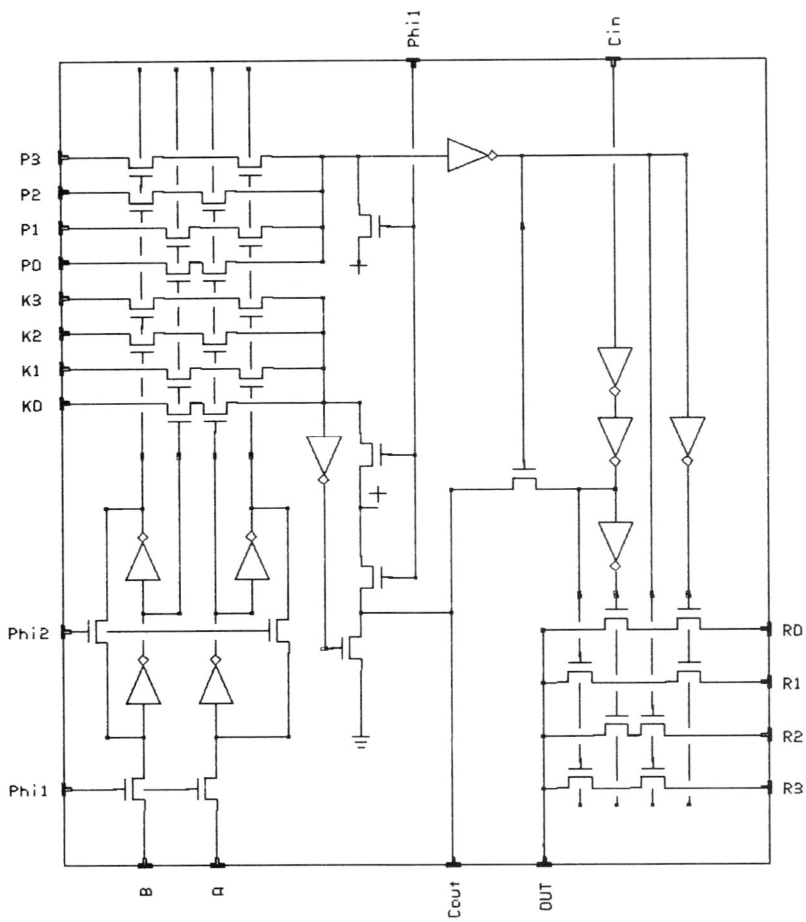

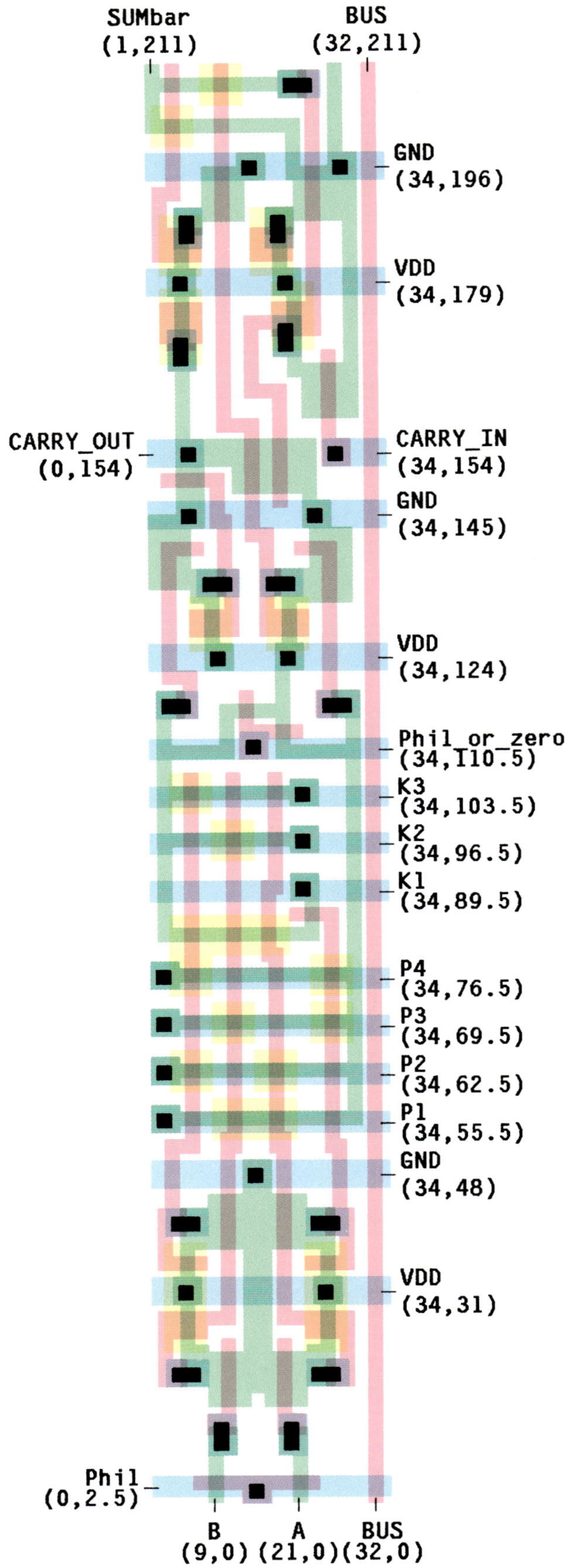
SUMbar
(1,211)
BUS
(32,211)
GND
(34,196)
VDD
(34,179)
CARRY_OUT
(0,154)
CARRY_IN
(34,154)
GND
(34,145)
VDD
(34,124)
Phi1_or_zero
(34,110.5)
K3
(34,103.5)
K2
(34,96.5)
K1
(34,89.5)
P4
(34,76.5)
P3
(34,69.5)
P2
(34,62.5)
P1
(34,55.5)
GND
(34,48)
VDD
(34,31)
Phi1
(0,2.5)
B
(9,0)
A
(21,0)
BUS
(32,0)

Name

StaticAdder—adder/subtractor

SYNOPSIS

An adder/subtractor based on the OM2 ALU. Identical in function to *Adder* but not precharged.

PROPERTIES

Size

34 x 211λ

Interface

A	input,	switched	5 □ C_g
B	input,	switched	5 □ C_g

P1 is typical of the P and K control lines.

P1	input,	switched	2 □ C_g
$\overline{\phi 1}$	input,	switch control	3 □ C_g
Carry-In	input,	switched	5 □ C_g
Carry-Out	output,	switch-logic	2 □ C_g
$\overline{SUM}$	output,	switch-logic	1 □ C_g

DC Current

0.3 mA

DESCRIPTION

StaticAdder is a functional equivalent to *Adder* that can be used as combinational logic. The carry chain is static, the function blocks need not be precharged (*i.e.,* ϕ1__or__zero may be tied low), and the inputs need not be latched (*i.e.,* ϕ1 may be tied high).

TESSELATION

Abuts vertically, without overlap.

ORIGIN

Stanford University (Linda Shwetz).

STATUS

Checked.

SEE ALSO

ALU, AddRestore, AddOut, AddClkOut, ALUBufPair.

LAYOUT SCALE

25 λ per inch.

CIF ID

175

LOGIC DIAGRAM

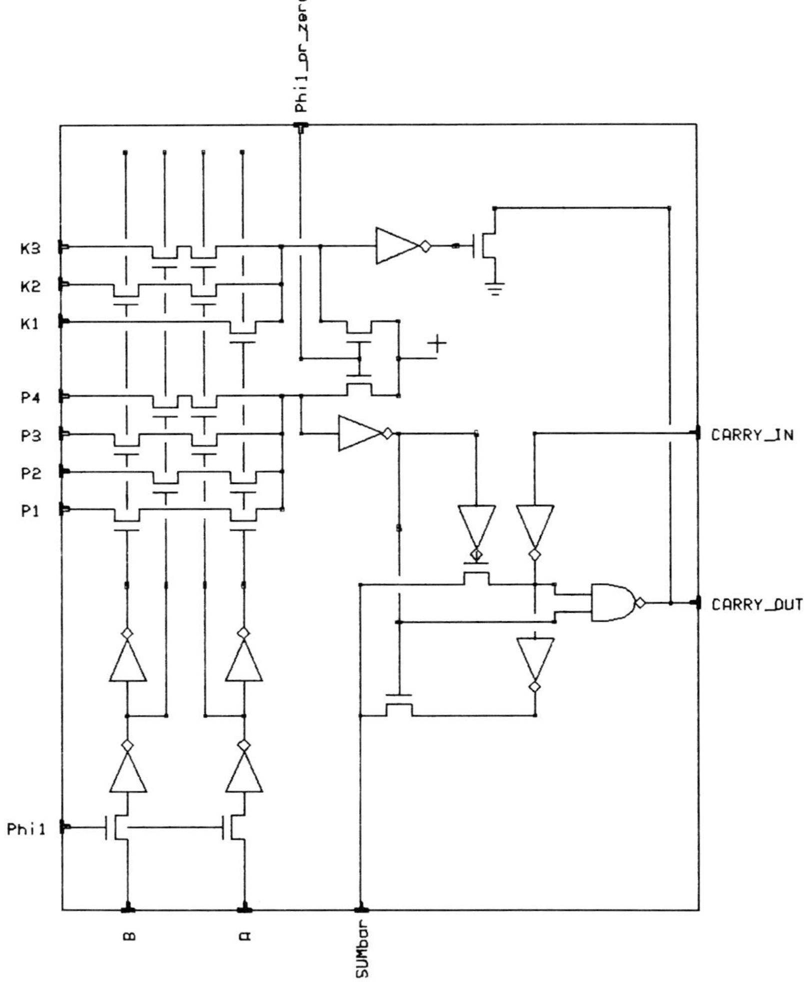

2 / Buffers

NAME

intro—introduction to superbuffers

SYNOPSIS

This chapter describes inverting and noninverting superbuffers, used to drive high capacitance lines. These superbuffers are meant to be used as clock and control line drivers. Unlike ordinary buffers the rise times on their outputs are only about twice their fall times, Mead and Conway to the contrary.

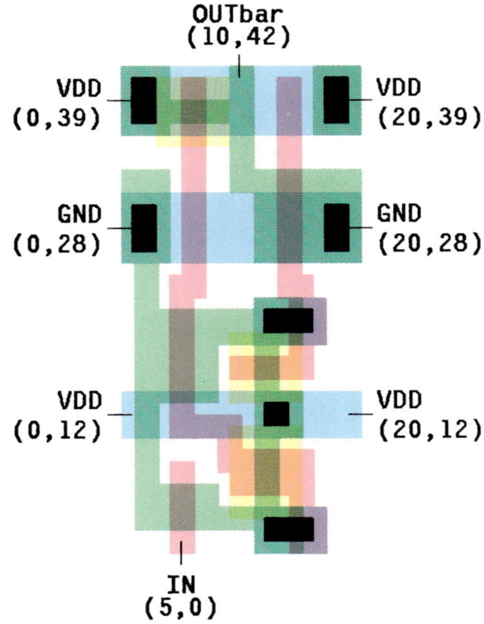
OUTbar
(10,42)
VDD
(0,39)
VDD
(20,39)
GND
(0,28)
GND
(20,28)
VDD
(0,12)
VDD
(20,12)
IN
(5,0)

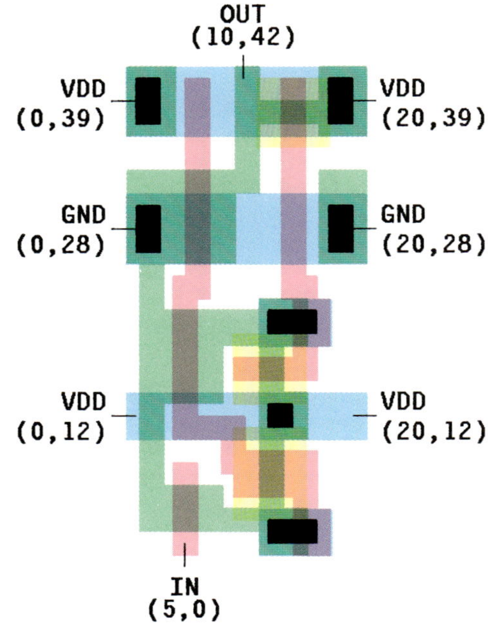
OUT
(10,42)
VDD
(0,39)
VDD
(20,39)
GND
(0,28)
GND
(20,28)
VDD
(0,12)
VDD
(20,12)
IN
(5,0)

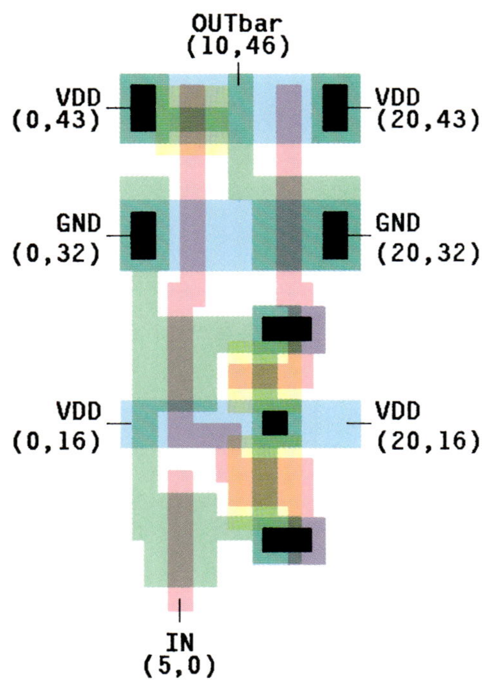
OUTbar
(10,46)
VDD
(0,43)
VDD
(20,43)
GND
(0,32)
GND
(20,32)
VDD
(0,16)
VDD
(20,16)
IN
(5,0)

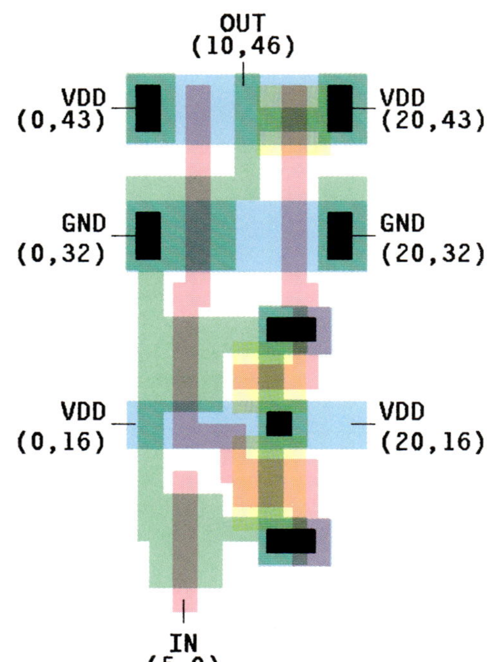
OUT
(10,46)
VDD
(0,43)
VDD
(20,43)
GND
(0,32)
GND
(20,32)
VDD
(0,16)
VDD
(20,16)
IN
(5,0)

NAME

superbuffers—superbuffers

SYNOPSIS

Inverting or non-inverting superbuffers with 4:1 or 8:1 input ratios.

PROPERTIES

Size

20 x 46 λ for SB8 versions, and
20 x 42 λ for SB4 versions.

Interface

For the SB4 cells:

IN	input,	4:1 ratio	3 □ C_g	

For the SB8 cells:

IN	input,	8:1 ratio	5 □ C_g	

For all of them:

OUT	output,	super-buffered	2 □ C_g	[1 / ¼] □ *R*

DC current

0.3 mA per bit.

DESCRIPTION

These cells are composed of 2 inverters in series, with a superbuffer output stage. The cells whose names end in "4" have 4:1 input ratios, and those ending in "8" have 8:1 input ratios.

TESSELATION

Designed to stack horizontally, mirroring in *y* and overlapping 4 λ, giving outputs on a 16 λ pitch. An iterated, mirrored pair is 36 λ wide.

ORIGIN

Xerox PARC (Dick Lyon).

STATUS

Generally used.

LAYOUT SCALE

8 λ per inch.

CIF ID

InvSB4:	234
NonInvSB4:	236
InvSB8:	251
NonInvSB8:	253

LOGIC DIAGRAM

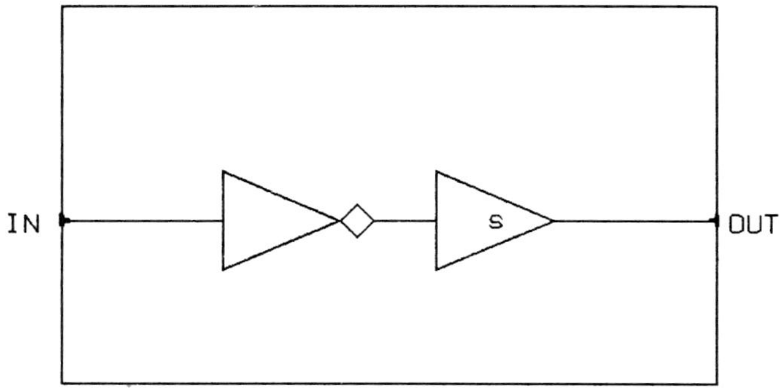

(Inverting Superbuffer)

LOGIC DIAGRAM

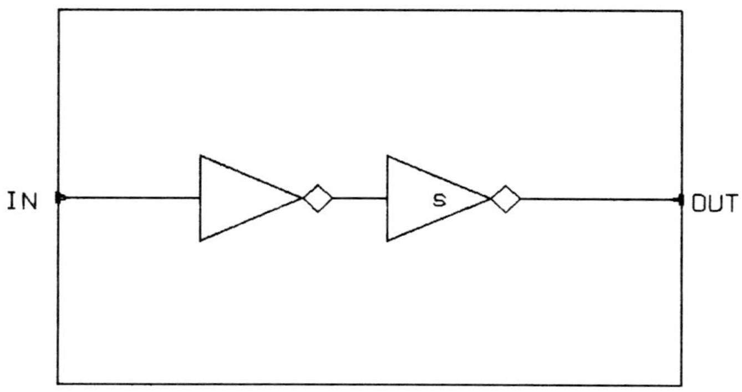

(NonInverting Superbuffer)

3 / Clocking Cells

NAME

intro—introduction to timing control circuits

SYNOPSIS

This section has circuits for asynchronous communication, and will be expanded in the future with other related timing control cells.

Currently it includes cells for a clock generator and for a synchronizer.

DESCRIPTION

The clock (*Vsclk*) is a compact design that provides adjustable speed and a mechanism for requesting stretching of a phase. It generates 2 non-overlapping phases.

The synchronizer (*Sync*) is used to receive external asynchronous signals, without risk of synchronization failure. Infrequently it may request the stretching of a clock phase.

ORIGIN

Stanford University (Daniel Chapiro and Robert Mathews).

SEE ALSO

Chapter 7 of Mead & Conway on System Timing, by Charles Seitz.

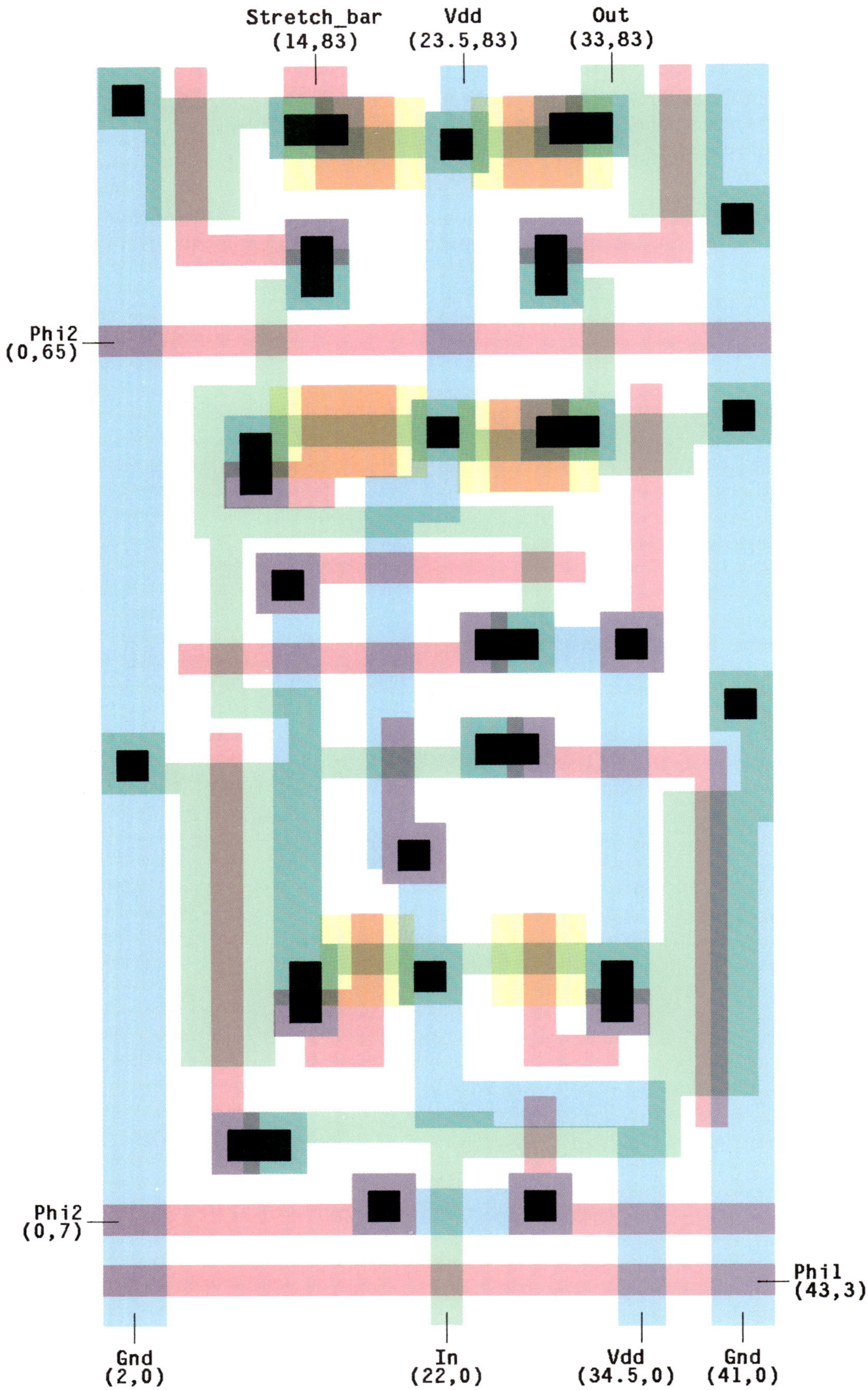
Stretch_bar
(14,83)
Vdd
(23.5,83)
Out
(33,83)
Phi2
(0,65)
Phi2
(0,7)
Phi1
(43,3)
Gnd
(2,0)
In
(22,0)
Vdd
(34.5,0)
Gnd
(41,0)

NAME

Sync—synchronizer for stoppable clock

SYNOPSIS

Sync uses the local clock to sample asynchronous data, and outputs the data synchronized to the clock. It differs from plain synchronizers in that it cannot have synchronization failures. It is designed to control a stoppable clock.

PROPERTIES

Size

43 x 83 λ

Interface

IN	input,		1 □ C_g	
STRETCH-BAR	output,	gate	4 □ C_g	[2 / ¼] □ *R*
OUT	output,	gate	4 □ C_g	[2 / ¼] □ *R*
$\phi 1$	input,	switch control	1 □ C_g	
$\phi 2$	input,	switch control	3 □ C_g	

DC current

0.57 mA.

Performance

Operates over the range of 25 Hz to 10 MHz.

DESCRIPTION

Sync is a flip-flop-like circuit for synchronizing an external signal to a local clock. It samples the signal, then stretches the affected clock cycle (via the STRETCH output) if its internal flip-flop is metastable. Hence, data is always sampled in a single clock, but with some small probability, the corresponding clock phase gets stretched for an unbounded, but normally small, period of time.

Sync samples the data during $\phi 1$, and stores it in a static storage cell. An analog XNOR gate is used to detect non-digital values and to control the stretching of $\phi 2$. For $\phi 2$ actually to be stretched, IN must have changed precisely when $\phi 1$ was going low, and the non-digitally-defined value loaded in the storage cell must have caused the cell to remain in a metastable condition during the $\phi 1/\phi 2$ gap and $\phi 2$. The throughput is almost unaffected by these occasional phase stretchings because they occur with a small probability, and this probability decreases exponentially with the length of the stretching period.

The synchronized data will appear in the output during $\phi 2$, but $\phi 2$ may end immediately after the output appears. This may occur independently of how slow the clock is running because the synchronizer will stop stretching $\phi 2$ immediately after the latched data becomes defined. Hence, the output is only stable during $\phi 1$.

Sync has the following pins:

IN: An asynchronous external signal. It is gated with $\phi 1$.

STRETCH:
A signal that infrequently requests a stretching of $\phi 2$.

CLOCK:
A two-phase, non-overlapping clock ($\phi 1$ and $\phi 2$).

OUT: The synchronized signal (stable $\phi 1$, not valid $\phi 2$).

LOGIC DIAGRAM

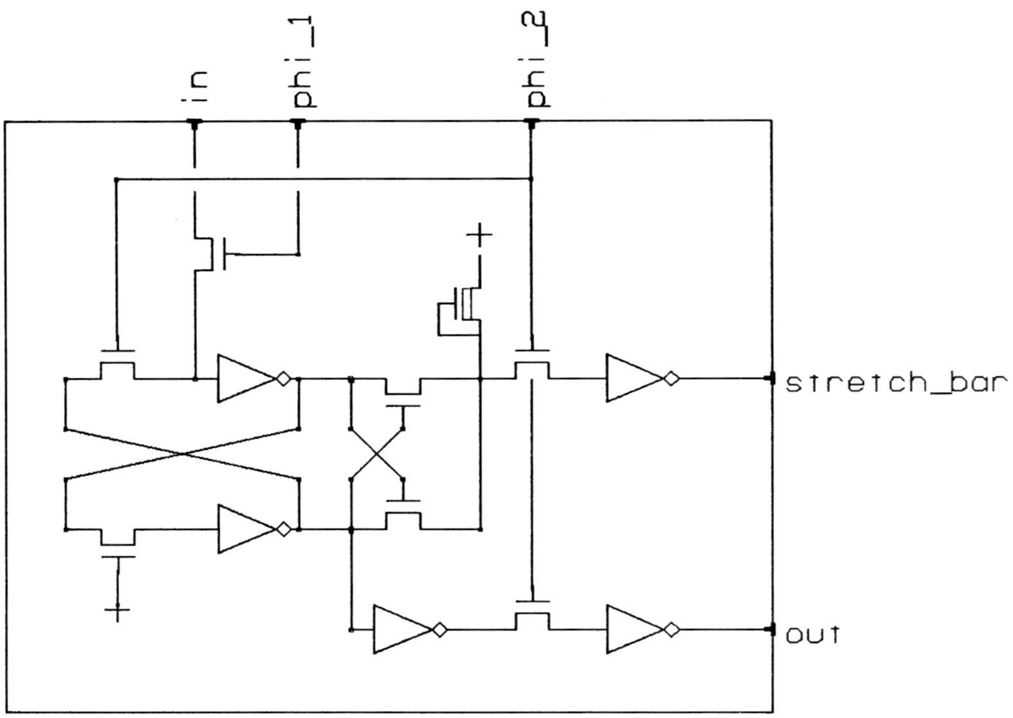

Synchronization of multiple signals

If several asynchronous signals are received, one synchronizer should be used for each one, and the NAND of all the STRETCH_BARs should be sent to the stretch control of the stoppable clock.

ORIGIN

Stanford University (Daniel Chapiro and Robert Mathews).

STATUS

Checked.

SEE ALSO

Vsclk (CLOCK).

LAYOUT SCALE

10 λ per inch.

CIF ID

521

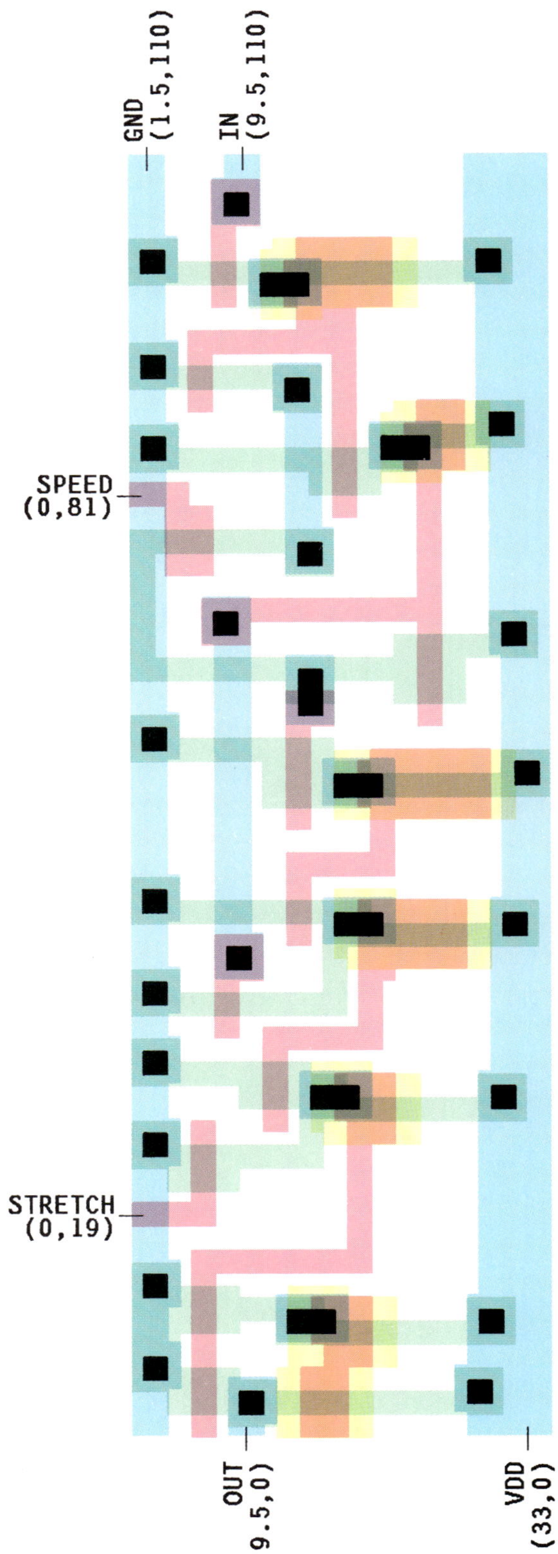
GND (1.5,110)
IN (9.5,110)
SPEED (0,81)
STRETCH (0,19)
OUT (9.5,0)
VDD (33,0)

NAME

Vsclk—variable speed stoppable clock

SYNOPSIS

Vsclk is used to build a 2-phase clock, with analog inputs that permit controlling the width of each phase and gap independently. It can also stretch any phase or gap for arbitrarily long periods. This permits stopping the clock synchronously and restarting it asynchronously.

PROPERTIES

Size

35 x 110 λ

Interface

IN	input,	switch control	1 □ C_g	
STRETCH	input,	switch control	2 □ C_g	
SPEED	input,	analog control	0–5V	
OUT	output,	super-buffered	2 □ C_g	[2 / 1] □ *R*

DC current

Up to 1.7ma.

Performance

Operates over the range of 0.3 MHz to 10 MHz.

DESCRIPTION

A clock is composed of 4 *vsckl*s, which resemble one-shot circuits. The pins of each *vsclk* behave as follows:

IN: A pulse of arbitrary width, exceeding a fabrication-dependent minimum.

STRETCH:
Forces OUT high as long as it is high, independently of IN. If it appears while OUT is high, and goes low before OUT would have gone low by itself, it will not alter the normal width of the OUT pulse.

SPEED:
Sets the normal width of the OUT pulse.

OUT: A pulse that is emitted immediately after the IN pulse goes low, and whose normal width is set by SPEED, unless it is extended by STRETCH.

To make a 2-phase non-overlapping clock, put 4 vsclks in a ring tying the OUT of each one of the IN of the next one. Numbering these from 1 to 4, make $\phi 1 = OUT_2$ and $\phi 2 = OUT_4$. Connect all the 4 SPEED lines together, or provide different voltages to each phase and gap, to optimize the design. Each OUT line should not drive more than 8 □ C_g for the clock to operate correctly.

CONTROL

To start the clock, one of the STRETCH lines has to be pulsed, possibly with the system reset. The clock will start as soon as the STRETCH line goes down, with the following stage. Hence if we want to have $\phi 1$ and $\phi 2$low on RESET and start operating with $\phi 1$, RESET should be connected to $STRETCH_1$.

LOGIC DIAGRAM

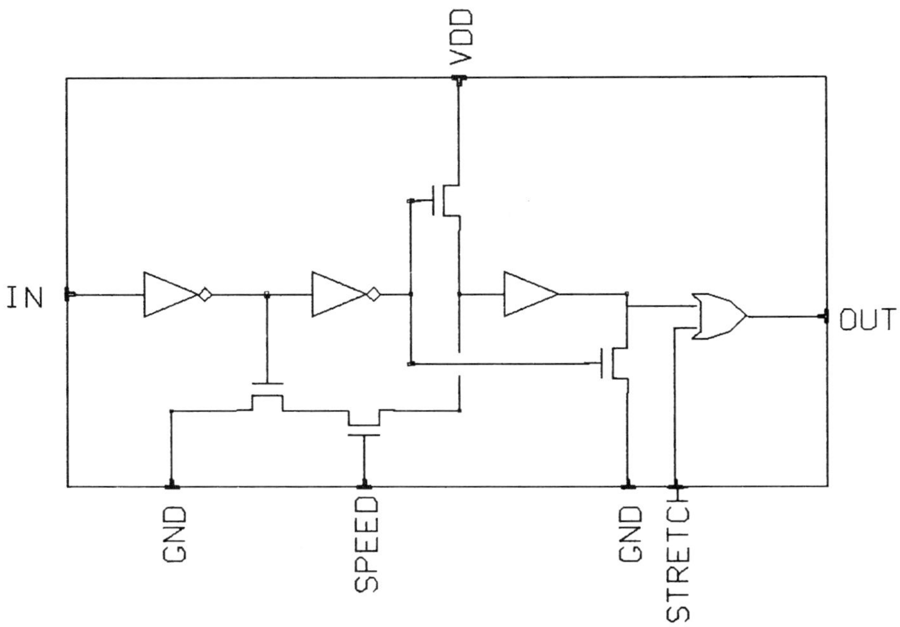

Once the clock is running, any of the four stages can be stretched by raising its corresponding STRETCH line. However, a stretch line should be asserted only while the corresponding OUT is high.

Running the clock too fast will make the clock go away. To restart it you need to lower the control voltage (SPEED) and pulse a STRETCH line again. Running it too slow will cause jitter and other unpredictable behavior.

TESSELATION

Abuts vertically. It also abuts laterally after a 180° rotation (with a 2λ overlap), sharing the VDD line.

ORIGIN

Stanford University (Daniel Chapiro).

STATUS

Tested. The operating frequency at each control voltage was the same for chips within the same batch, but differed by a factor of two in between chips from different batches. The faster chips run at 0.6MHz for SPEED=1V, and 10MHz for SPEED=2.25V. The frequency sensitivity to temperature is in the order of −0.1% per °F, with ambient temperature between 32°F and 150°F. The sensitivity to VDD fluctuations is in the order of −0.025% per mV for VDD between 4V and 7V.

SEE ALSO

Sync (CLOCK).

LAYOUT SCALE

16 λ per inch.

CIF ID

520

4 / Contexts

NAME

intro—introduction to design contexts

SYNOPSIS

This chapter describes frames that support a design project and interface it to the outside world.

DESCRIPTION

A design context provides a design environment in which a user project is implemented. A context is directly analogous to operating system software, in which the details of I/O interfacing, system-provided functions, and device drivers are invisible to the user. A minimal context (*e.g.*, BasicFrame, following) provides system clocks and simple I/O for the user circuit. Contexts also define standard interfaces transparent to the casual user, such as PC board layout or bonding diagrams. The context described is useful both in the construction of first-time projects (often games), and in breadboarding simple circuits.

ORIGIN

Xerox PARC (Al Paeth) University of North Carolina at Chapel Hill (Henry Fuchs).

STATUS

Tested, works.

SEE ALSO

A Basic Design Context for VLSI Systems, Xerox PARC Technical Report VLSI-83-xx (in press).

A Standard Design Frame for VLSI Circuit Prototyping, University of Wisconsin Computer Sciences Technical Report #485.

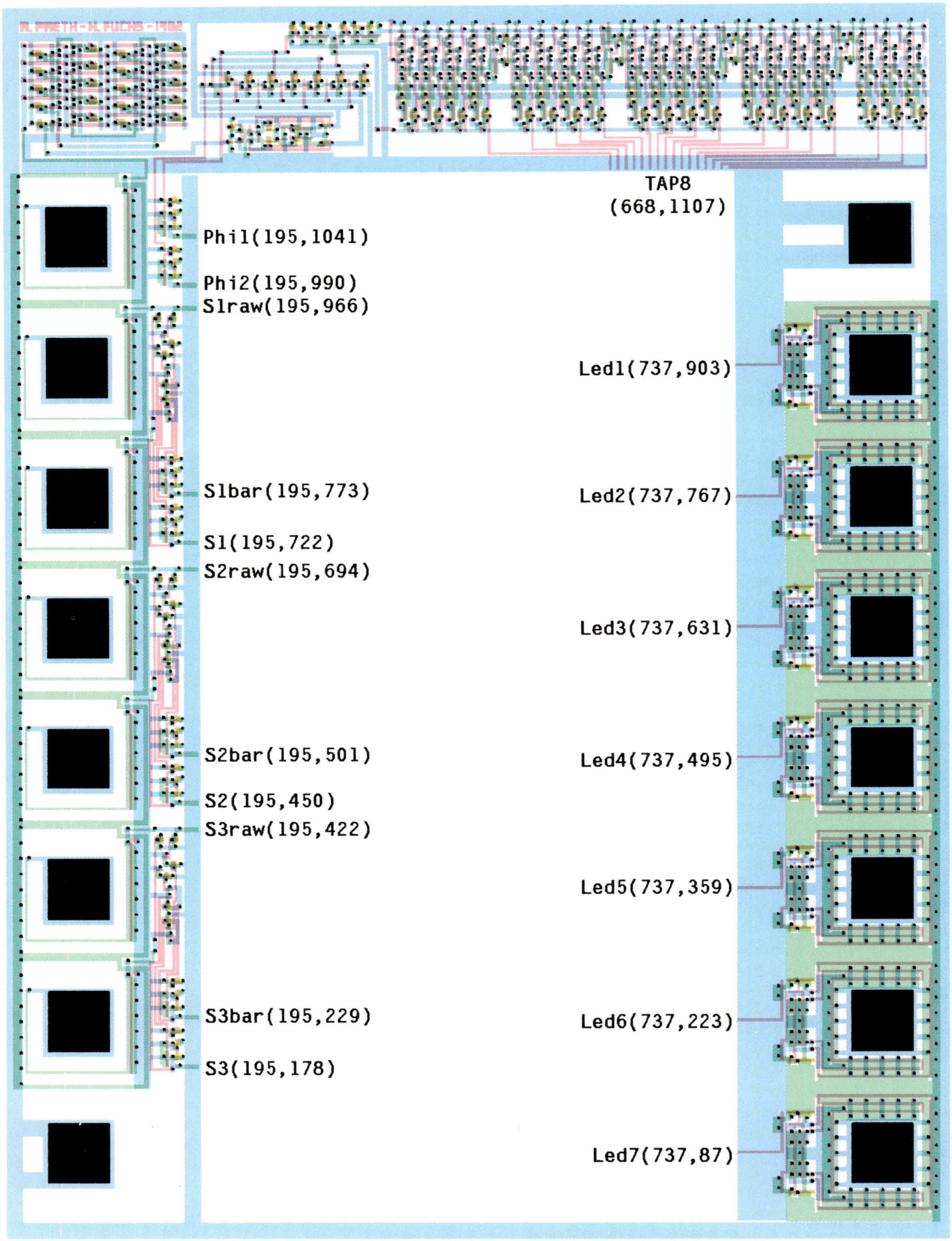
TAP8
(668,1107)
Phi1(195,1041)
Phi2(195,990)
S1raw(195,966)
S1bar(195,773)
S1(195,722)
S2raw(195,694)
S2bar(195,501)
S2(195,450)
S3raw(195,422)
S3bar(195,229)
S3(195,178)
Led1(737,903)
Led2(737,767)
Led3(737,631)
Led4(737,495)
Led5(737,359)
Led6(737,223)
Led7(737,87)

NAME

BasicFrame—design context for simple VLSI projects

SYNOPSIS

BasicFrame provides pushbutton or toggleswitch input, LED output, and system clocking.

PROPERTIES

Size

542 x 1091 λ (user cavity)
950 x 1279 λ (system)

Interface

$\phi 1$	output,	super-buffered	[½ / ⅛]	□ *R*
$\phi 2$	output,	super-buffered	[½ / ⅛]	□ *R*
Siraw	output,	pad with pullup	[2 / —]	□ *R*
Sibar	output,	super-buffered	[½ / ⅛]	□ *R*
Si	output,	super-buffered	[½ / ⅛]	□ *R*
TAPj	output,	gate	[3 / 2⁄11]	□ *R*
LEDk	input,	4:1 ratio		

DC current

2.4 mA internal + 210 mA LEDs (worst case) @5.0 VDC.

DESCRIPTION

BasicFrame is an elementary design context, targeted at a novice attempting a first project in VLSI design. It is particularly useful in creating custom, hand-held games. The frame provides three sets of debounced output signals, seven input lines controlling LEDs, taps spanning a wide range of frequencies, and non-overlapping clocks. The frame is self-supporting; no additional solid-state components are needed to complete the system.

The project cavity is bounded by VDD to the left and bottom, and GND to the top and right. All interface signals appear flush with these power rails. Left edge outputs are of 4 λ diffusion, clock taps and pad inputs of 2 λ polysilicon. Switch signals are debounced, non-overlapping complementary signal pairs and "raw" pad signals. System clocks ($\phi 1$, $\phi 2$) are 25% duty-cycle and non-overlapping. TAP*j* are 50% duty-cycle taps, which span 16 octaves. TAP*j* oscillates at 2^j Hz and is a level restored logical signal. An external trimmer fine-tunes both the system clock and the octave taps. The lamp inputs control high-current inverting output pads, so asserting input line LED*k* grounds the LED cathode and lights the device.

This part uses no butting or buried contacts. It is pad bondable at values of both $\lambda = 2.0\ \mu$ and $\lambda = 2.5\ \mu$.

TESSELATION

Packs in 2x3 or 3x4 arrays to form multi-project chips.

ORIGIN

Xerox PARC (Al Paeth). University of North Carolina at Chapel Hill (Henry Fuchs).

STATUS

Tested, works.

SEE ALSO

A Basic Design Context for VLSI Systems, Xerox PARC Technical Report VLSI-83-xx (in press).

LAYOUT SCALE

150 λ per inch.

CIF ID

944

Chip Bonding

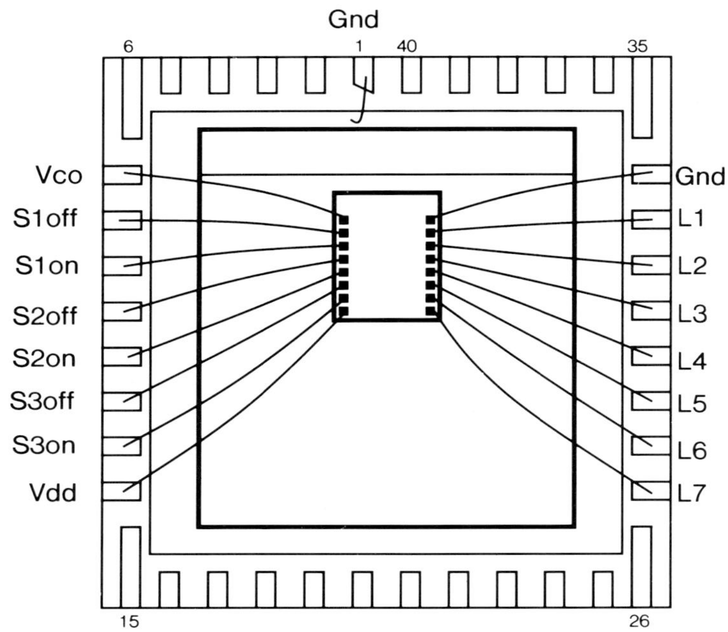

PC Layout

(scale 1.0x)

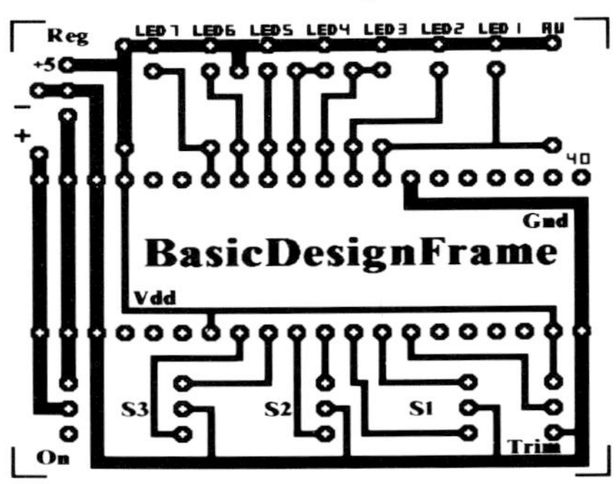

Chip Packaging

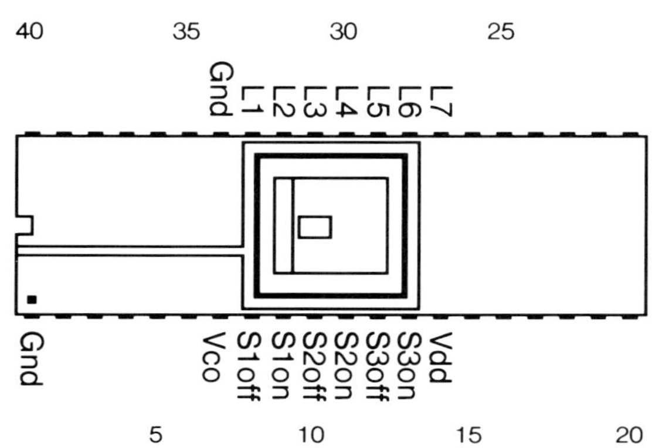

NAME

BasicFramePC—basic design context P.C. board

SYNOPSIS

BasicFramePC is a P.C. board layout that connects a *BasicFrame* chip and external components together in a system.

PROPERTIES

Size

2.1" x 1.8" (2100 x 1800 λ, λ = 254.00 microns =10 mils)

Interface

VCO	input,	Gnd $\leq$ VCO $\leq$ Vdd,	VCO sinks < 10μA	
L*k*	output,	push-pull,	~35mA pulldown,	~25mA pullup
S*i*on	input,	switch,	on resistance must be < 10KΩ	
S*i*off	input,	switch,	on resistance must be < 10KΩ	

DC current

2.4 mA internal + 210 mA LEDs (worst case) @5.0 VDC.

DESCRIPTION

This companion PC board provides a second (board-level) interface between the chip and the user hardware. Left to the user is creating the silicon design, choosing toggle or push-button switches, and choosing a 7-segment LED display or single lamps of various colors. These, together with a plastic package, trim-pot, voltage regulator, and 9-volt transistor battery complete the entire system.

The metal layer defines P.C. board traces and the poly layer defines components. The wiring is single-sided and planar; no jumpers are necessary. The components are on a standard .1" grid, with λ = 10 mils. The board provides both electrical and mechanical support for the user hardware, and is mounted in a 3"x2"x1" box, with room left for a battery.

A spare set of output lugs are provided in parallel with signal L1, together with some unused board space, for adding a sub-miniature phone jack or other external output. This output can directly drive a small speaker or a TTL load for integration into a larger system.

ADDITIONAL PARTS

DESCRIPTION	MAKE
9V Transistor Battery	standard
9V Battery Clip	standard
40-pin IC socket	Mouser 562-ICI-640T
+5V Regulator	Motorola MC78M05CT, National 340-5, LM321, LM323
20K Trimpot	Bournes #3299V-20K, Beckmann 68WR, Spectrol 64W
Slide switch	Mouser 10SP001
Pushbutton switches	Alco TPF11-FG-PC
Toggle switches	Mouser 10TC410
Discrete LEDs	any miniature(T-1 package)
7-seg LED	Monsanto MAN-1 (common anode)

ORIGIN

Xerox PARC (A1 Paeth).

University of North Carolina at Chapel Hill (Henry Fuchs).

STATUS

Tested, works.

SEE ALSO

A Basic Design Context for VLSI Systems, Xerox PARC Technical Report VLSI-83-xx (in press).

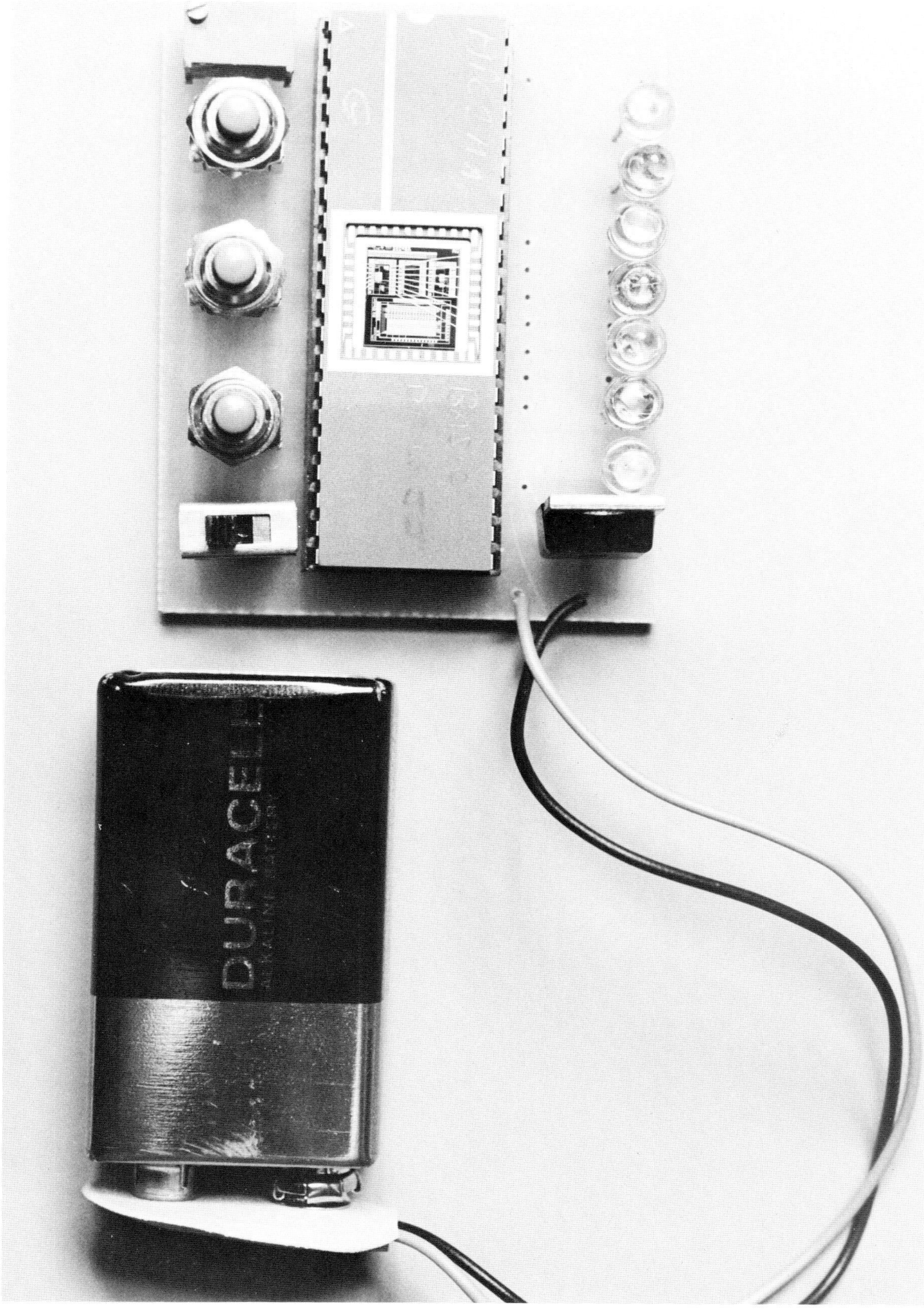
DURACELL

5 / Counters

NAME

intro—introduction to counter cells

SYNOPSIS

The cells described in this chapter can be used to construct binary counters.

DESCRIPTION

There are four main cells, *Cnt, CntRestore, CntUD* and *CntUDRestore*, which are designed to piece together to form up and up/down counters of arbitrary bit size.

Cnt contains the circuitry for one bit of a ripple-carry counter. *CntRestore* contains one bit of a counter and a buffer for restoring the signal levels in the carry chain of pass transistors. It should be put in every fourth bit. The increment count input should be put into CarryIn and $\overline{\text{CarryIn}}$ of the least significant bit.

Up/down counters can be built out of *CntUD* and *CntUDRestore*. They are used just like the regular counters except that two extra inputs, UP and $\overline{\text{UP}}$, must be supplied to indicate the direction of the count. When UP is high (and $\overline{\text{UP}}$ low), the cell counts upward. When UP is low (and $\overline{\text{UP}}$ high), the cell counts downward. *CntUDRestore* buffers the carry lines from *CntUD*, just as *CntRestore* does with *Cnt*, and should also be used every four bits.

The basic cell operates by either propagating or killing CarryIn and $\overline{\text{CarryIn}}$, depending on the "inputs" IN and $\overline{\text{IN}}$ that it gets as state information from the driver circuit in the lower part of the cell. With IN low (and $\overline{\text{IN}}$ high), there is a 1 square pass transistor from VDD to $\overline{\text{CarryOut}}$ and another from GND to CarryOut. Thus, $\overline{\text{CarryIn}}$ is high and CarryIn is pulled to ground. When IN is high (and $\overline{\text{IN}}$ is low), the CarryIn and $\overline{\text{CarryIn}}$ signals are passed through unchanged. Since the delay through the carry chain is ultimately quadratic, buffers need to be inserted to keep the delay reasonable.

A PLA-type structure can be attached on top if decoding of the counter's state is desired. The counter can be reset by attaching a transistor to ground at the bottom. This, in effect, makes the output inverter a NOR gate with the reset signal as one of its inputs. Proper ratios should be observed.

Cnt - 155

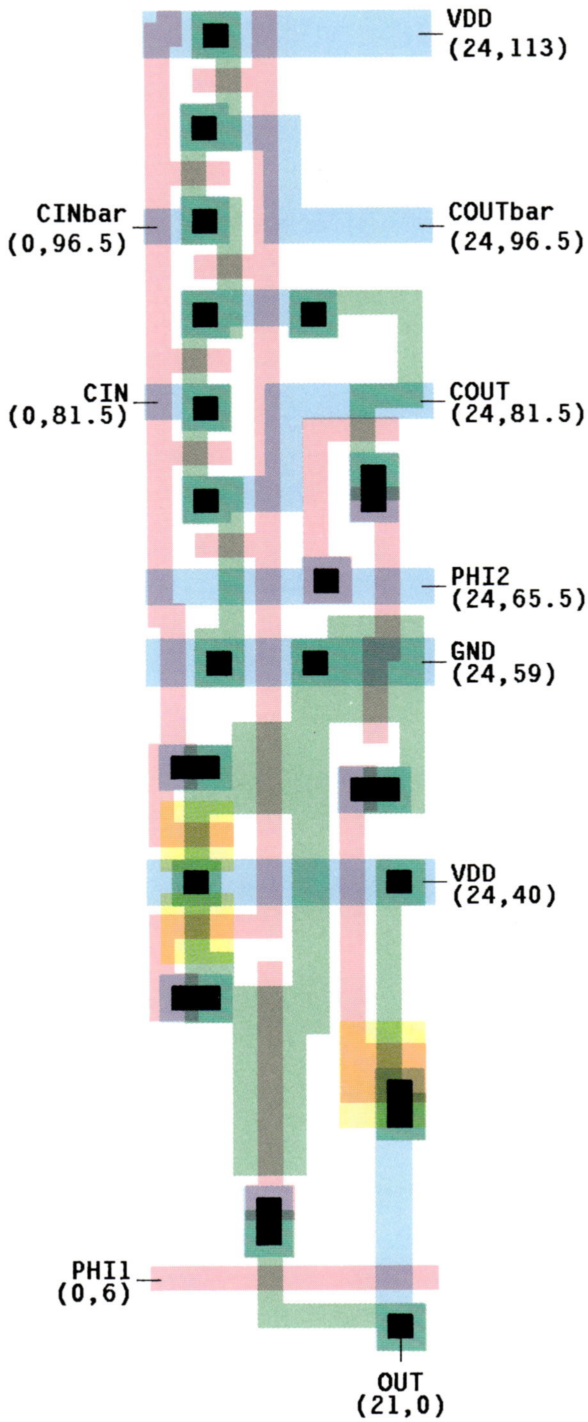
VDD
(24,113)
CINbar
(0,96.5)
COUTbar
(24,96.5)
CIN
(0,81.5)
COUT
(24,81.5)
PHI2
(24,65.5)
GND
(24,59)
VDD
(24,40)
PHI1
(0,6)
OUT
(21,0)

NAME
Cnt—counter cell

SYNOPSIS
Cnt contains the input driver, gating, and output driver for one bit of a ripple-carry counter.

PROPERTIES
Size

24 x 115 λ

Interface

Signal	Type	Kind	Load	Drive
CIN	input,	switched	1.5 □ C_g	
$\overline{CIN}$	input,	switched	1.5 □ C_g	
$\phi 1$	input,	switch control	2 □ C_g	
$\phi 2$	input,	switch control	3.25 □ C_g	
COUT	output,	switch-logic	2.85 □ C_g	
$\overline{COUT}$	output,	switch-logic	2.85 □ C_g	
OUT	output,	gate	11.75 □ C_g	[2 / ¼] □ *R*

DC current

0.3 mA.

DESCRIPTION
Cnt contains the circuitry for a single bit of a ripple-carry counter. The signal that increments the counter should be input as CIN **and** $\overline{\text{CIN}}$ in the least significant bit.

TESSELATION
Abuts horizontally, with no overlap.

ORIGIN
Stanford University (John Redford).

STATUS
Generally used.

SEE ALSO
CntRestore, CntUD, CntUDRestore.

LAYOUT SCALE
16 λ per inch.

CIF ID
155

LOGIC DIAGRAM

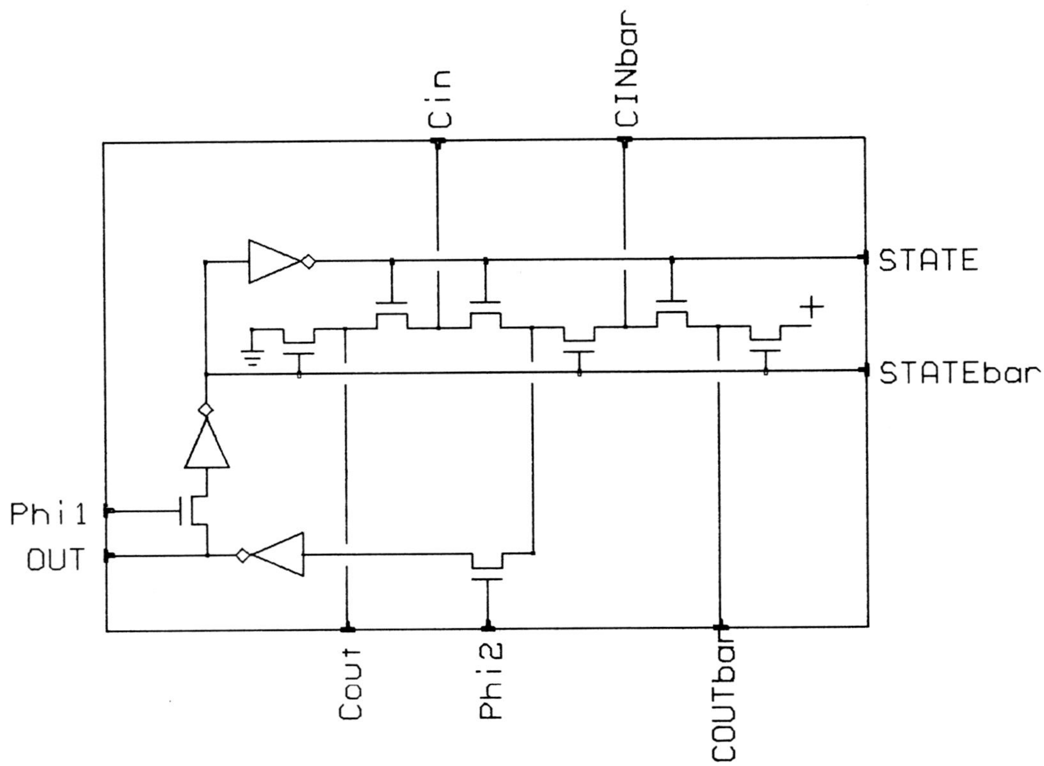

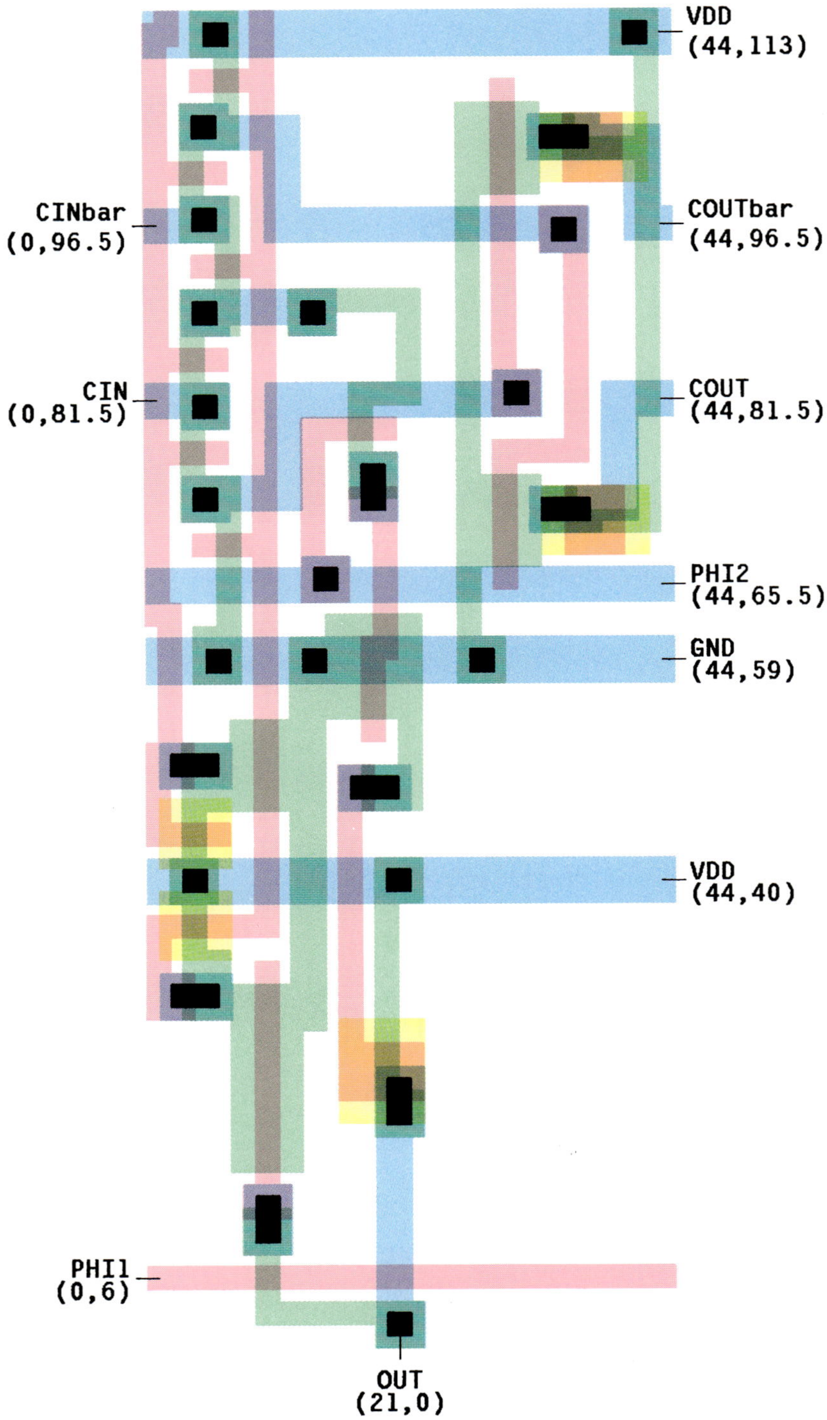
VDD
(44,113)
CINbar
(0,96.5)
COUTbar
(44,96.5)
CIN
(0,81.5)
COUT
(44,81.5)
PHI2
(44,65.5)
GND
(44,59)
VDD
(44,40)
PHI1
(0,6)
OUT
(21,0)

NAME

CntRestore—carry buffer

SYNOPSIS

CntRestore is used to buffer the carry lines.

PROPERTIES

Size

44 x 115 λ

Interface

CIN	input,	switched	1.5 □ C_g	
$\overline{CIN}$	input,	switched	1.5 □ C_g	
ϕ1	input,	switch control	3 □ C_g	
ϕ2	input,	switch control	4.5 □ C_g	
$\overline{COUT}$	output,	gate	5.5 □ C_g	[2 / ¼] □ *R*
$\overline{COUT}$	output,	gate	5.5 □ C_g	[2 / ¼] □ *R*
OUT	output,	gate	11.75 □ C_g	[2 / ¼] □ *R*

DC current

0.4 mA.

DESCRIPTION

The four counter cells can be used to construct both ordinary binary counters and up/down counters. *CntRestore* is used to buffer the CIN lines and contains the gating for one bit of a counter. It should be used every 4 bits or so.

TESSELATION

Abuts horizontally, without overlap.

ORIGIN

Stanford University (John Redford).

STATUS

Generally used.

SEE ALSO

Cnt, CntUD, CntUDRestore.

LAYOUT SCALE

16 λ per inch.

CIF ID

162

LOGIC DIAGRAM

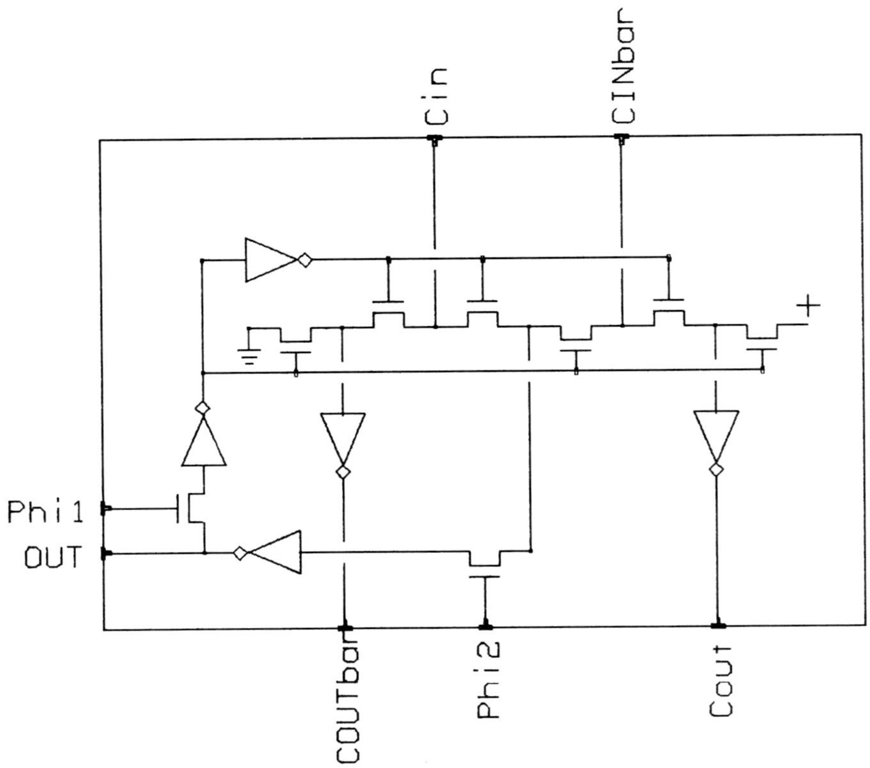

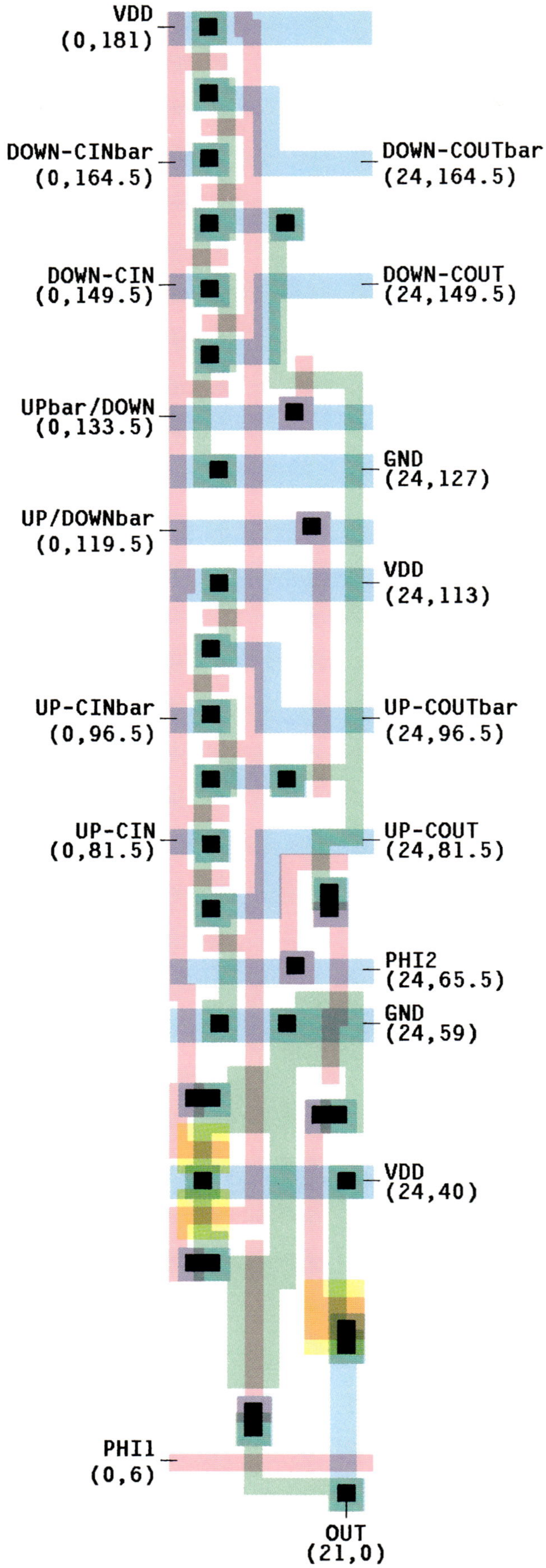
VDD
(0,181)
DOWN-CINbar
(0,164.5)
DOWN-COUTbar
(24,164.5)
DOWN-CIN
(0,149.5)
DOWN-COUT
(24,149.5)
UPbar/DOWN
(0,133.5)
GND
(24,127)
UP/DOWNbar
(0,119.5)
VDD
(24,113)
UP-CINbar
(0,96.5)
UP-COUTbar
(24,96.5)
UP-CIN
(0,81.5)
UP-COUT
(24,81.5)
PHI2
(24,65.5)
GND
(24,59)
VDD
(24,40)
PHI1
(0,6)
OUT
(21,0)

NAME

CntUD—up/down counter cell

SYNOPSIS

CntUD contains the input driver, gating, up/down control logic, and output driver for one bit of an up/down ripple carry counter.

PROPERTIES

Size

24 x 183 λ

Interface

CIN and COUT lines are all switched logic with about 1 □ C_g of capacitance. The control inputs $\phi 1$ and $\phi 2$ and UP and $\overline{\text{UP}}$ are all switch control with about 2 □ C_g of capacitance.

OUT	output, gate	4 □ C_g	[2 / ¼] □ *R*

DC current

0.3 mA.

Performance

An eight bit version operated over a range of 3 Hz to 6 MHz.

DESCRIPTION

Up/down counters can be made out of *CntUD*, *CntUDRestore*. *CntUD* is used exactly like *Cnt* except that two extra inputs, UP and $\overline{\text{UP}}$, must be supplied to determine whether the counter counts up or down: with UP high ($\overline{\text{UP}}$ low) the cell counts up; with UP low ($\overline{\text{UP}}$ high) the cell counts down. The signal that increments the counter should be input to UP-CIN and $\overline{\text{UP-CIN}}$ when UP is high, and to DOWN-CIN and $\overline{\text{DOWN-CIN}}$ when UP is low.

TESSELATION

Abuts horizontally, without overlap.

ORIGIN

Stanford University (John Redford).

STATUS

Generally used.

SEE ALSO

Cnt, CntUDRestore.

LAYOUT SCALE

20 λ per inch.

CIF ID

161

LOGIC DIAGRAM

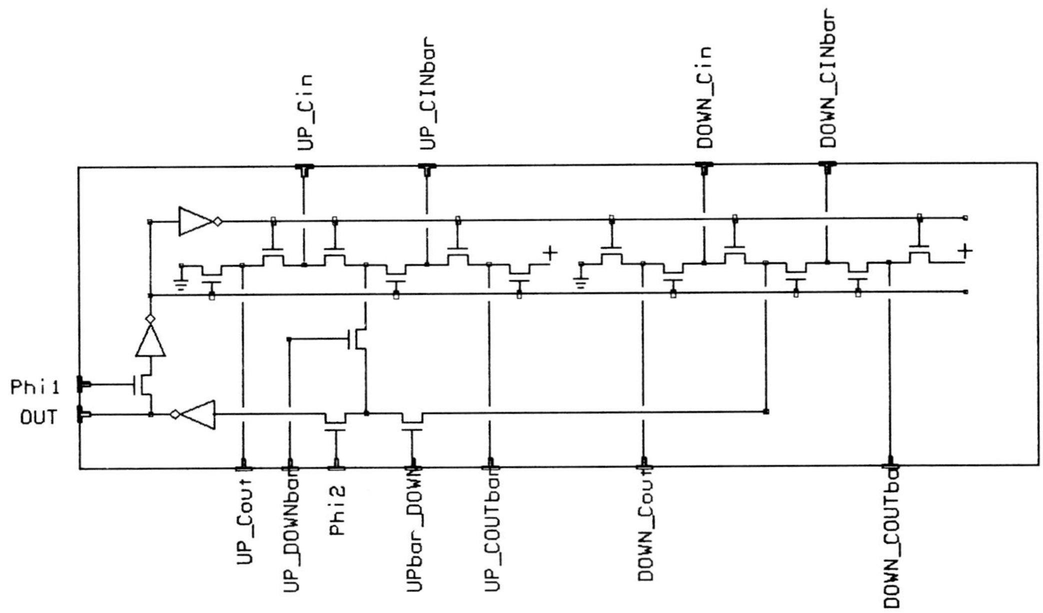

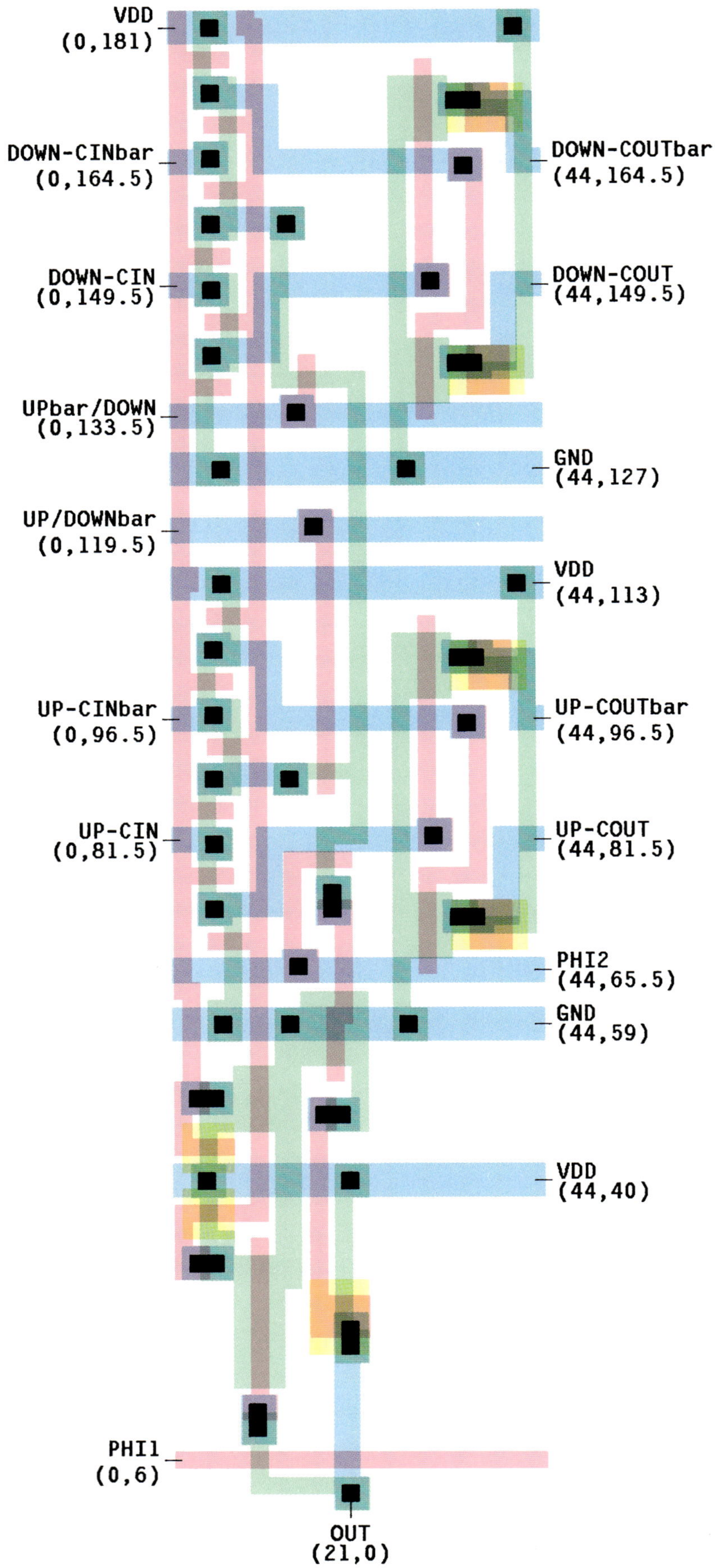
VDD
(0,181)
DOWN-CINbar
(0,164.5)
DOWN-COUTbar
(44,164.5)
DOWN-CIN
(0,149.5)
DOWN-COUT
(44,149.5)
UPbar/DOWN
(0,133.5)
GND
(44,127)
UP/DOWNbar
(0,119.5)
VDD
(44,113)
UP-CINbar
(0,96.5)
UP-COUTbar
(44,96.5)
UP-CIN
(0,81.5)
UP-COUT
(44,81.5)
PHI2
(44,65.5)
GND
(44,59)
VDD
(44,40)
PHI1
(0,6)
OUT
(21,0)

NAME

CntUDRestore—buffered up/down counter cell

SYNOPSIS

CntUDRestore is used to buffer the carry lines between bits in an up/down counter.

PROPERTIES

Size

44 x 183 λ

Interface

Like CntUD, except that the carry-out lines are all like:

COUT output, gate 5.5 □ C_g [2 / ¼] □ *R*

DC current

0.5 mA.

Performance

An eight bit version operated over a range of 3 Hz to 6 MHz.

DESCRIPTION

CntUDRestore and *CntUD* can be used to construct up/down binary counters. *CntUDRestore* buffers the carry lines from *CntUD*, just as *CntRestore* does with *Cnt*. With *CntUDRestore*, however, there are true and complement lines for UP-CIN and DOWN-CIN. As with *CntRestore*, *CntUDRestore* should be used every four bits to keep the propagation delay reasonable.

TESSELATION

Abuts horizontally, with no overlap.

ORIGIN

Stanford University (John Redford).

STATUS

Generally used.

SEE ALSO

CntUD, CntRestore.

LAYOUT SCALE

20 λ per inch.

CIF ID

163

LOGIC DIAGRAM

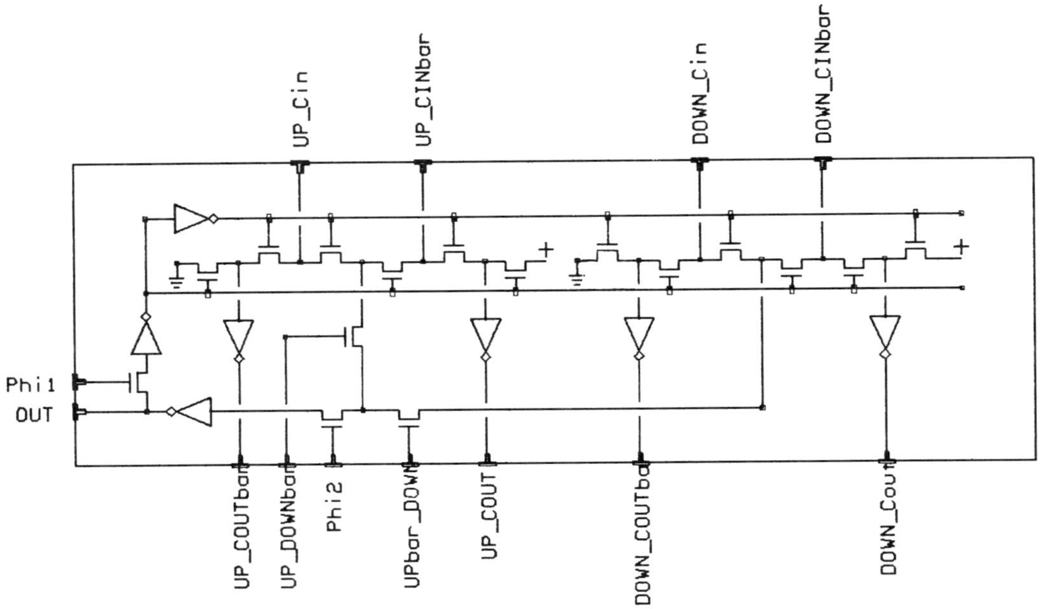

6 / Memory

An example memory subsystem

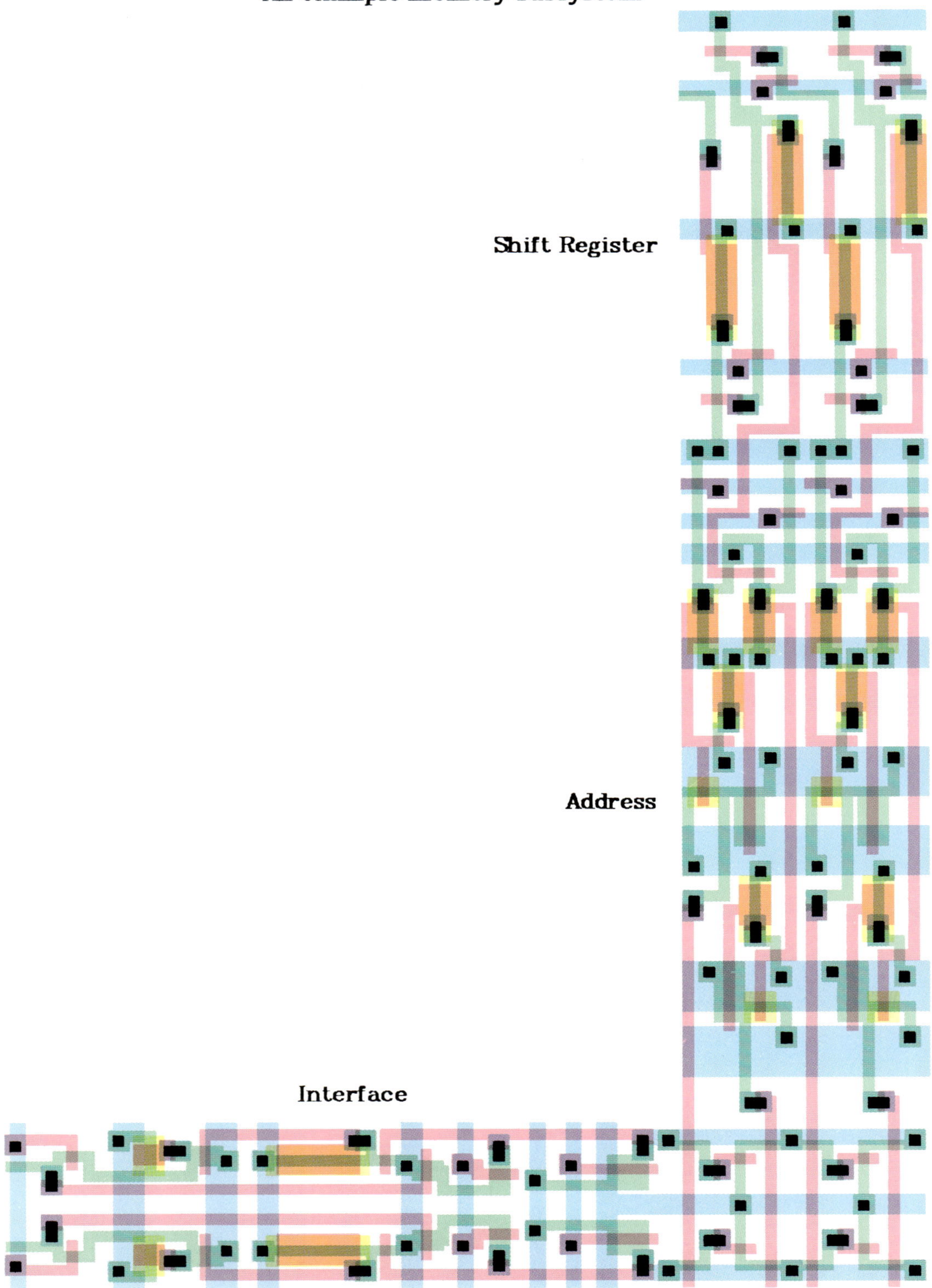

NAME

intro—introduction to 3-transistor memory subsystem

SYNOPSIS

This chapter describes a set of cells that can be used to construct a single-bus, precharged Random Access Memory using the 3-T RAM cells.

DESCRIPTION

The RAM subsystem is intended to provide a compact, dynamic, random-access memory. The most important cell herein is, of course, the 3-T RAM cell itself. In addition, there is an *Interface* to link the array of memory cells with the outside world (by providing storage and inversion), and an *Address* cell to drive the appropriate Read and Write signals to the memory elements. Finally, there is a simple shift register that is useful when generating sequential addresses.

LOGIC DIAGRAM

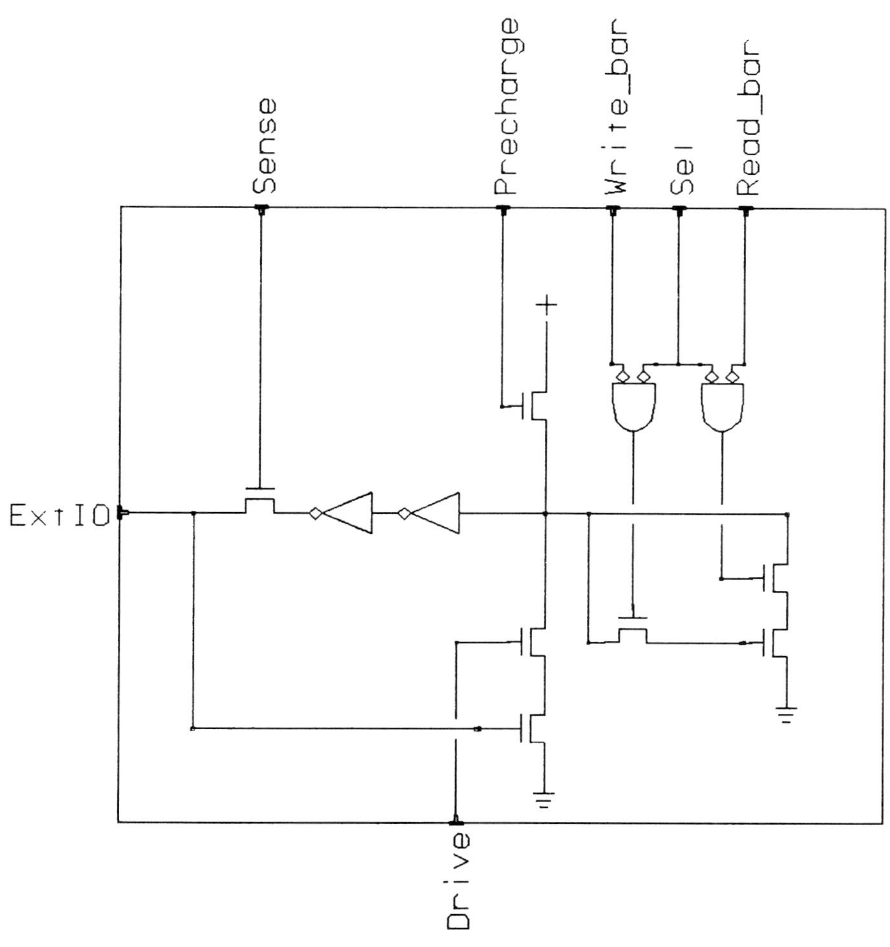

OVERVIEW

The overall scheme is as shown in the figure on the previous page. Appropriate clocking is:

Signal	Clocking type
Precharge	qϕ1
Read	qϕ2 or sϕ2
Write	qϕ2
Sel	qϕ2 or sϕ2
Sense	qϕ2
Drive	qϕ2
Extio(read)	sϕ1
Extio(write)	sϕ2

(The notation of qϕ_1 means a qualified ϕ1 clock; sϕ1, a data signal stable throughout ϕ11; and vϕ1, a data signal valid by the falling edge of ϕ1.)

This "safe" clocking scheme is not completely safe: we have assumed that the capacitance of the bit line is much greater than the capacitance of the node between the Drive- and EXTio-controlled pulldown transistors in the interface cell.

No addressing scheme has been built into these cells. It has, however, been recognized that most efficient addressing schemes would do part of their decoding on the input and part on the output side of the memory. While this poses no problem when reading, to write we will need to read out the entire column, modify the one bit we are interested in, and restore the whole column. The facility for doing this buffering has been built into the *Interface* cell.

ORIGIN

Stanford University (Rob Mathews, John Newkirk, Tim Saxe).

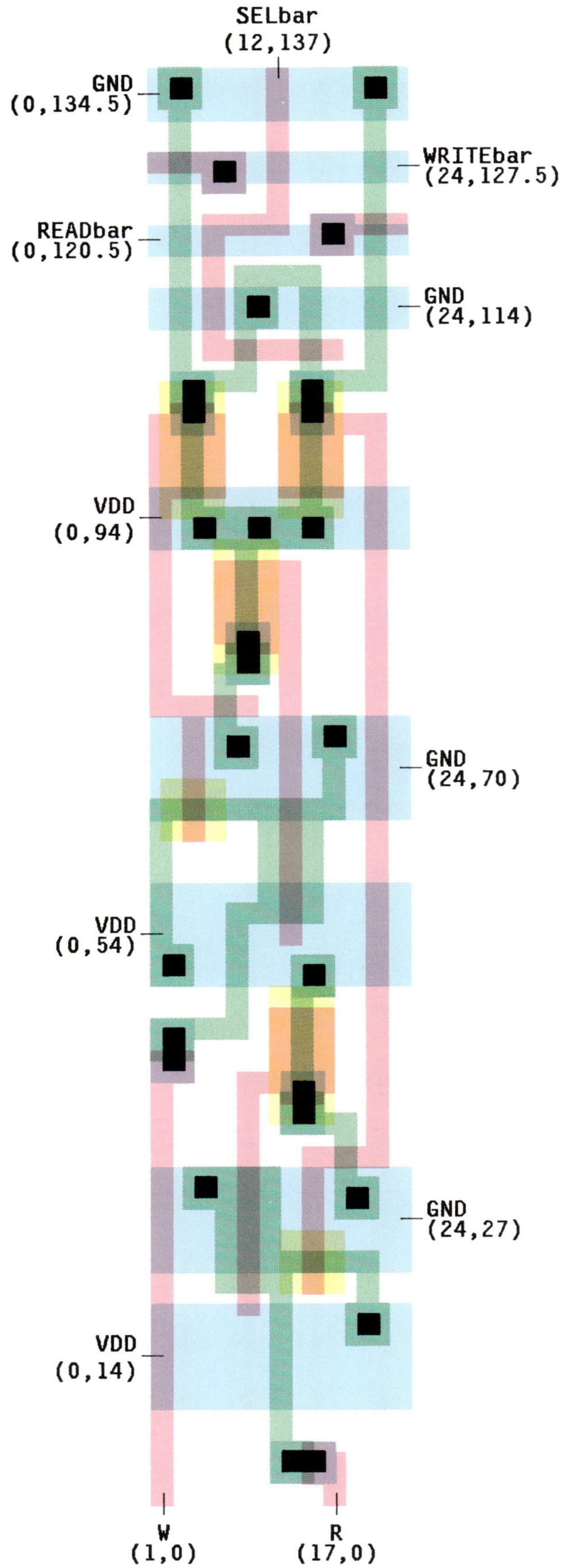
SELbar
(12,137)
GND
(0,134.5)
WRITEbar
(24,127.5)
READbar
(0,120.5)
GND
(24,114)
VDD
(0,94)
GND
(24,70)
VDD
(0,54)
GND
(24,27)
VDD
(0,14)
W
(1,0)
R
(17,0)

NAME

Address—RAM read/write line driver

SYNOPSIS

Driver for the Read and Write control lines

PROPERTIES

Size

24 x 137λ

Interface

R	output,	super-buffered	8.2 □ C_g	[8 / 1] □ *R*
W	output,	super-buffered	4.9 □ C_g	[8 / 1] □ *R*
$\overline{Read}$	input,	switch control	2.5 □ C_g	
$\overline{Write}$	input,	switch control	2.5 □ C_g	
$\overline{Sel}$	input,	switch control	5.3 □ C_g	

DC Current

0.4 mA

DESCRIPTION

Address qualifies the $\overline{\text{Read}}$ and $\overline{\text{Write}}$ clocks with the $\overline{\text{Sel}}$ line and provides superbuffered drive for the R and W word lines crossing a memory array.

TESSELATION

Abuts horizontally with no overlap. Abuts *Ram* and *RamPair* with optional 0.5λ overlap. Note that there is a 5λ displacement between a row of *Address* cells and the *Ram* array it connects to.

LOGIC DIAGRAM

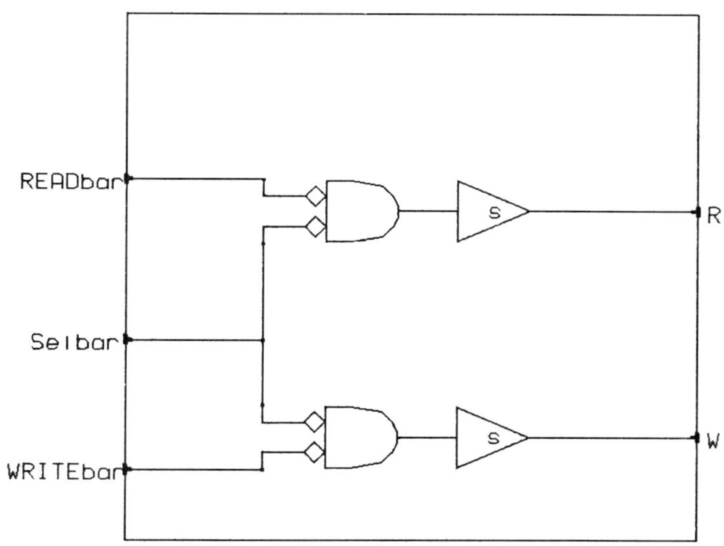

ORIGIN

Stanford University (Rob Mathews, John Newkirk, Tim Saxe; modified by H. V. Jagadish).

STATUS

Generally used.

SEE ALSO

Ram, Interface, ShiftReg, RamPair, InterfacePair.

LAYOUT SCALE

16 λ per inch.

CIF ID

503

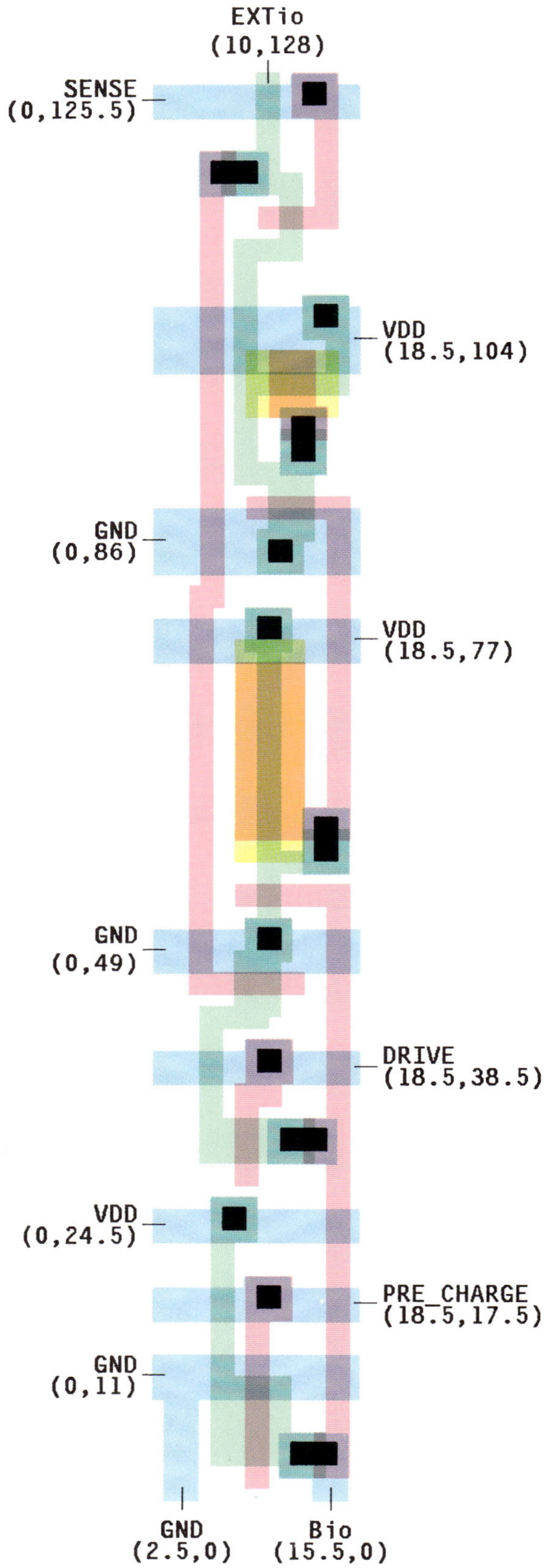
EXTio
(10,128)
SENSE
(0,125.5)
VDD
(18.5,104)
GND
(0,86)
VDD
(18.5,77)
GND
(0,49)
DRIVE
(18.5,38.5)
VDD
(0,24.5)
PRE_CHARGE
(18.5,17.5)
GND
(0,11)
GND
(2.5,0)
Bio
(15.5,0)

NAME

Interface—RAM interface cell

SYNOPSIS

This cell provides the circuitry necessary for the outside world to access the RAM cell.

PROPERTIES

Size

18.5 x 128 λ

Interface

EXTio	input,	8:1 ratio	8.3 □ C_g
EXTio	output,	switch-logic	5.7 □ C_g
Bio	input,	4:1 ratio	6.7 □ C_g
Bio	output,	precharged	6.7 □ C_g
SENSE	input,	switch control	2.9 □ C_g
DRIVE	input,	switch control	5.6 □ C_g

DC Current

.1 mA

DESCRIPTION

Interface precharges **Bio** during $\phi 1$. It reads during $\phi 2$ when **Sense** is clocked, and writes during $\phi 2$ when **Drive** is clocked. **EXTio** must be stable during $\phi 2$ for writing. Notice that the **EXTio** datum is inverted to compensate for the net inversion of *Ram*.

TESSELATION

Abuts *Ram* vertically with no overlap. (Horizontally same as Ram, except see *InterfacePair.)*

ORIGIN

Stanford University (Rob Mathews, John Newkirk, Tim Saxe; modified by H. V. Jagadish).

STATUS

Generally used.

SEE ALSO

Ram, Address, ShiftReg, InterfacePair.

LAYOUT SCALE

16 λ per inch.

CIF ID

502

LOGIC DIAGRAM

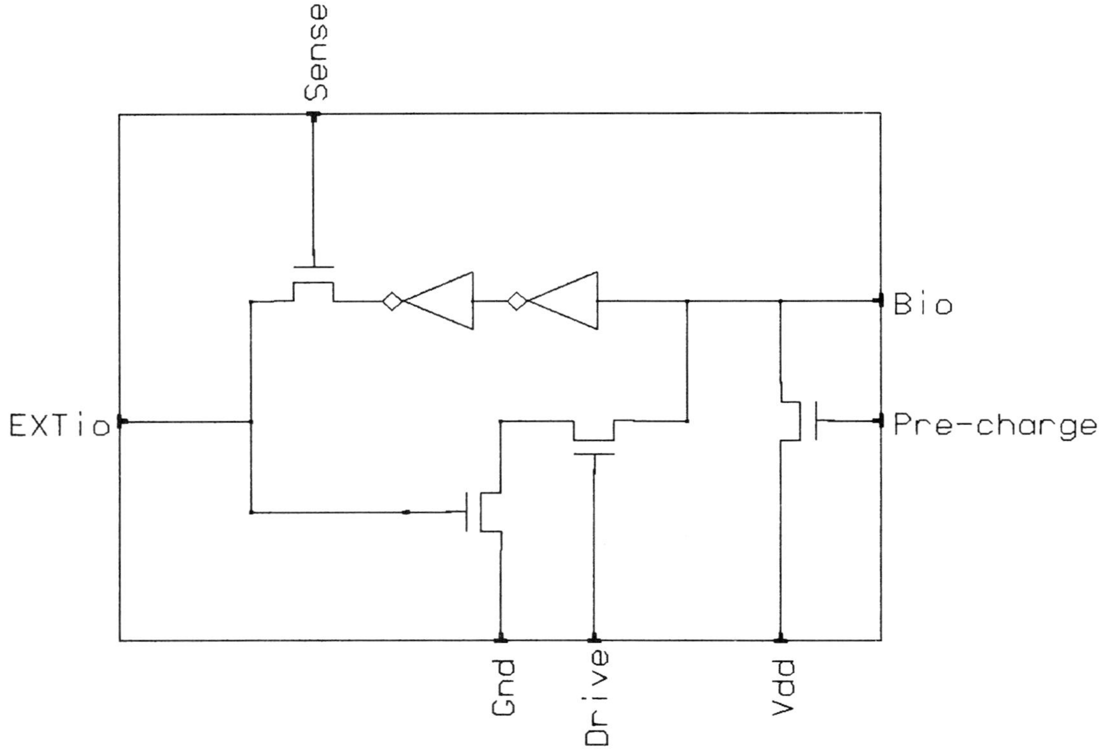

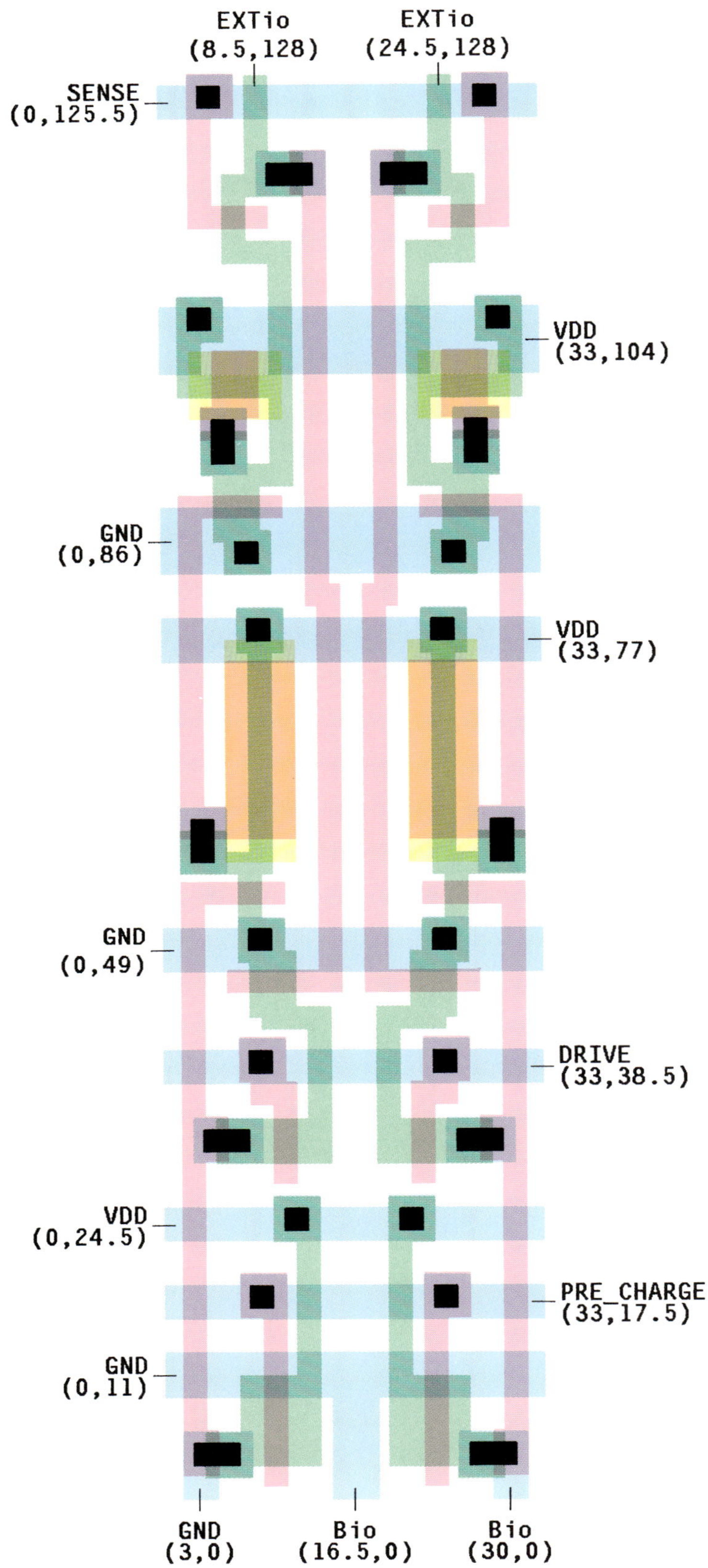
EXTio
(8.5,128)
EXTio
(24.5,128)
SENSE
(0,125.5)
VDD
(33,104)
GND
(0,86)
VDD
(33,77)
GND
(0,49)
DRIVE
(33,38.5)
VDD
(0,24.5)
PRE_CHARGE
(33,17.5)
GND
(0,11)
GND
(3,0)
Bio
(16.5,0)
Bio
(30,0)

NAME

InterfacePair—pair of memory interface cells

SYNOPSIS

A mirrored pair of *Interfaces.*

PROPERTIES

Size

33 x 128λ

Interface

EXTio	input,	8:1 ratio	8.3 □ C_g
EXTio	output,	switch-logic	5.7 □ C_g
Bio	input,	4:1 ratio	6.7 □ C_g
Bio	output,	precharged	6.7 □ C_g
SENSE	input,	switch control	5.5 □ C_g
DRIVE	input,	switch control	10.9 □ C_g

DC Current

.2 mA

DESCRIPTION

See *Interface.*

TESSELATION

Abuts horizontally with no overlap. Abuts the *RamPair* cell with no overlap.

ORIGIN

Stanford University (Rob Mathews, John Newkirk, Tim Saxe; modified by H. V. Jagadish).

STATUS

Generally used.

SEE ALSO

Rampair, Address, ShiftReg, Interface.

LAYOUT SCALE

16 λ per inch.

CIF ID

506

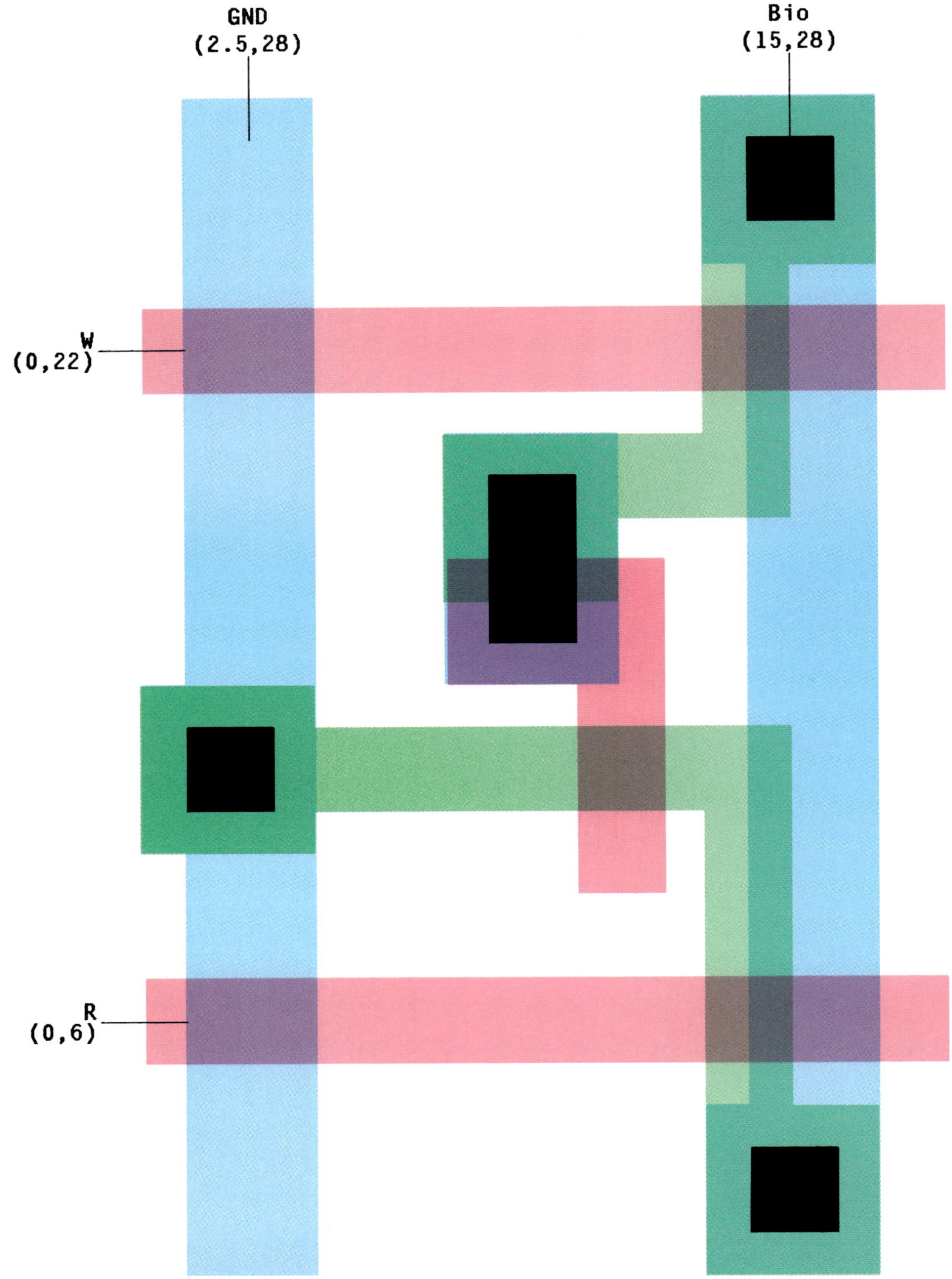
GND
(2.5,28)
Bio
(15,28)
W
(0,22)
R
(0,6)

NAME
Ram—3-transistor RAM cell

SYNOPSIS
The basic 3-T RAM cell.

PROPERTIES
Size

18.5 x 28 λ

Interface

R	input,	switch control	1.8 □ C_g
W	input,	switch control	1.8 □ C_g
Bio	input/output, switch		2.5 □ C_g

DC Current

None

DESCRIPTION
Ram is a 3-T RAM cell. To write into the cell, Bio is driven to the desired value and W is clocked. To read from it, Bio is precharged (or conceivably, pulled high with a static pullup), R is clocked, and the inverted value appears on Bio. Both reading and writing entail some charge-sharing from Bio, so be careful if you perform global operations on an array of cells, *e.g.*, writing 1s into all cells, or if the bit I/O line is very low-capacitance.

TESSELATION
Abuts vertically, with 4λ overlap. Forms mirrored pairs horizontally, overlapping 4λ to share ground; pairs then abut horizontally. (See *RamPair.*)

ORIGIN
Stanford University (Rob Mathews, John Newkirk, Tim Saxe; modified by H. V. Jagadish).

STATUS
Generally used.

SEE ALSO
Interface, Address, ShiftReg, RamPair.

LAYOUT SCALE
4 λ per inch.

CIF ID
501

LOGIC DIAGRAM

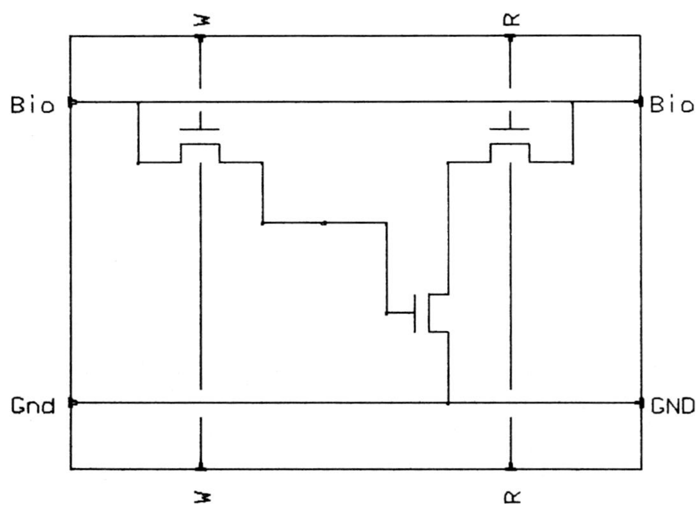

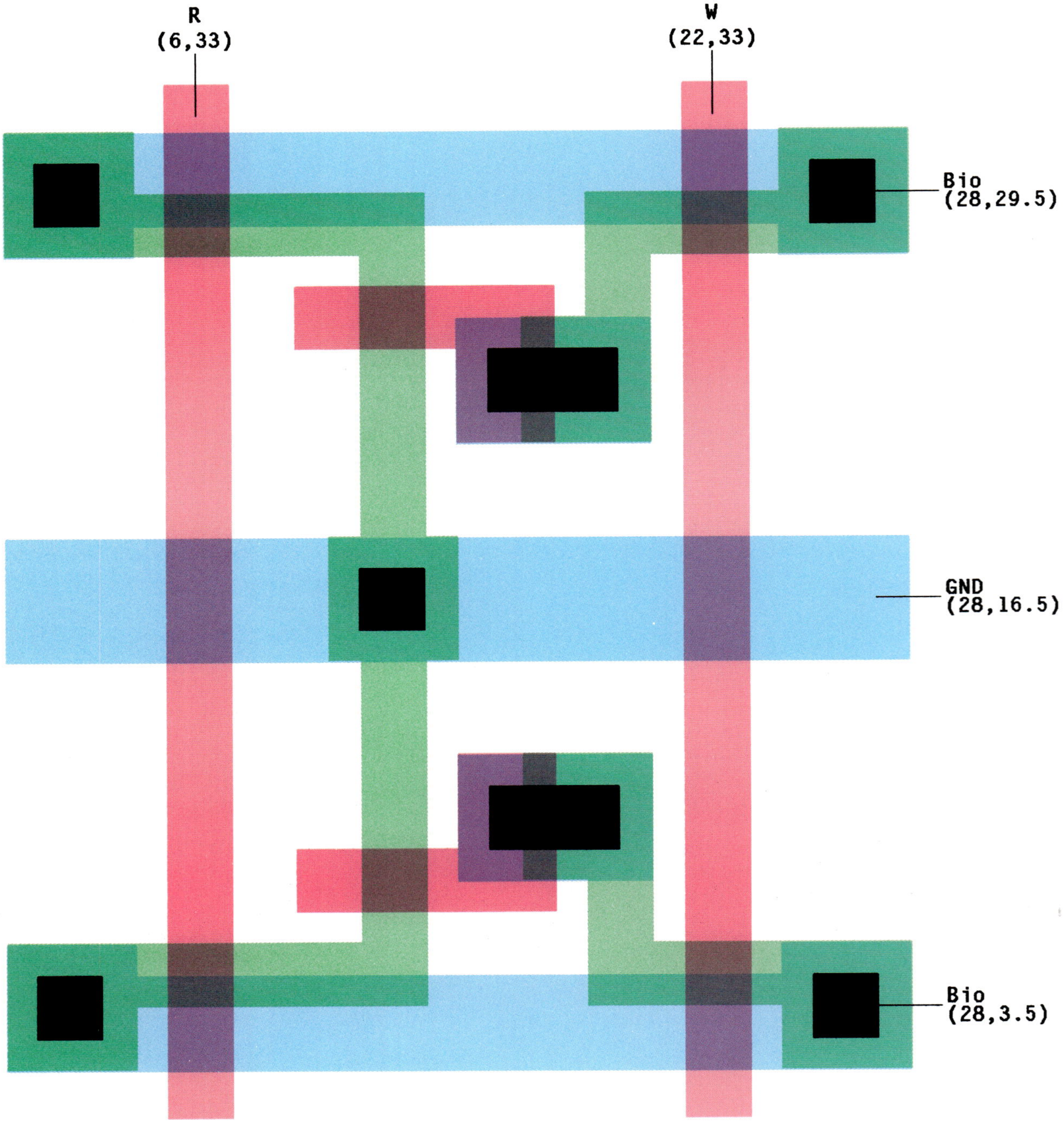
R
(6,33)
W
(22,33)
Bio
(28,29.5)
GND
(28,16.5)
Bio
(28,3.5)

NAME

RamPair—6-transistor RAM cell pair

SYNOPSIS

A mirrored pair of *Rams*.

PROPERTIES

Size

28 x 33λ

Interface

R	input,	switch control	3.5 □ C_g
W	input,	switch control	3.5 □ C_g
Bio	input/output, switch		

DC Current

None

DESCRIPTION

See *Ram*.

TESSELATION

Abuts horizontally, with 4λ overlap. Abuts vertically, with no overlap.

ORIGIN

Stanford University (Rob Mathews, John Newkirk, Tim Saxe; modified by H. V. Jagadish).

STATUS

Generally used.

SEE ALSO

Ram, Interface, Address, ShiftReg, InterfacePair.

LAYOUT SCALE

5 λ per inch.

CIF ID

505

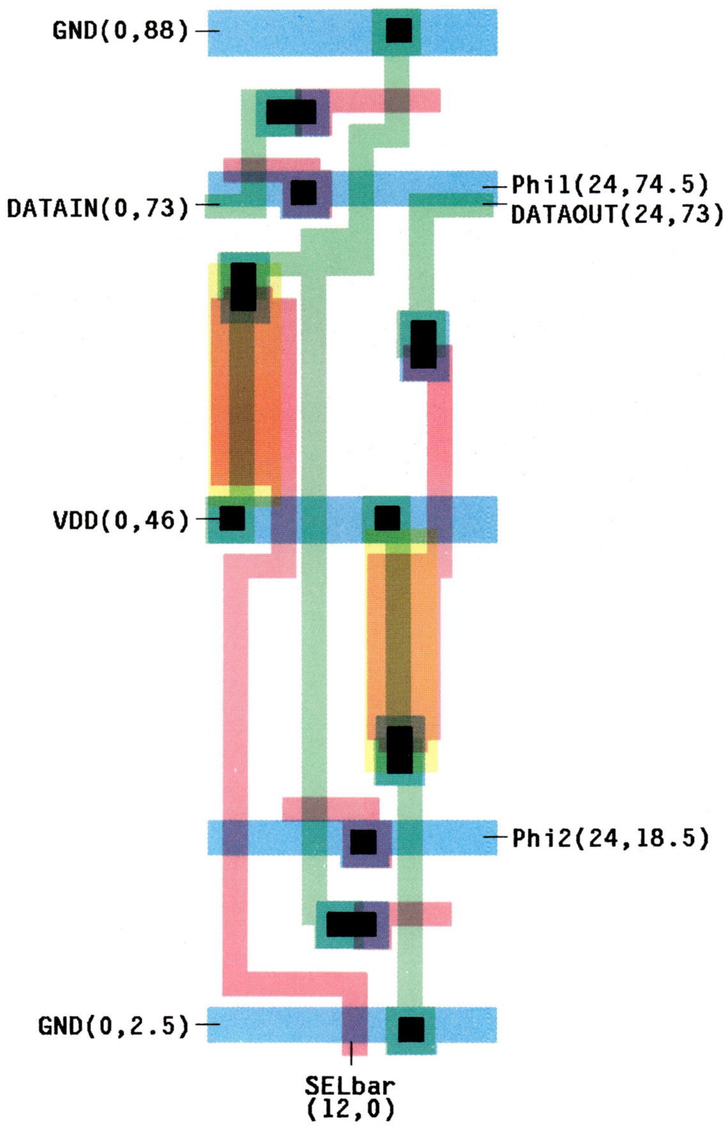
GND(0,88)
Phi1(24,74.5)
DATAIN(0,73)
DATAOUT(24,73)
VDD(0,46)
Phi2(24,18.5)
GND(0,2.5)
SELbar
(12,0)

NAME

ShReg—Shift Register

SYNOPSIS

A shift register pitch-matched to the 3-T RAM.

PROPERTIES

Size

24 x 90λ

Interface

Out	output,	gate	16.2 □ C_g	[8 / 1] □ *R*
In	input,	4:1 ratio	24 □ C_g	
$\overline{Sel}$	output,	gate	24 □ C_g	[8 / 1] □ *R*
$\phi 1$	input,	switch control	2.8 □ C_g	
$\phi 2$	input,	switch control	2.8 □ C_g	

DC Current

0.1 mA

DESCRIPTION

ShReg is a simple shift register with an inverted output tap, $\overline{\text{Sel}}$. It has been pitch-matched to *Address* to permit a Ram array to be sequentially addressed, forming a power- and area-efficient shift-register array.

TESSELATION

Abuts horizontally, with no overlap. Abuts *Address* vertically, with a 4λ overlap.

LOGIC DIAGRAM

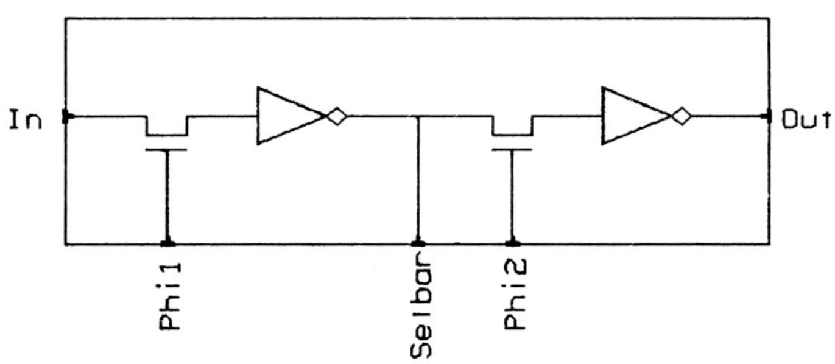

ORIGIN

Stanford University (Rob Mathews, John Newkirk, Tim Saxe; modified by H. V. Jagadish).

STATUS

Tested.

SEE ALSO

Ram, Interface, Address, RamPair, InterfacePair.

LAYOUT SCALE

16 λ per inch.

CIF ID

504

7 / Pads—Input/Output

NAME

intro—introduction to bonding pads

SYNOPSIS

This chapter describes a variety of bonding pads.

DESCRIPTION

These pads are exactly 100 λ wide, so that automatic probing on metric machines is easy. They have wide (16 λ) GND lines, which makes it possible to put up to fourteen *Tristate* pads in a row, or to use this GND line to conveniently supply a connection to some of the components on chip. The VDD lines can be widened by running more metal around the outside of the chip.

The pads described in this section are: *PadBlank, PadIn4, PadIn8, PadOut4, PadOut8, PadClkOut, PadTriState4, PadTriState8, PadClock, PadVdd, PadGnd,* and *PadMux*. A suffix of "4" (as in 4:1 ratio) means that the pad expects gate-logic (restored) levels if it is to function properly; "8" mean switch-logic levels are acceptable. The clock pad, *PadClk*, generates true versions of $\phi 1$ and $\phi 2$.

Be reasonable in your placement of pads. The chip carrier with *n* pins typically has a square cavity with *n/4* pin connections on each side. If you put more than *n/4* pads on one edge of the chip, bonding wires may leave some pads at such an obtuse angle that they will be very prone to shorting. Pads should only be placed around the perimeter of a design, since interior pads are hard to bond.

For a metal bonding blank of 130 μ (see PadBlank), there must be a separation of 20 μ between the bonding blank and any other unconnected metal to assure that no shorts occur when the pad is bonded.

Pads intrinsically do not scale well. These were designed for $\lambda = 2.5\mu$, but have proved to be satifactory for $\lambda = 2.0\mu$. Bonding is more difficult, though—check with your fabrication service about the smaller λ.

GND
(0,98)
VDD
(0,4)

NAME

PadBlank—blank pad

SYNOPSIS

PadBlank is the standard bonding pad structure used by all other pads except the VDD pad.

PROPERTIES

Size

100 x 106 λ.

DESCRIPTION

PadBlank is used by all other pads except *PadVdd*. It consists of a 130 μ (53 λ) metal square with a 115 μ (46 λ) square overglassing window in a 100 by 106 λ square area. A 16 λ horizontal metal line forms the top edge; and 8 λ line forms the bottom edge. The bottom metal line is used for a VDD connection. The top metal line is used for ground.

TESSELATION

Abuts horizontally, with no overlap.

ORIGIN

Xerox PARC (Dick Lyon), modified at Stanford University (John Redford).

STATUS

Generally used.

LAYOUT SCALE

16 λ per inch.

CIF ID

202

BUGS

Not scalable; designed for $\lambda = 2.5\mu$.

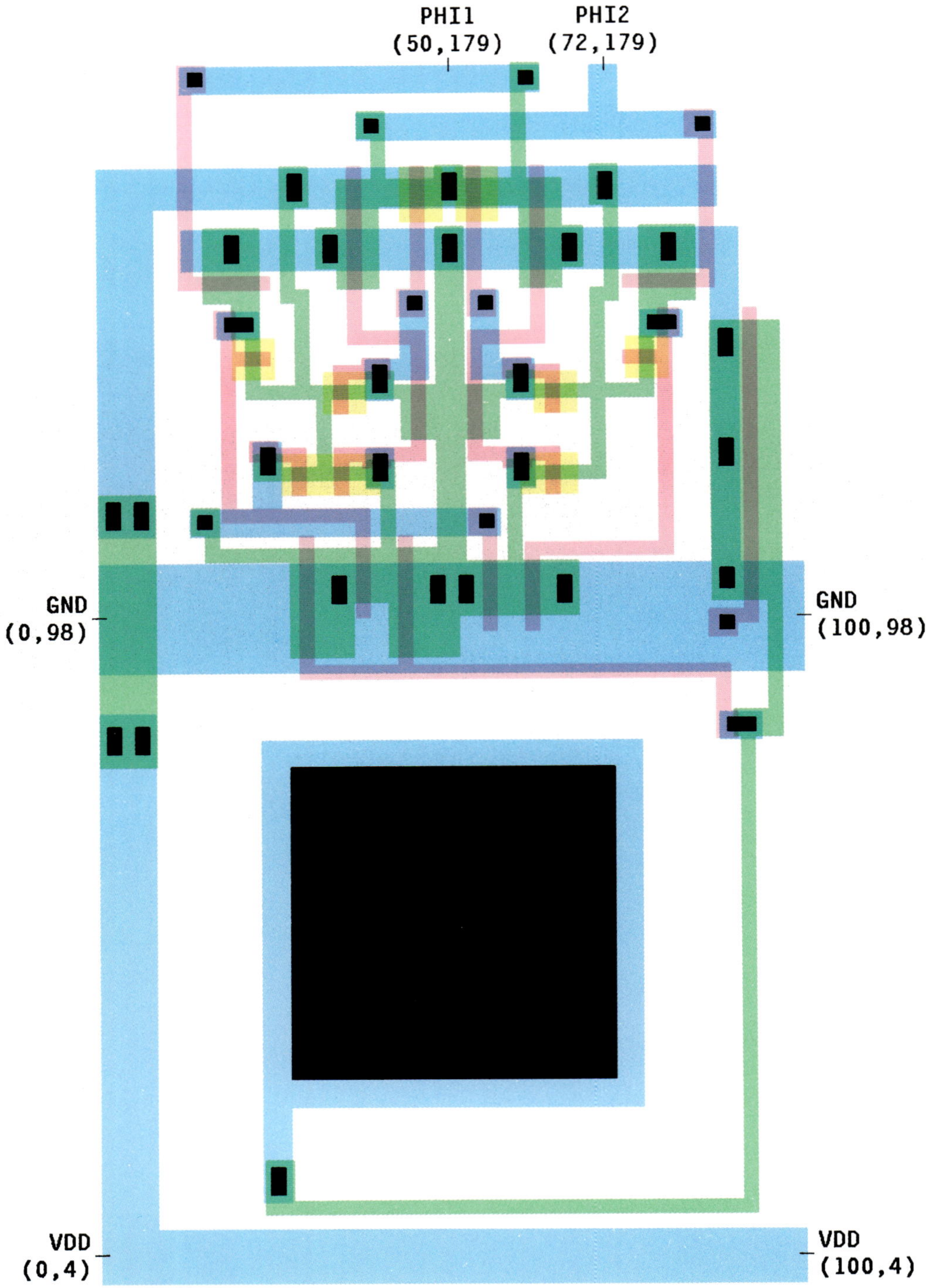
PHI1
(50,179)
PHI2
(72,179)
GND
(0,98)
GND
(100,98)
VDD
(0,4)
VDD
(100,4)

NAME

PadClock—clock pad

SYNOPSIS

PadClock generates a two-phase, nonoverlapping clock signal from a single-phase, TTL-compatible input.

PROPERTIES

Size

100 x 179 λ.

Interface

$\phi 1$	output,	super-buffered	15 □ C_g	[1/8 / 1/8] □ *R*
$\phi 2$	output,	super-buffered	15 □ C_g	[1/8 / 1/8] □ *R*

DC current

1.2 mA (without load).

DESCRIPTION

PadClock converts its single-phase input ϕ into a 2-phase clock, where $\phi 1$ follows ϕ and $\phi 2$ follows $\overline{\phi}$. The two output phases are guaranteed never to overlap. These signals are driven by superbuffers for distribution around the chip.

NOTE: on a prototype chip, it is often best to bring the two clock phases into the project using separate pads to increase testing flexibility. As a case in point, this cell may still cause overlapping clock phases because of qualified clocks.

TESSELATION

Abuts horizontally, with no overlap.

ORIGIN

Xerox PARC (Dick Lyon), modified at Stanford University (John Redford, C.Y. Chu).

STATUS

Checked.

SEE ALSO

BUFFER

LAYOUT SCALE

25 λ per inch.

CIF ID

209

BUGS

Not scalable; designed for $\lambda = 2.5\mu$.

LOGIC DIAGRAM

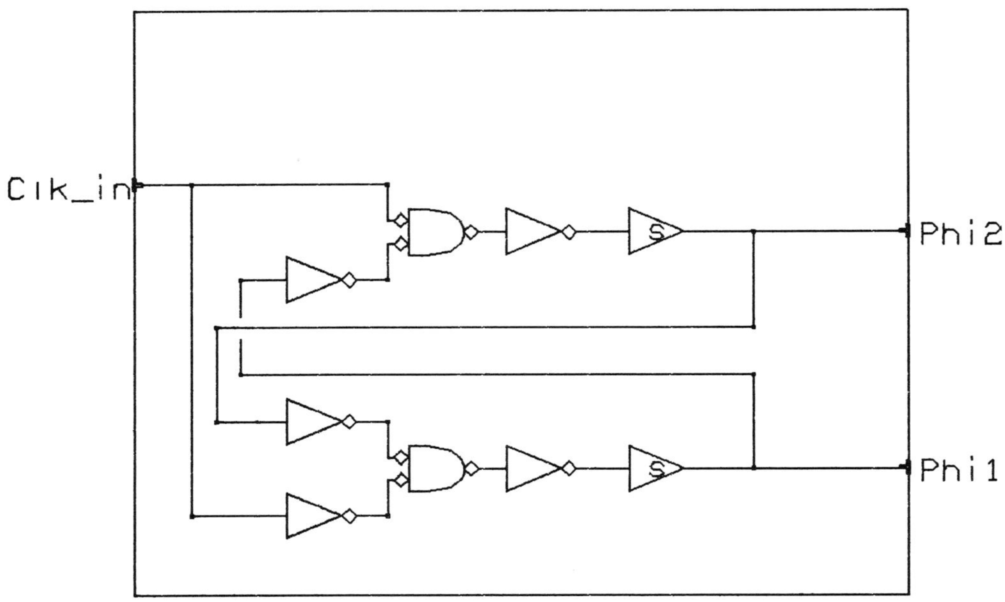

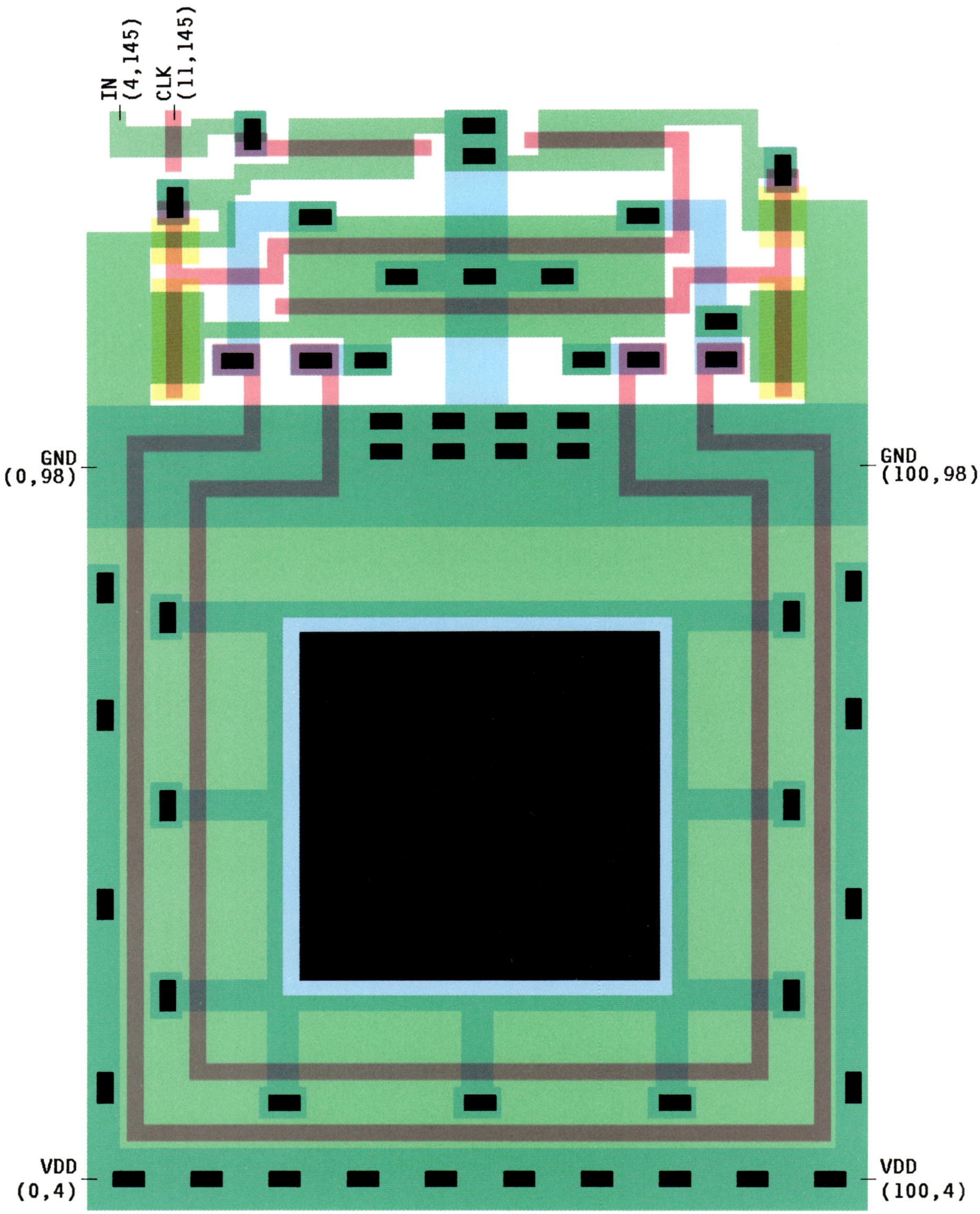
IN
(4,145)
CLK
(11,145)
GND
(0,98)
GND
(100,98)
VDD
(0,4)
VDD
(100,4)

NAME

PadClockOut—clocked output pad

SYNOPSIS

PadClockOut is a standard output pad with a clocked input.

PROPERTIES

Size

100 x 145 λ.

Interface

IN	input,	switched	10 □ C_g
CLK	input,	switch control	2 □ C_g

DC current

1.6 mA (without external load).

DESCRIPTION

Same as *PadOut4*, but with a clocked input.

TESSELATION

Abuts horizontally, with no overlap.

ORIGIN

Xerox PARC (Dick Lyon), modified at Stanford University (John Redford).

STATUS

Generally used.

SEE ALSO

PadOut4, PadTriState4.

LAYOUT SCALE

20 λ per inch.

CIF ID

210

BUGS

Not scalable; designed for $\lambda = 2.5\mu$.

LOGIC DIAGRAM

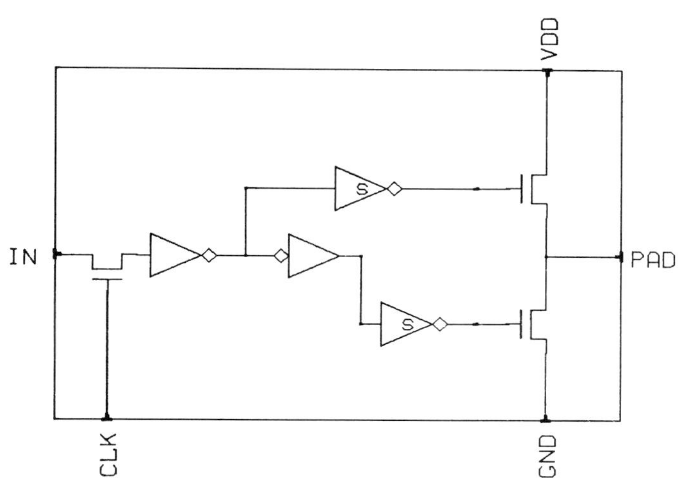

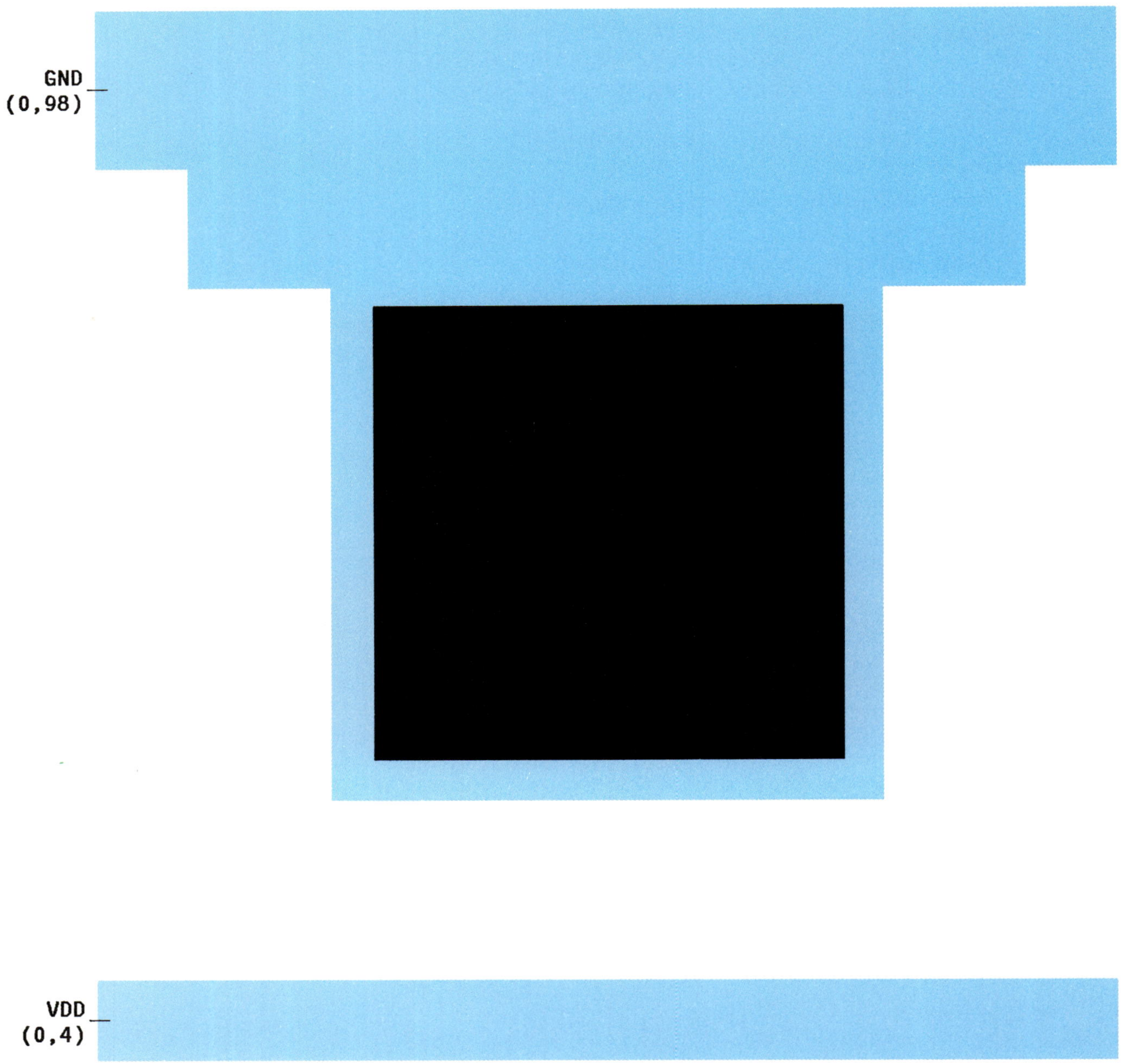
GND
(0,98)
VDD
(0,4)

NAME

PadGnd—GND pad

SYNOPSIS

PadGnd is a bonding pad for connecting to the inner, GND ring.

PROPERTIES

Size

100 x 106 λ.

DC current

maximum current in 16 λ metal is 40 mA (with $\lambda = 2.5\mu$).

DESCRIPTION

PadGnd has the standard pad configuration with a substantial metal strip connecting the bonding pad to the metal line which forms the inner GND ring of the chip.

ORIGIN

Xerox PARC (Dick Lyon), modified at Stanford University (John Redford).

STATUS

Generally used.

SEE ALSO

PadBlank.

LAYOUT SCALE

16 λ per inch.

CIF ID

206

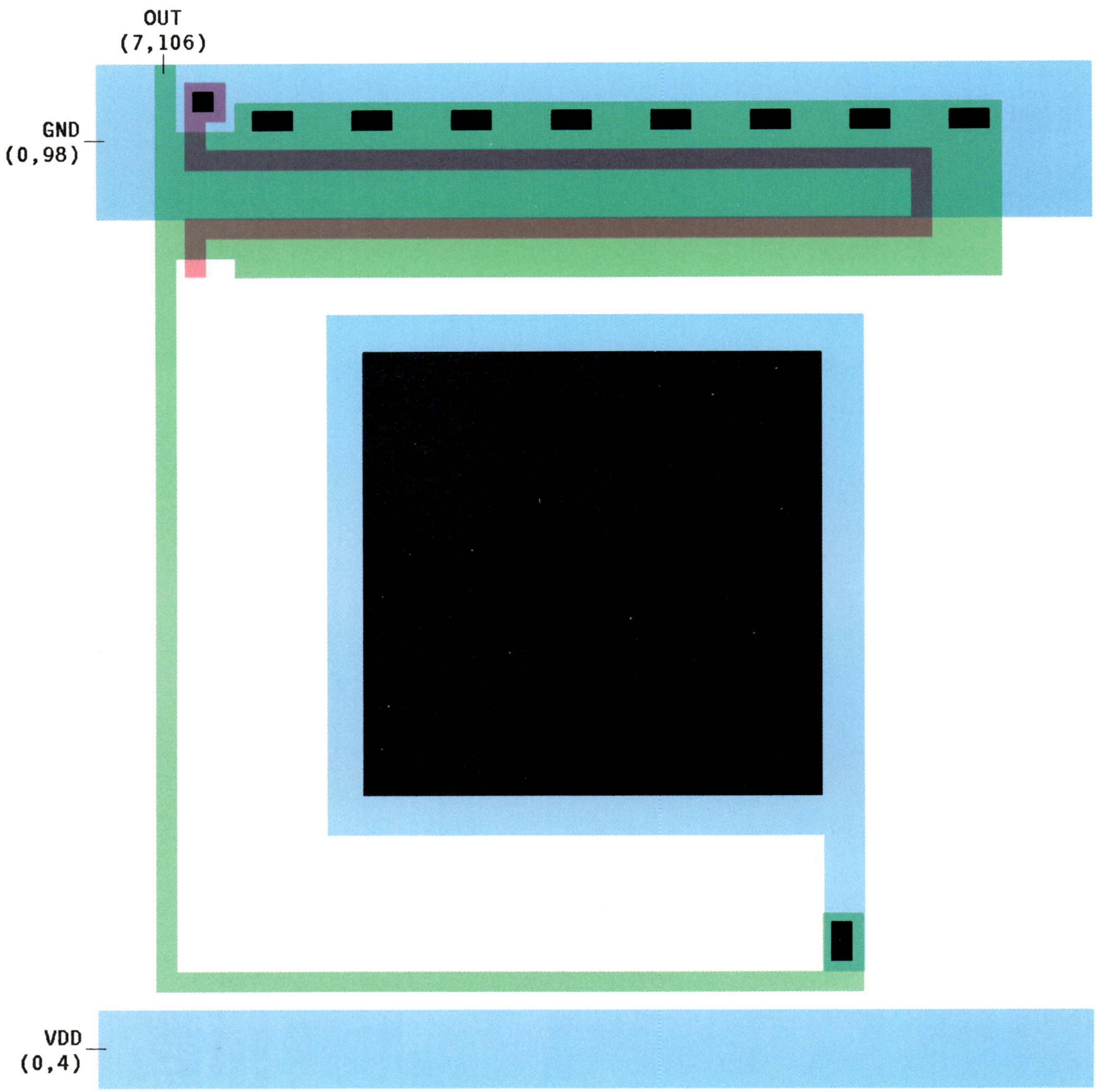
OUT
(7,106)
GND
(0,98)
VDD
(0,4)

NAME
PadIn4—input pad

SYNOPSIS
PadIn4 is the standard input pad.

PROPERTIES
Size
100 x 106 λ.

Interface
OUT output, switch-logic 101 □ C_g

DC current
none

DESCRIPTION
PadIn4 has the standard pad configuration with a lightning arrestor connected from the pad to the ground line. This pad does no level restoration, so if external inputs are to be TTL compatible, then this pad should only connect to gate logic with a ratio of *8:1;* it should not connect to switch logic.

NOTE: The signal path is in diffusion.

TESSELATION
Abuts horizontally, with no overlap.

ORIGIN
Xerox PARC (Dick Lyon), modified at Stanford University (John Redford).

STATUS
Generally used.

SEE ALSO
PadBlank, PadIn8.

LAYOUT SCALE
16 λ per inch.

CIF ID
205

BUGS
Not scalable; designed for $\lambda = 2.5\mu$.

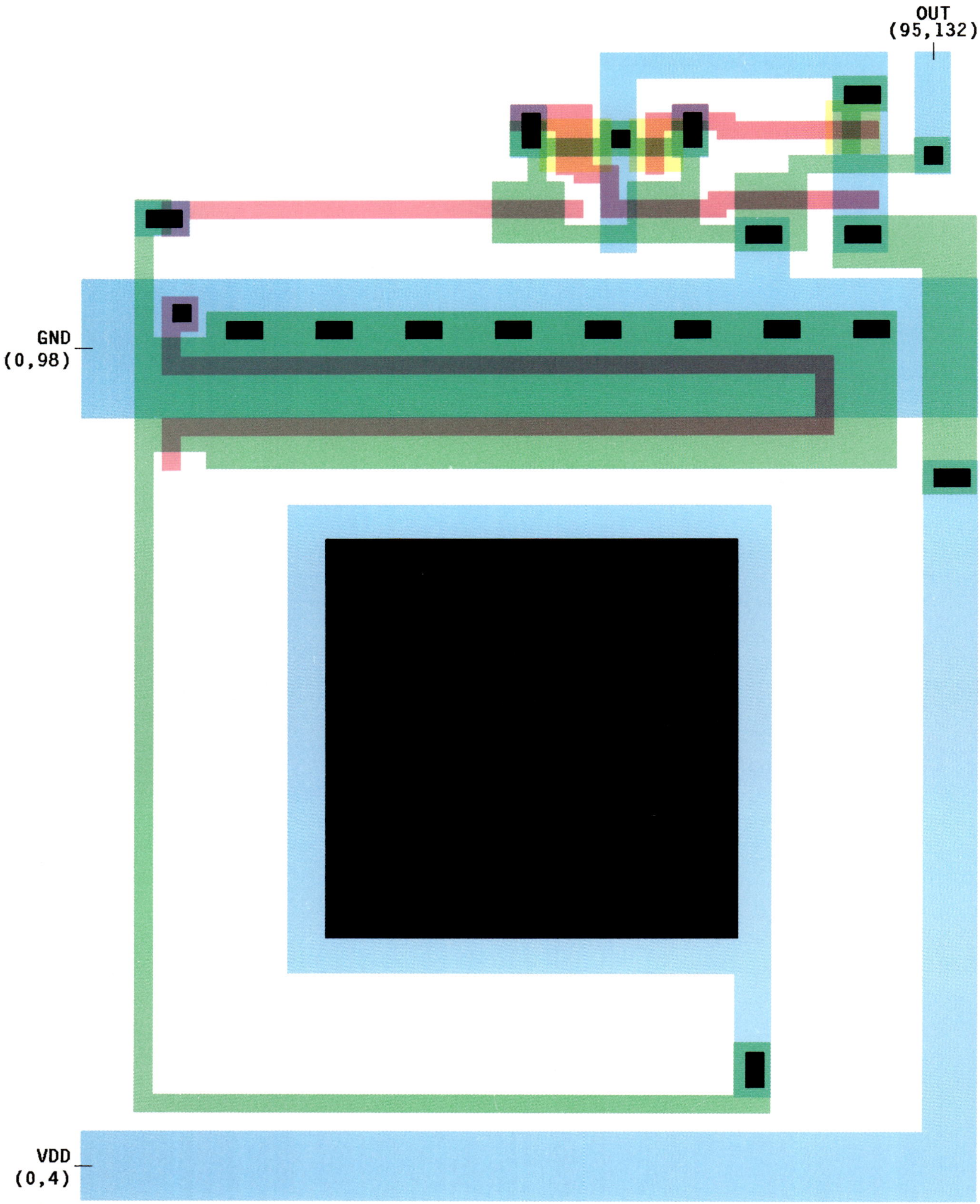
OUT
(95,132)
GND
(0,98)
VDD
(0,4)

NAME

PadIn8—superbuffered input pad

SYNOPSIS

PadIn8 is an input pad with a superbuffer output.

PROPERTIES

Size

100 x 132 λ.

Interface

OUT output, super-buffered 4.75 □ C_g [1 / ¼] □ *R*

DC Current

0.3 mA.

DESCRIPTION

PadIn8 has the standard pad configuration with a lightning arrestor connected from the pad to the ground line. In addition, it has a non-inverting superbuffer output stage, so that even with TTL-compatible signals driving the pad, the output can be used directly to drive switch logic.

TESSELATION

Abuts horizontally, with no overlap.

ORIGIN

Stanford University (John Redford).

STATUS

Generally used.

SEE ALSO

PadIn4, BUFFER.

LAYOUT SCALE

16 λ per inch.

CIF ID

280

BUGS

Not scalable; designed for λ=2.5μ.

LOGIC DIAGRAM

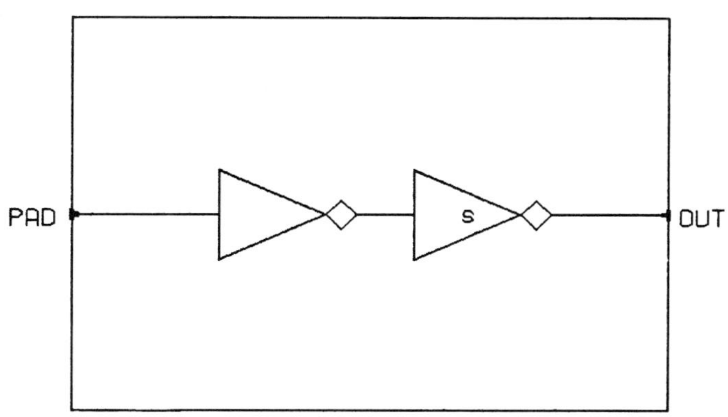

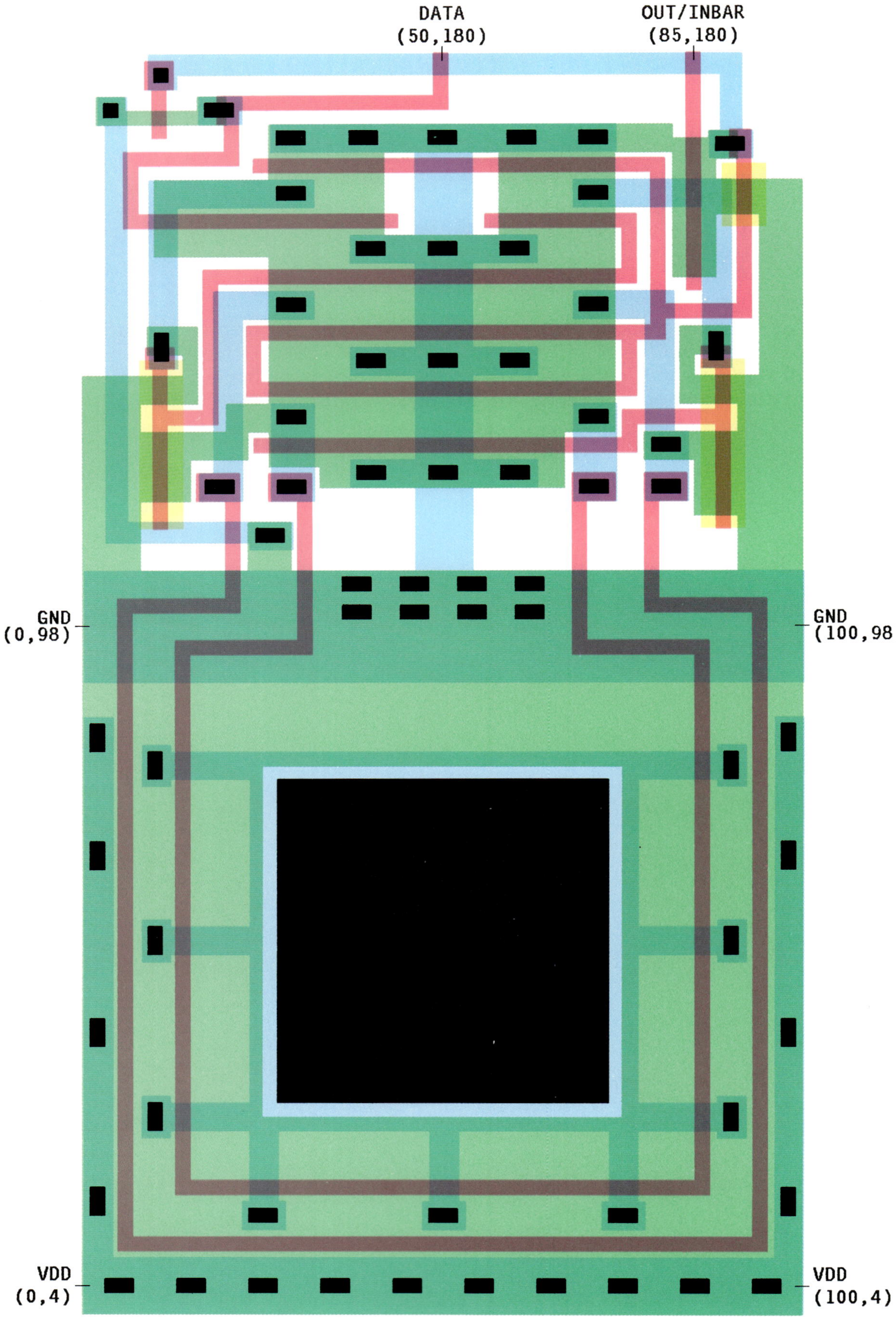
DATA
(50,180)
OUT/INBAR
(85,180)
GND
(0,98)
GND
(100,98)
VDD
(0,4)
VDD
(100,4)

NAME

PadMux—tristate pad with bidirectional input

SYNOPSIS

PadMux acts like an input pad when $OUT/\overline{IN}$ is low, and like an output pad when $OUT/\overline{IN}$ is high.

PROPERTIES

Size

100 x 180 λ.

Interface

DATA	input,	8:1 ratio	19 □ C_g
$OUT/\overline{IN}$	input,	4:1 ratio	10 □ C_g

DC current

1.6 mA with $OUT/\overline{IN}$ high (without external load).
3.2 mA with $OUT/\overline{IN}$ low.

DESCRIPTION

PadMux is similar to *TriState8* except that the *OUT* and *IN* lines have been combined into a single *DATA* line. A pass transistor is used to break the feedback loop when the pad is used for output.

LOGIC DIAGRAM

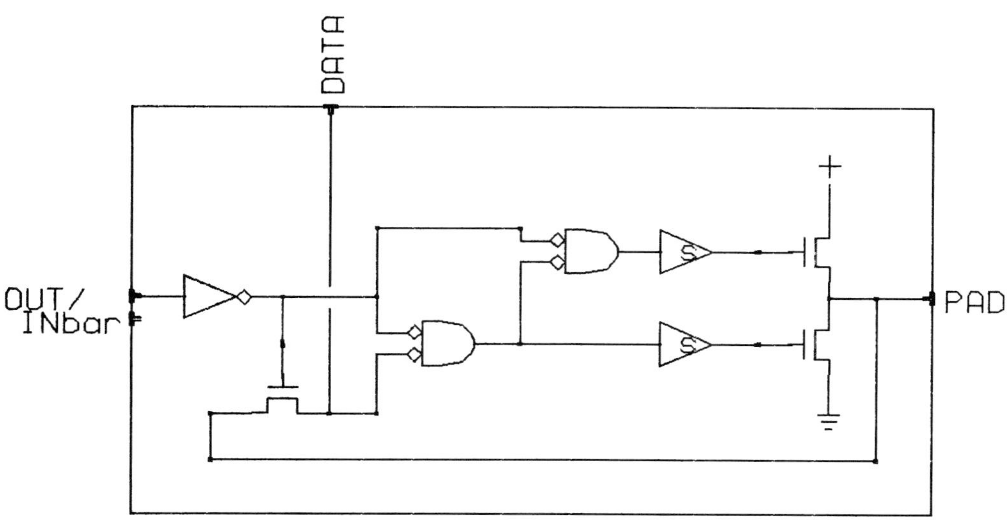

TESSELATION

Abuts horizontally, with no overlap.

ORIGIN

Xerox PARC (Dick Lyon), modified at Stanford University (Tim Saxe).

STATUS

Tested.

SEE ALSO

PadOut8, PadIn4, PadTriState8.

LAYOUT SCALE

20 λ per inch.

CIF ID

282

BUGS

Not scalable; designed for $\lambda=2.5\mu$.

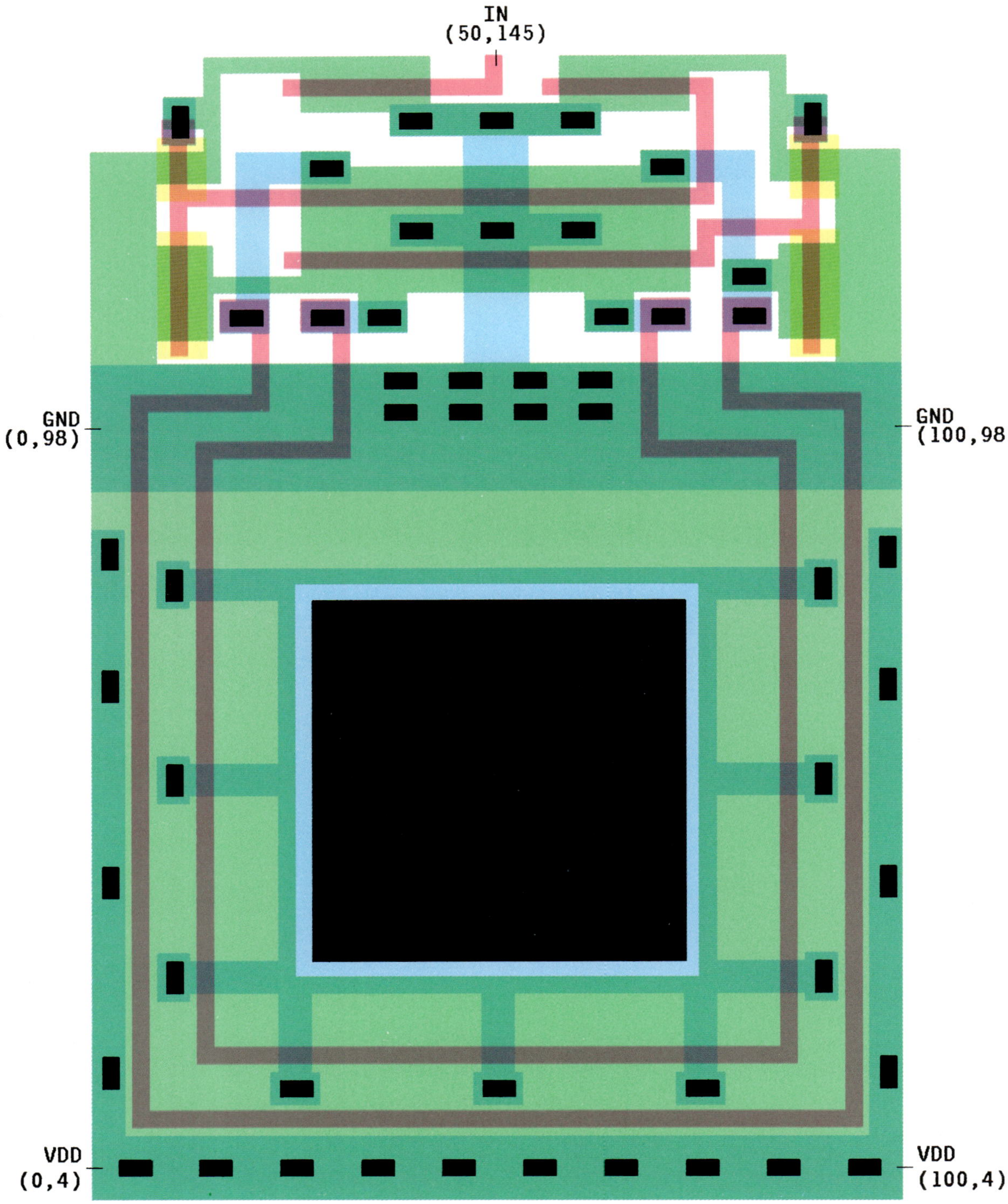
IN
(50,145)
GND
(0,98)
GND
(100,98)
VDD
(0,4)
VDD
(100,4)

NAME

PadOut4—output pad

SYNOPSIS

PadOut4 is the standard output pad with 4:1 input ratio.

PROPERTIES

Size

100 x 145 λ.

Interface

IN	input,	4:1 ratio	8 □ C_g

DC current

1.6 mA (without external load).

Performance

About 13 ns delay with a 50 pF load.

DESCRIPTION

PadOut4 uses an enhancement-mode output stage to provide TTL-compatible output levels. The IN signal drives both inverting and non-inverting superbuffers whose respective outputs connect the pad to power and to ground via pass transistors. *PadOut4* provides a non-inverted signal at its output.

TESSELATION

Abuts horizontally, with no overlap.

LOGIC DIAGRAM

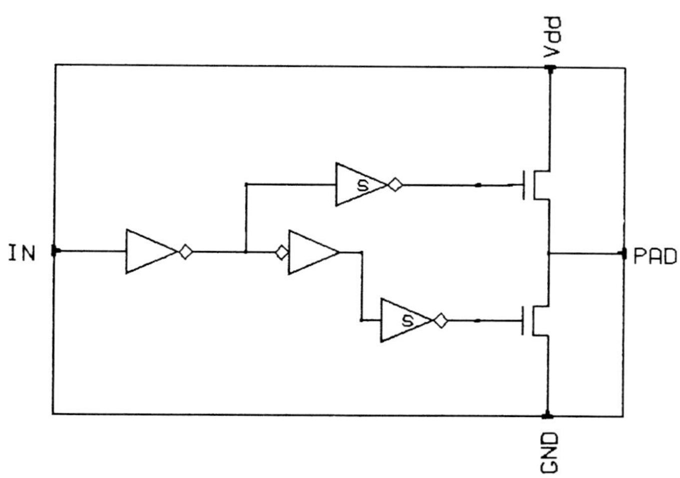

ORIGIN

Xerox PARC (Dick Lyon), modified at Stanford University (John Redford).

STATUS

Generally used.

SEE ALSO

PadOut8.

LAYOUT SCALE

20 λ per inch.

CIF ID

204

BUGS

Not scalable; designed for $\lambda=2.5\mu$.

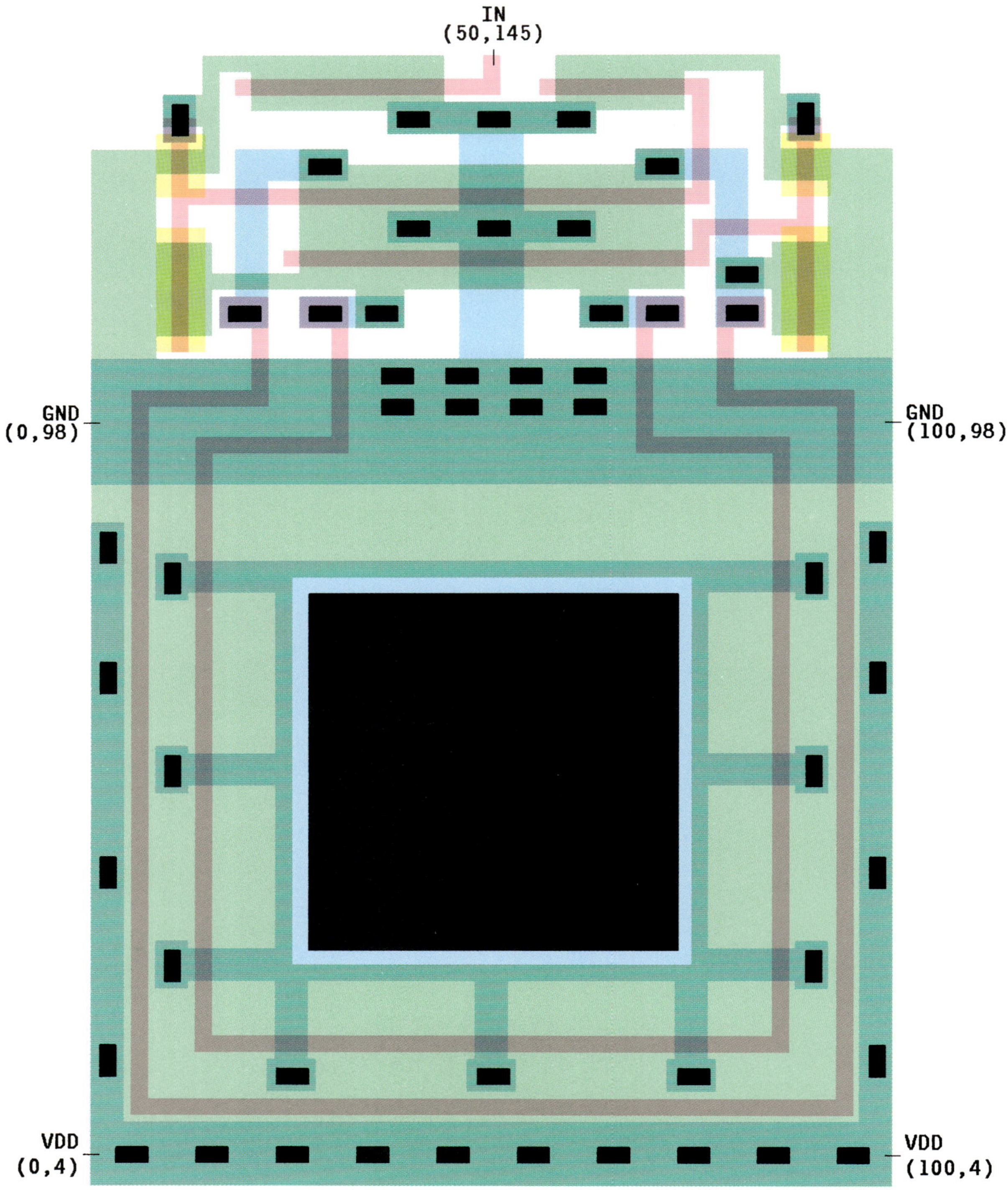
IN
(50,145)
GND
(0,98)
GND
(100,98)
VDD
(0,4)
VDD
(100,4)

NAME

PadOut8—output pad

SYNOPSIS

PadOut8 is the standard output pad with 8:1 input ratio.

PROPERTIES

Size

100 x 145 λ.

Interface

IN input, 8:1 ratio 16 □ C_g

DC current

1.6 mA (without external load).

DESCRIPTION

PadOut8 uses an enhancement-mode output stage to provide TTL-compatible output levels. The IN signal drives both inverting and non-inverting superbuffers whose respective outputs connect the pad to power and to ground via pass transistors. *PadOut8* provides a non-inverted signal at its output.

TESSELATION

Abuts horizontally, with no overlap.

ORIGIN

Xerox PARC (Dick Lyon), modified at Stanford University (John Redford, C.Y. Chu).

STATUS

Generally used.

SEE ALSO

PadOut4.

LAYOUT SCALE

20 λ per inch.

CIF ID

250

BUGS

Not scalable; designed for $\lambda=2.5\mu$.

LOGIC DIAGRAM

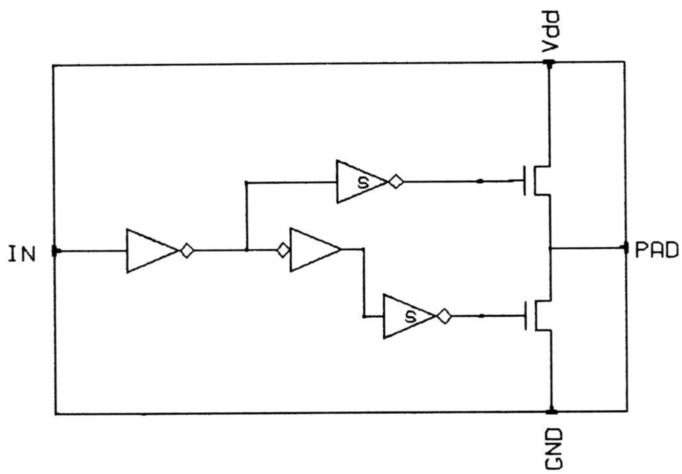

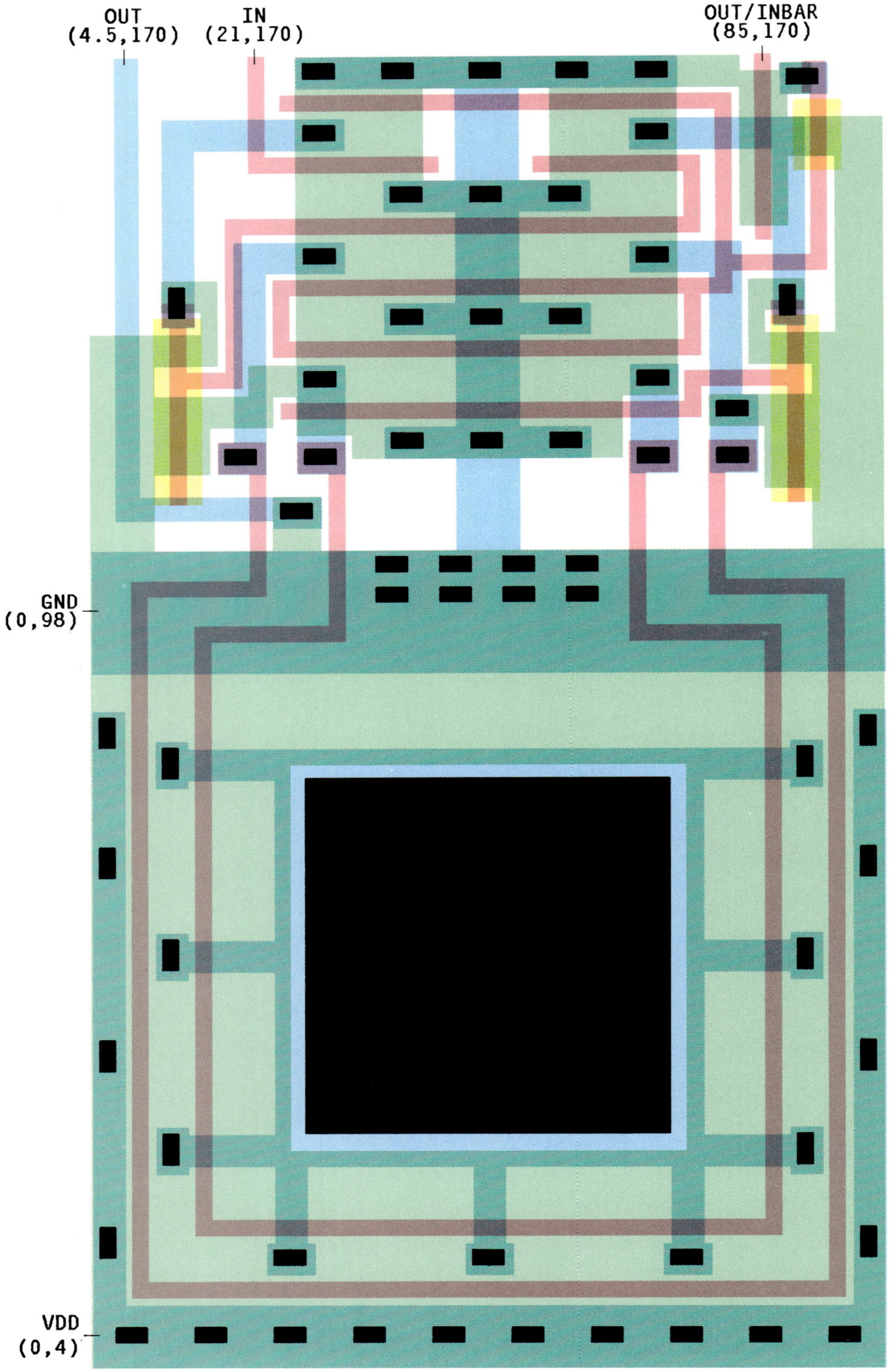
OUT
(4.5,170)
IN
(21,170)
OUT/INBAR
(85,170)
GND
(0,98)
VDD
(0,4)

NAME

PadTriState4—tristate pad with 4-to-1 input

SYNOPSIS

PadTriState4 acts like an input pad when $OUT/\overline{IN}$ is low, and like an output pad when $OUT/\overline{IN}$ is high.

PROPERTIES

Size

100 x 170 λ.

Interface

IN	input,	4:1 ratio	9 □ C_g
OUT	output,	switch-logic	177 □ C_g
$OUT/\overline{IN}$	input,	4:1 ratio	10 □ C_g

DC current

1.6 mA with $OUT/\overline{IN}$ high (without external load).
3.2 mA with $OUT/\overline{IN}$ low.

DESCRIPTION

PadTriState4 is similar to *PadOut4* except for some additional logic that disables the output circuitry when $OUT/\overline{IN}$ is low. In this state the pad can be used as an input pad, since the pad will see infinite impedance when looking into the output circuitry. *PadTriState4* provides a non-inverted signal at its output.

NOTE: as with *PadIn4*, when *PadTriState4* serves as an input pad there is no level restoration. Assuming TTL-compatible input signals, *OUT* should be taken into gate logic with an input ratio of at least *8:1*.

TESSELATION

Abuts horizontally, with no overlap.

ORIGIN

Xerox PARC (Dick Lyon), modified at Stanford University (C. Y. Chu).

STATUS

Revised to repair a long-standing ratio error.

SEE ALSO

PadOut4, PadIn4.

LAYOUT SCALE

20 λ per inch.

CIF ID

208

BUGS

Not scalable; designed for $\lambda = 2.5\mu$.

LOGIC DIAGRAM

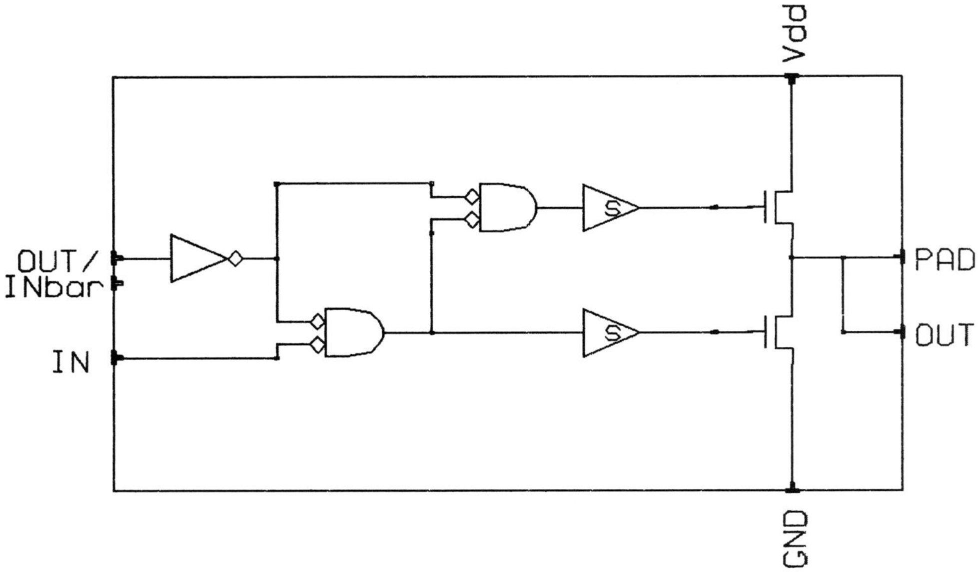

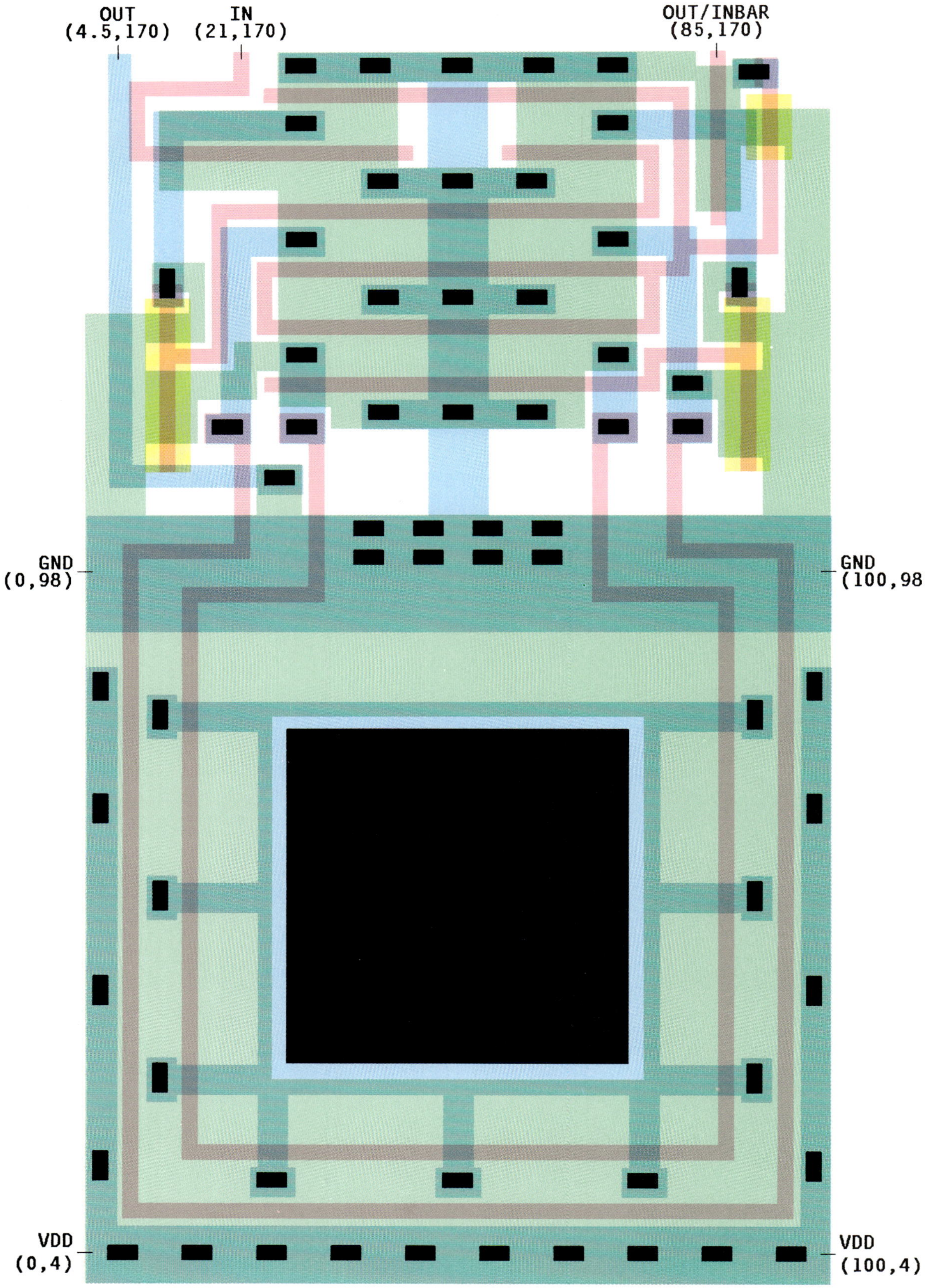

OUT
(4.5,170)
IN
(21,170)
OUT/INBAR
(85,170)
GND
(0,98)
GND
(100,98)
VDD
(0,4)
VDD
(100,4)

NAME

PadTriState8—tristate pad with 8-to-1 input

SYNOPSIS

PadTriState8 acts like an input pad when $OUT/\overline{IN}$ is low, and like an output pad when $OUT/\overline{IN}$ is high.

PROPERTIES

Size

100 x 170 λ.

Interface

IN	input,	8:1 ratio	17.5 □ C_g
OUT	output,	switch-logic	177 □ C_g
$OUT/\overline{IN}$	input,		10 □ C_g

DC current

1.6 mA with $OUT/\overline{IN}$ high (without external load).
3.2 mA with $OUT/\overline{IN}$ low.

DESCRIPTION

PadTriState8 is similar to *PadOut8* except for some additional logic that disables the output circuitry when $OUT/\overline{IN}$ is low. In this state the pad can be used as an input pad, since the pad will see infinite impedance when looking into the output circuitry. *PadTriState8* provides a non-inverted signal at its output.

NOTE: as with *PadIn4*, when *PadTriState8* serves as an input pad there is no level restoration. Assuming TTL-compatible input signals, *OUT* should be taken into gate logic with an input ratio of at least *8:1*.

TESSELATION

Abuts horizontally, with no overlap.

ORIGIN

Xerox PARC (Dick Lyon), modified at Stanford University (C.Y. Chu).

STATUS

Revised to remove a long-standing ratio error.

SEE ALSO

PadOut8, PadIn4.

LAYOUT SCALE

20 λ per inch.

CIF ID

281

BUGS

Not scalable; designed for $\lambda=2.5\mu$.

LOGIC DIAGRAM

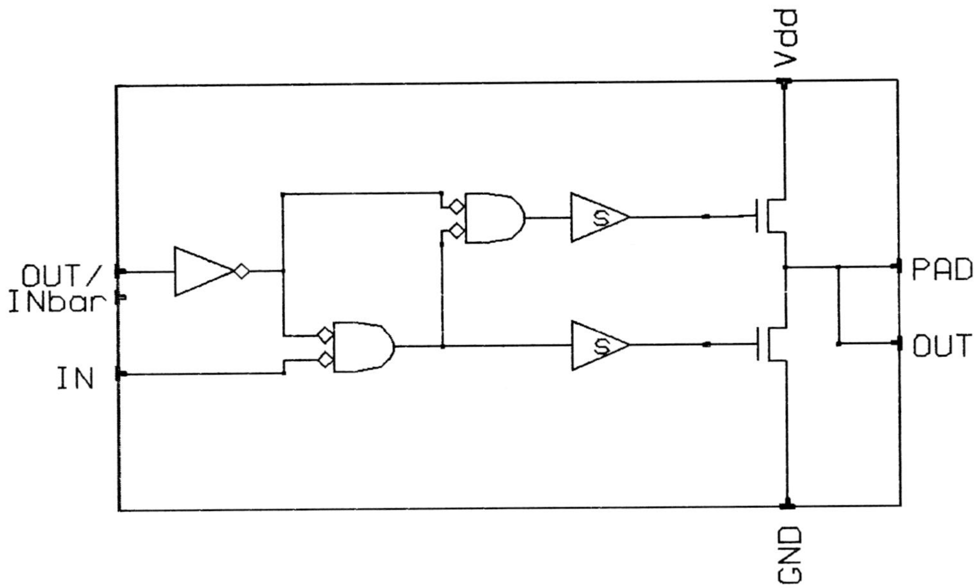

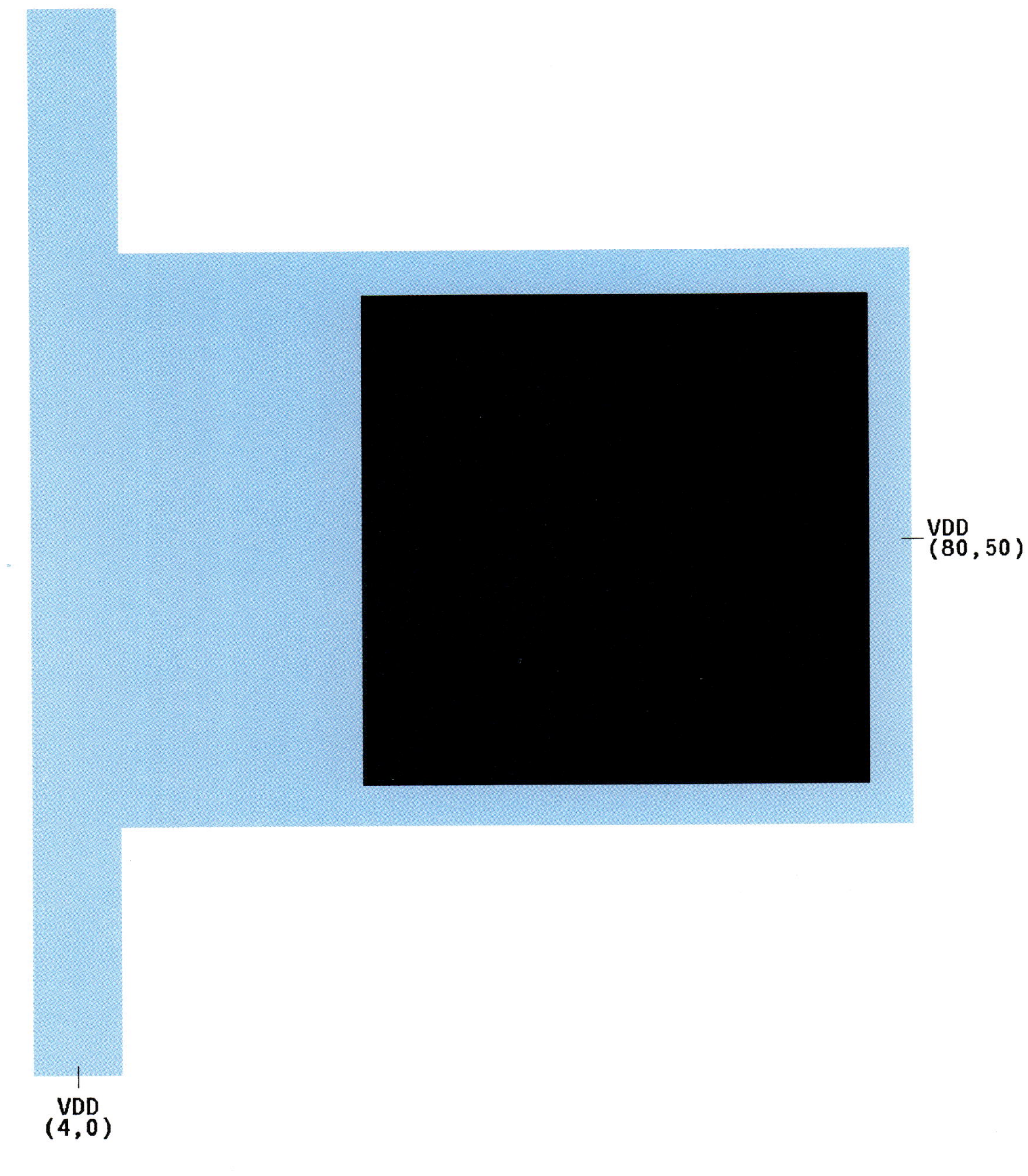
VDD
(80,50)
VDD
(4,0)

NAME

PadVdd—power pad

SYNOPSIS

PadVdd is a bonding pad for connecting to the outer, VDD ring.

PROPERTIES

Size

80 x 100 λ.

DC current

maximum current in 8 λ metal is 20 mA (with $\lambda = 2.5\mu$).

DESCRIPTION

PadVdd is a nonstandard pad in that the GND metal line is omitted. Thus, there is a gap in the GND ring, permitting power to be brought into the project.

ORIGIN

Xerox PARC (Dick Lyon), modified at Stanford University (John Redford).

STATUS

Generally used.

SEE ALSO

PadBlank.

LAYOUT SCALE

16 λ per inch.

CIF ID

207

BUGS

Not scalable; designed for $\lambda=2.5\mu$.

8/ PLA Drivers, Buffers and Pieces

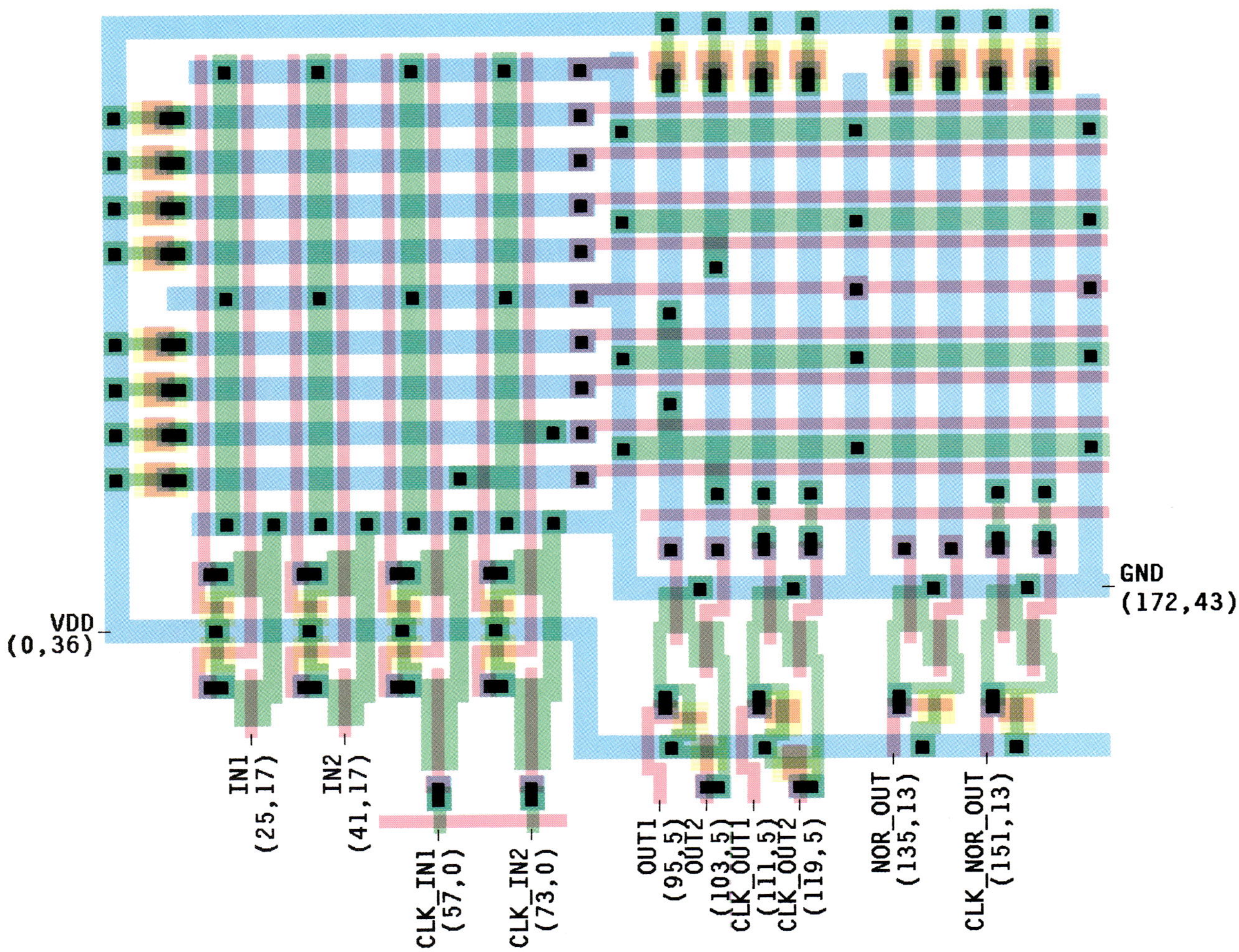
VDD
(0,36)
GND
(172,43)
IN1
(25,17)
IN2
(41,17)
CLK_IN1
(57,0)
CLK_IN2
(73,0)
OUT1
(95,5)
OUT2
(103,5)
CLK_OUT1
(111,5)
CLK_OUT2
(119,5)
NOR_OUT
(135,13)
CLK_NOR_OUT
(151,13)

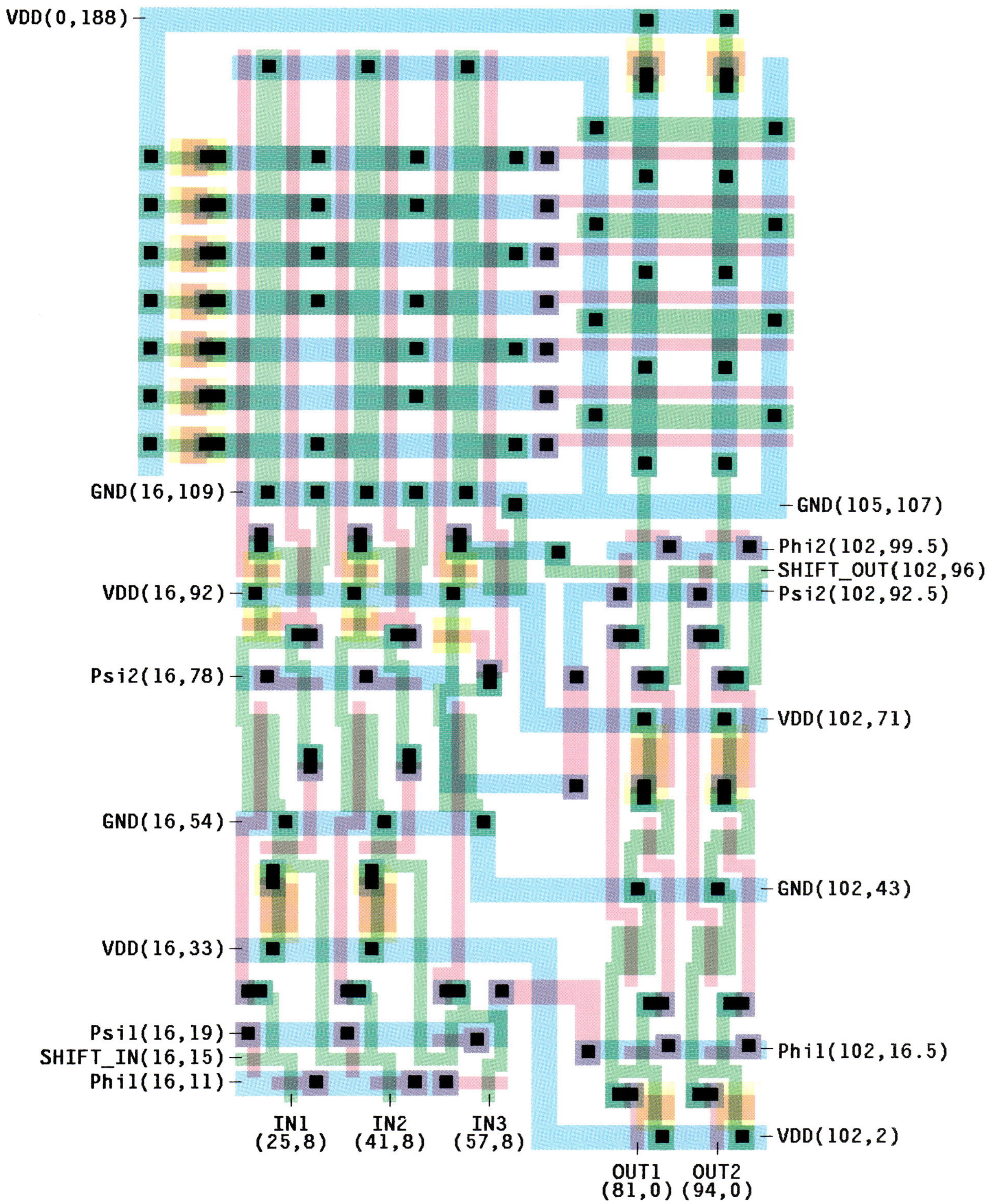
VDD(0,188)
GND(16,109)
VDD(16,92)
Psi2(16,78)
GND(16,54)
VDD(16,33)
Psi1(16,19)
SHIFT_IN(16,15)
Phi1(16,11)
IN1
(25,8)
IN2
(41,8)
IN3
(57,8)
GND(105,107)
Phi2(102,99.5)
SHIFT_OUT(102,96)
Psi2(102,92.5)
VDD(102,71)
GND(102,43)
Phi1(102,16.5)
VDD(102,2)
OUT1
(81,0)
OUT2
(94,0)

NAME

intro—introduction to PLAs

SYNOPSIS

This chapter contains an overview of Programmed Logic Arrays and a collection of input and output buffers for customizing your PLA.

DESCRIPTION

A PLA is a regular implementation for arbitrary multi-input, multi-output combinational logic functions. There are 4 logic stages : input inversion and buffering, a NOR array used for ANDing to form product terms, a NOR array used for ORing together the product terms, and a final inversion to produce positive-logic outputs. A combinational function expressed as a set of sums of products of input literals maps directly into the PLA structure.

PLAs are also convenient for mechanizing finite-state machines (FSMs). State storage for the FSM is provided by suitable input and output buffers, and an encoded transition table maps directly into the PLA planes. State feedback is provided by connecting the next-state outputs back to the state inputs.

PLA planes are built out of primitive cells (see PlaPieces and the preceding example). The details, including programming the planes, are best handled by a PLA-generating program. The planes are simply large NOR gates, identical except for a 90-degree rotation.

The PLA buffers in this section are of two genera, simple and scan-path. The scan-path genus contains a dynamic and a static species; the simple genus is a single species. Buffers within a species fit together, but do not in general mate with other buffers.

The simple genus contains unclocked and clocked output buffers that can be used to do one bit of output decoding when the PLA is used as a ROM. Simple buffers are ideal for combinational PLAs and for very simple FSMs; for more complex FSMs, scan-path buffers are superior, although larger.

Each scan-path species has three types of input buffers: a clocked LSSD input cell (*ScanClkIn* and *StScanClkIn*), an unclocked LSSD input cell for PLA inputs that have already been clocked before reaching the cell (*ScanUnClkIn* and *StScanUnClkIn*), and an unclocked cell that does not become part of the LSSD chain (*ScanIn* and *StScanIn*). There is also a cell for joining each species of input buffer to the output buffers (*ScanJoin* and *StScanJoin). The scan-path output buffers (ScanOut* and *ScanClkOut*) match either input species.

Unfortunately, scan-path buffers are generally wider than their simple, 16 λ counterparts. Static input cells are 19 λ wide (+3 λ); output cells are 13 λ wide *per* (+10 λ per plane cell). The pitches of the planes must change accordingly.

So why use a scan path? LSSD (Level Sensitive Scan Design) is a testability technique due to IBM wherein all bits of the internal state of the chip are linked into a shift register and read out for examination or modification. This scan path greatly increases the observability and controllability of the chip by providing access to signals that would otherwise be invisible to the outside world without multiplexing schemes or large numbers of extra pins. IBM's primary motivation for using LSSD is to make automatic generation of manufacturing tests possible. For our designers, the primary motivation is greatly simplified debugging and a strong partitioning of the design. If a datapath is separated from its control FSM by a scan chain, then each can be tested independently, regardless of whether the other works, a valuable asset!

LSSD fits well with our clocking style. Two-phase, non-overlapping clocking provides level-sensitive designs. Four extra pins, two for the LSSD shift clocks and two for reading the data in and out, are needed. The data pins can sometimes be multiplexed onto other pins.

To illustrate the use of scan-path buffers consider a PLA FSM with its inputs clocked on ϕ_1 and its outputs clocked on ϕ_2. When the system ϕ_1 and ϕ_2 clocks are running, the buffers behave normally. Between clock phases, the input buffers hold the PLA input at the most recent ϕ_1 and the output buffers the PLA output at the most recent ϕ_2. Changing to the scan clock after ϕ_1 (...ϕ_1 ϕ_2 ϕ_1 ψ_2 ψ_1...) scans out the state of the input buffers; changing after ϕ_2 (...ϕ_1 ϕ_2 ψ_1 ψ_2...) scans out the state of the output buffers.

Setting the FSM state is similar. To set the input buffers' state, change after ψ_1 (...ψ_1 ψ_2 ψ_1 ϕ_2 ϕ_1...); the output state is, of course, overwritten by the PLA. Changing after ψ_2 (...ψ_1 ψ_2 ϕ_1 ϕ_2...) sets the output state.

The sequence ...ψ_1 ψ_2 ψ_1 ϕ_2 ψ_1 ψ_2... can shift in inputs to the PLA, let it compute its function, and shift out the result, independent of state feedback or external connections. In any event, the subscripts on the clocking must always alternate for correct operation.

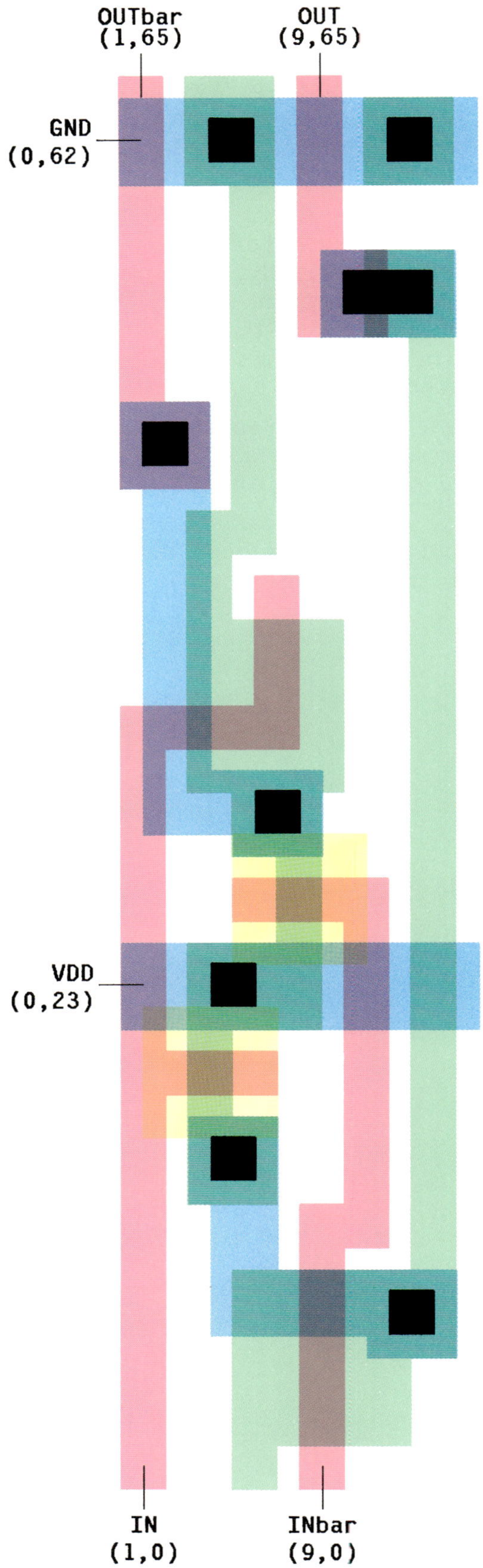
OUTbar
(1,65)
OUT
(9,65)
GND
(0,62)
VDD
(0,23)
IN
(1,0)
INbar
(9,0)

NAME
Afterburner—superbuffer for inputs

SYNOPSIS
Afterburner turns a PLA input driver into a superbuffer pair.

PROPERTIES
Size
16 x 65 λ.

Interface

IN	input,	4:1 ratio	7.5 □ C_g	
$\overline{IN}$	input,	4:1 ratio	6.25 □ C_g	
OUT	output,	switch-logic	10.5 □ C_g	[1 / ¼] □ *R*
$\overline{OUT}$	output,	switch-logic	7.5 □ C_g	[1 / ¼] □ *R*

DC current
0.2 mA.

DESCRIPTION
Large PLAs can be driven faster if they have superbuffer inputs. This cell can be tacked onto the top of any of the input drivers, thus boosting their speed.

NOTE: this cell reverses the locations of the true and complement outputs in going from input to output.

TESSELATION
Abuts horizontally and vertically, with no overlap.

ORIGIN
Stanford University (John Redford).

STATUS
Tested.

SEE ALSO
PlaIn, PlaClockIn, PlaStaticIn, PlaShiftIn.

LAYOUT SCALE
8 λ per inch.

CIF ID
82

LOGIC DIAGRAM

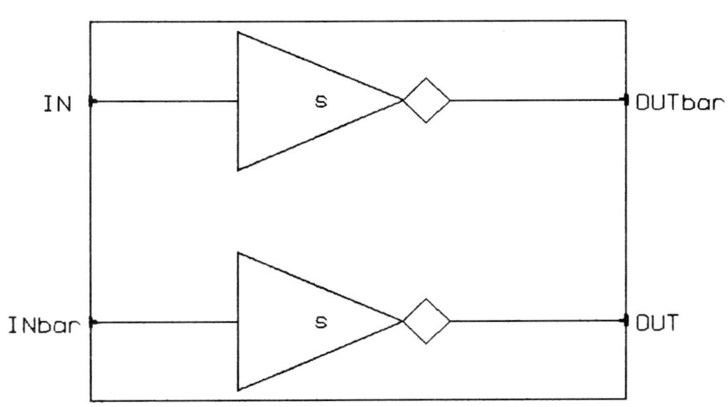

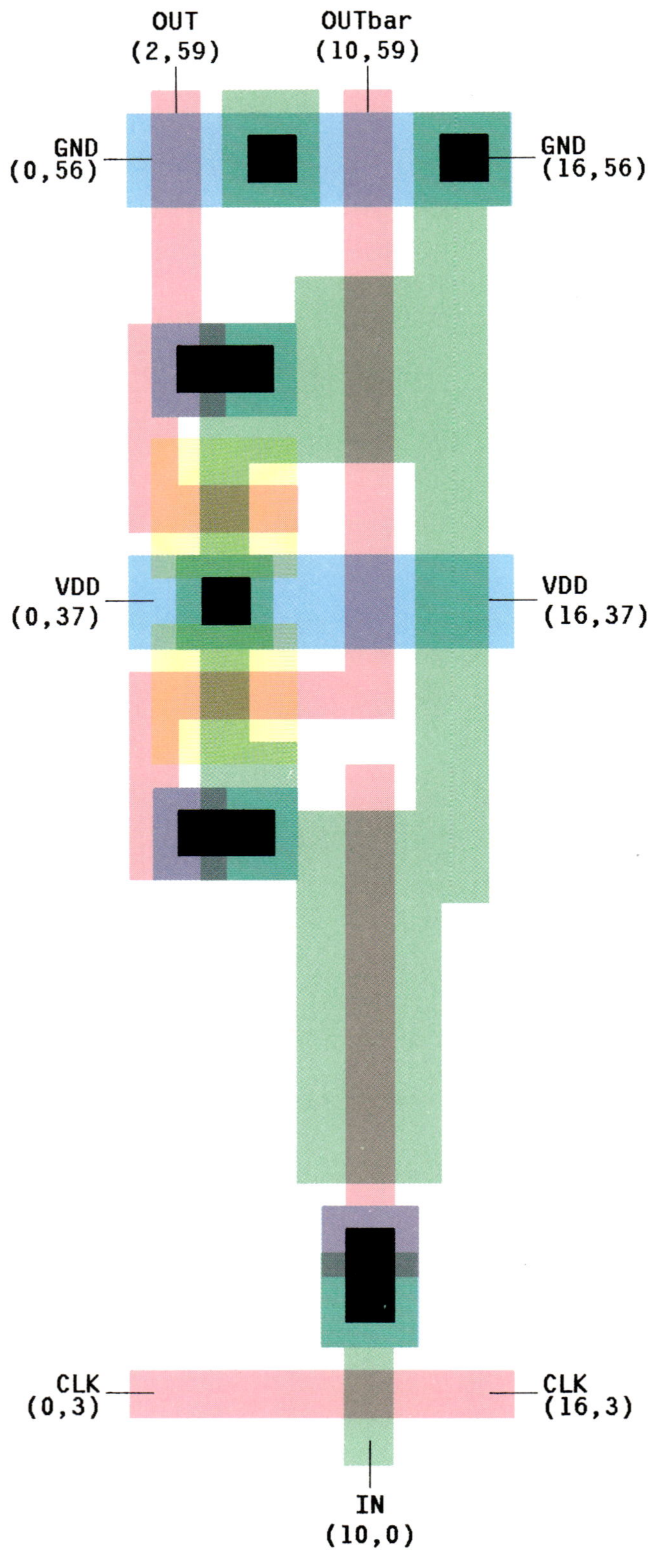
OUT
(2,59)
OUTbar
(10,59)
GND
(0,56)
GND
(16,56)
VDD
(0,37)
VDD
(16,37)
CLK
(0,3)
CLK
(16,3)
IN
(10,0)

NAME

PlaClockIn—clocked input driver

SYNOPSIS

PlaClockIn is a clocked input driver for PLAs.

PROPERTIES

Size

16 x 59 λ

Interface

IN	input,	8:1 ratio	10.5 □ C_g	
CLK	input,	switch control	1.8 □ C_g	
OUT	output,	gate	0.5 □ C_g	[1 / ¼] □ *R*
$\overline{OUT}$	output,	gate	0.5 □ C_g	[1 / ¼] □ *R*

DC current

0.2 mA

DESCRIPTION

PlaClockIn is a clocked buffer that consists of two inverters in series for producing both the true and complement versions of the input signal.

TESSELATION

Abuts horizontally and vertically, with no overlap.

ORIGIN

Xerox PARC (Dick Lyon).

STATUS

Generally used.

SEE ALSO

PlaIn, PlaStaticIn, PlaShiftIn, Afterburner.

LAYOUT SCALE

8 λ per inch.

CIF ID

20

LOGIC DIAGRAM

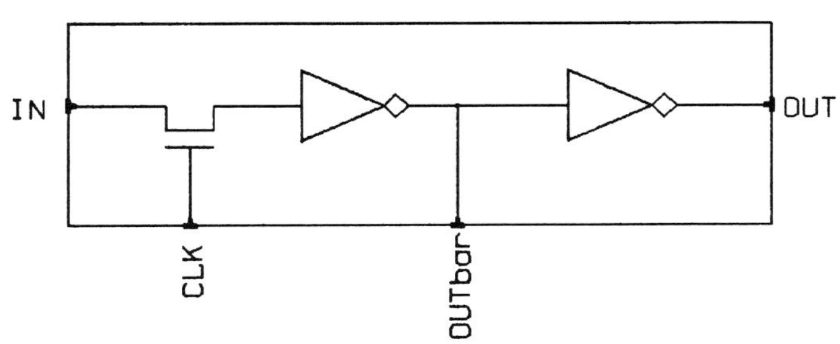

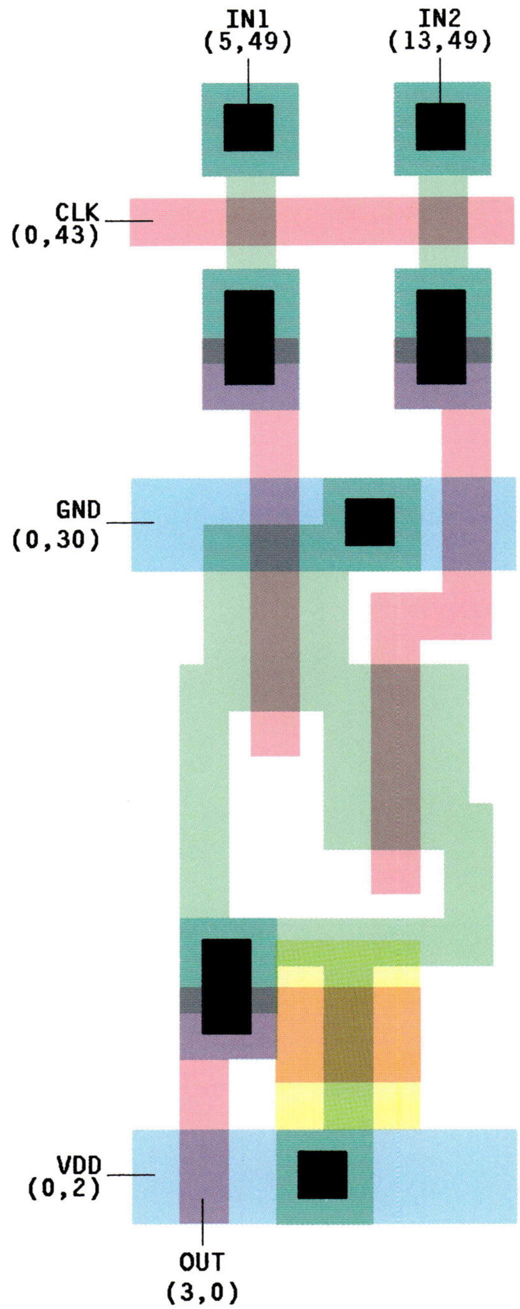
IN1
(5,49)
IN2
(13,49)
CLK
(0,43)
GND
(0,30)
VDD
(0,2)
OUT
(3,0)

NAME

PlaClockNorOut—clocked NOR of two outputs

SYNOPSIS

PlaClockNorOut is a clocked PLA output driver that joins the outputs into a NOR gate.

PROPERTIES

Size

16 x 49 λ.

Interface

IN1	input,	switched	7.5 □ C_g	
IN2	input,	switched	8 □ C_g	
CLK	input,	switch control	2.5 □ C_g	
OUT	output,	gate	2 □ C_g	[2 / ¼] □ *R*

DC current

0.1 mA (worst case).

DESCRIPTION

PlaClockNorOut attaches to two output lines of a PLA, joining them into a clocked NOR gate.

TESSELATION

Abuts horizontally, with no overlap; **vertically, must overlap the PLA OR-plane by 4λ.**

ORIGIN

Xerox PARC (Dick Lyon).

STATUS

Generally used.

SEE ALSO

PLA description.

LAYOUT SCALE

8 λ per inch.

CIF ID

31

LOGIC DIAGRAM

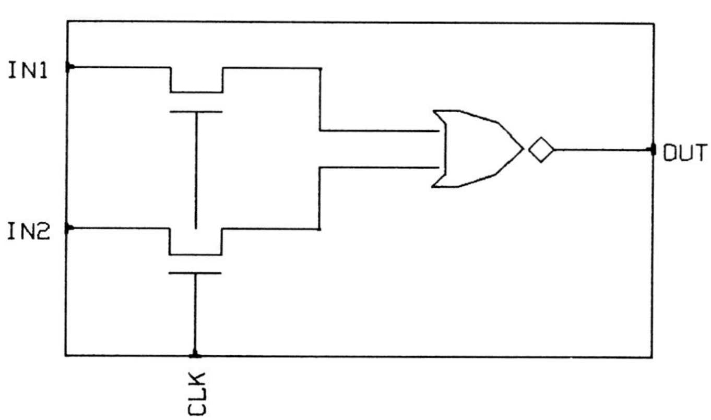

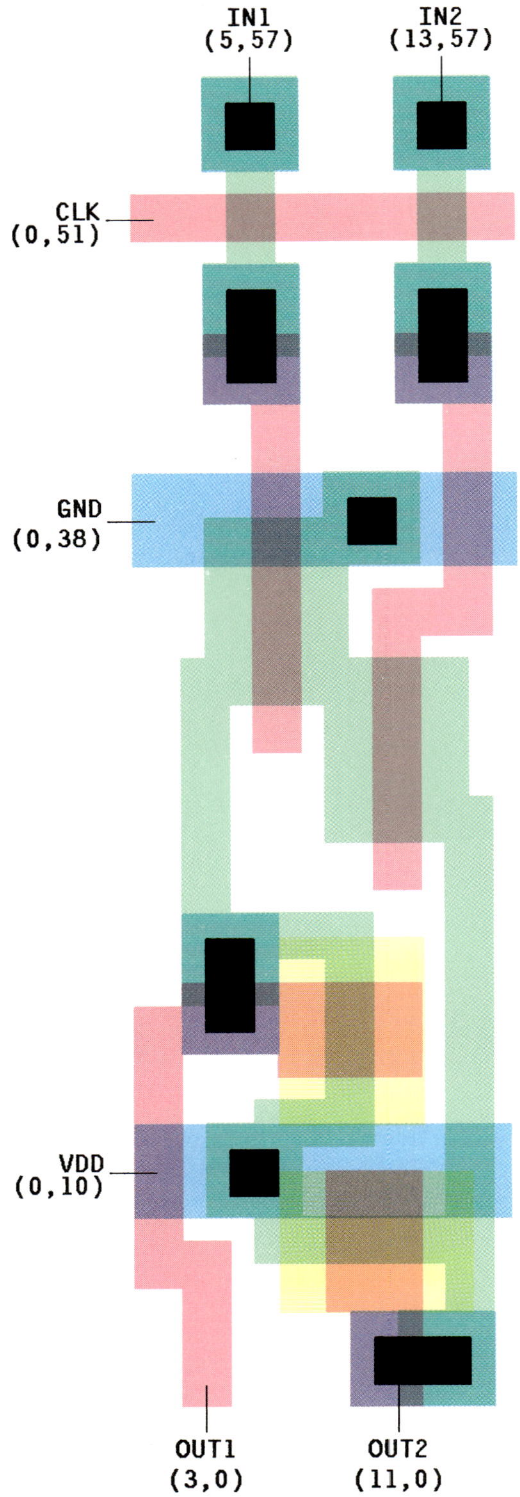
IN1
(5,57)
IN2
(13,57)
CLK
(0,51)
GND
(0,38)
VDD
(0,10)
OUT1
(3,0)
OUT2
(11,0)

NAME

PlaClockOut—clocked output buffer

SYNOPSIS

PlaClockOut is the standard clocked PLA output driver.

PROPERTIES

Size

16 x 57 λ.

Interface

IN1	input,	switched	6 □ C_g	
IN2	input,	switched	6 □ C_g	
CLK	input,	switch control	2.5 □ C_g	
OUT1	output,	gate	7 □ C_g	[2 / ¼] □ *R*
OUT2	output,	gate	7 □ C_g	[2 / ¼] □ *R*

DC current

0.2 mA

DESCRIPTION

PlaClockOut attaches to two output lines of a PLA. It clocks them in, before inverting and buffering them.

TESSELATION

Abuts horizontally, with no overlap; **vertically, must overlap the PLA OR-plane by** 4 λ.

ORIGIN

Xerox PARC (Dick Lyon).

STATUS

Generally used.

SEE ALSO

PLA description.

LAYOUT SCALE

8 λ per inch.

CIF ID

27

LOGIC DIAGRAM

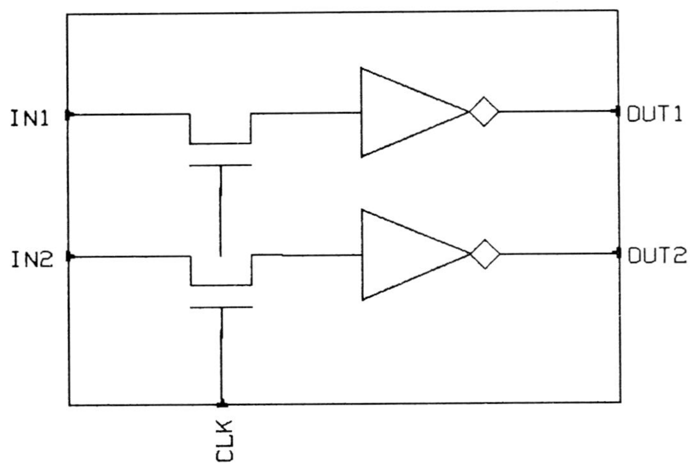

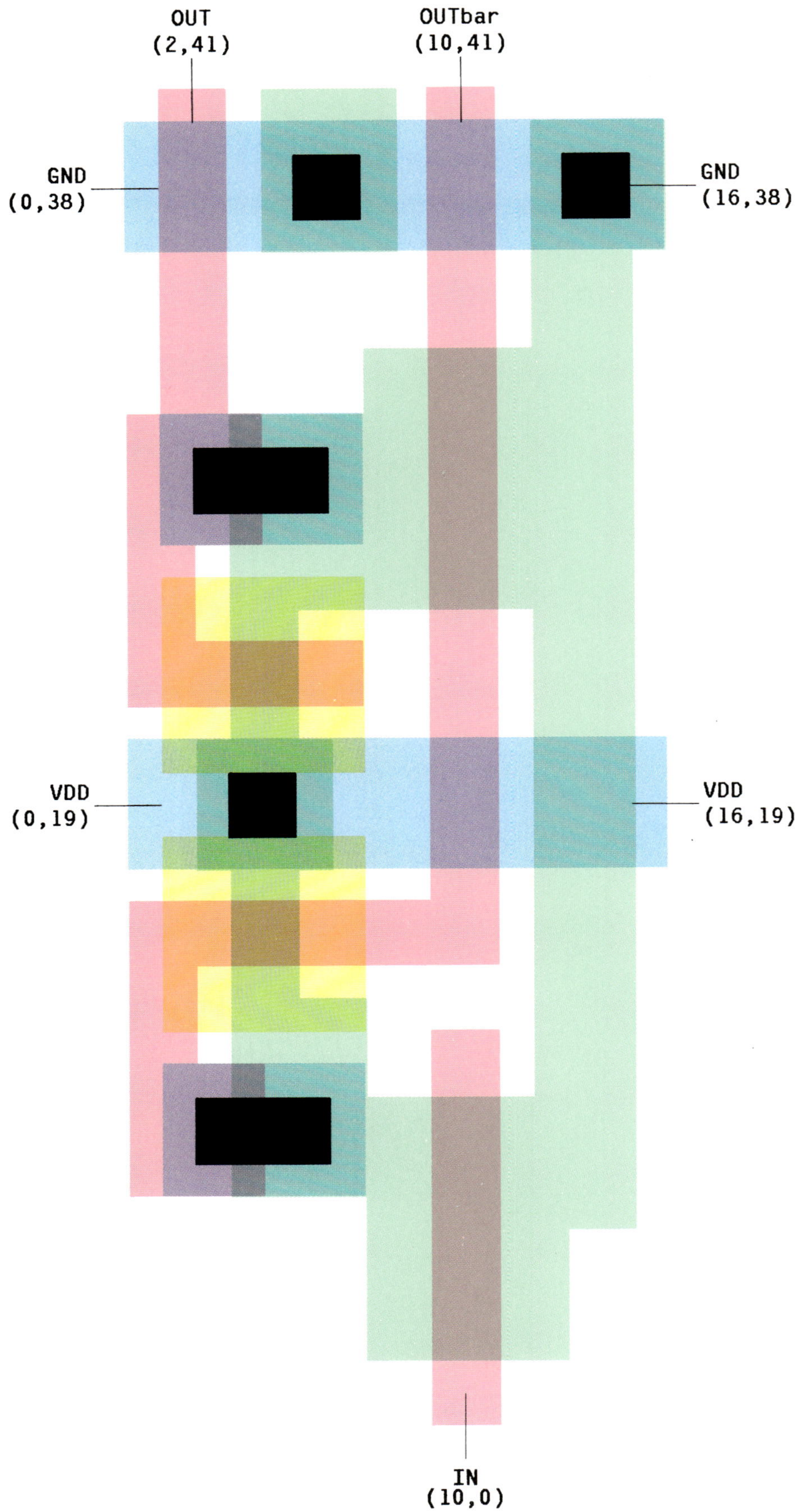
OUT
(2,41)
OUTbar
(10,41)
GND
(0,38)
GND
(16,38)
VDD
(0,19)
VDD
(16,19)
IN
(10,0)

NAME

PlaIn—input driver

SYNOPSIS

PlaIn is an unclocked input driver for PLAs.

PROPERTIES

Size

16 x 41 λ.

Interface

IN	input,	4:1 ratio	4.25 □ C_g	
OUT	output,	gate	4.75 □ C_g	[1 / ¼] □ *R*
$\overline{OUT}$	output,	gate	9.25 □ C_g	[1 / ¼] □ *R*

DC current

0.2 mA

DESCRIPTION

PlaIn is an unclocked buffer that consist of two inverters in series for producing both the true and complement versions of the input signal.

Note: If you are also using *PlaClockIn*, the clock lines will have to be provided below the *PlaIn* cell.

TESSELATION

Abuts horizontally and vertically, with no overlap.

ORIGIN

Xerox PARC (Dick Lyon).

STATUS

Generally used.

SEE ALSO

PlaClockIn, PlaStaticIn, PlaShiftIn, Afterburner.

LAYOUT SCALE

5 λ per inch.

CIF ID

19

LOGIC DIAGRAM

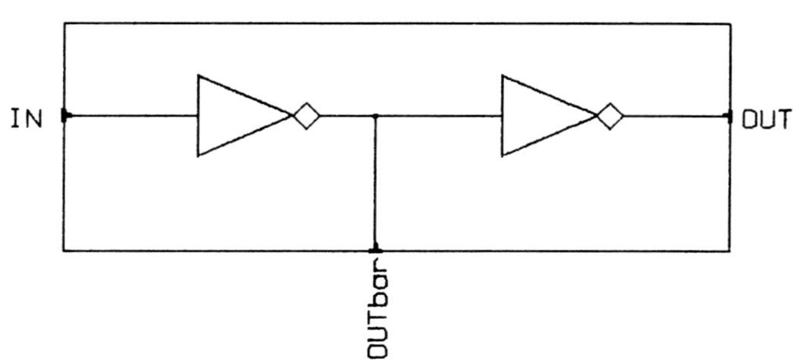

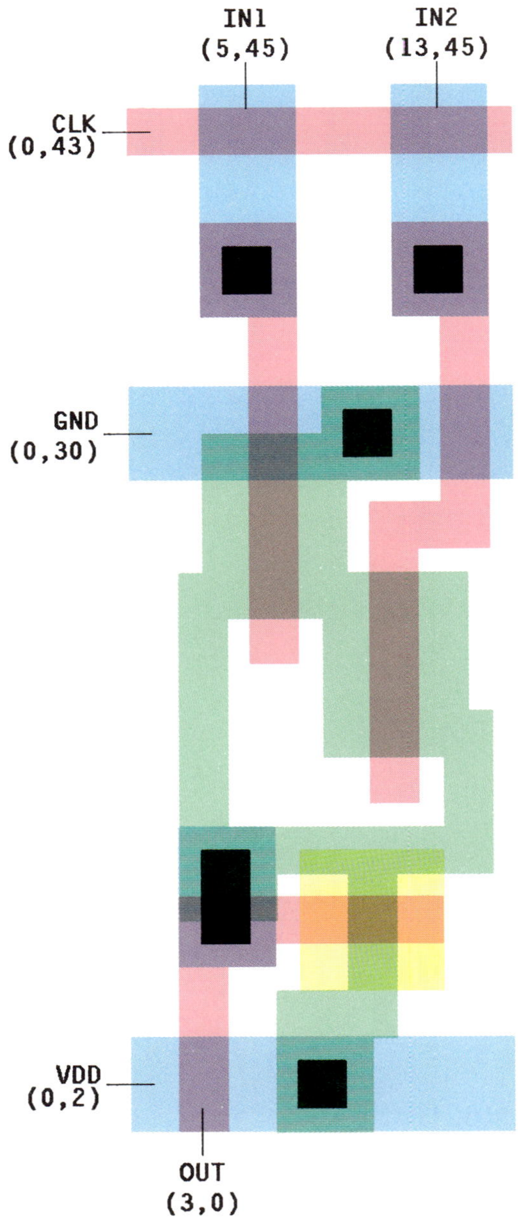
IN1
(5,45)
IN2
(13,45)
CLK
(0,43)
GND
(0,30)
VDD
(0,2)
OUT
(3,0)

NAME
PlaNorOut—NOR of two outputs

SYNOPSIS
PlaNorOut is an unclocked PLA output driver that joins two output terms in a NOR gate.

PROPERTIES
Size
16 x 45 λ.

Interface

IN1	input,	4:1 ratio	5 □ C_g	
IN2	input,	4:1 ratio	5.5 □ C_g	
OUT	output,	gate	2 □ C_g	[1 / ¼] □ *R*

DC current
0.2 mA

DESCRIPTION
PlaNorOut attaches to two output lines of a PLA, joining them into a NOR gate.

Note: the clock line provides continuity in case, for example, a *PlaClockNorOut* is being used.

TESSELATION
Abuts horizontally, with no overlap.

ORIGIN
Xerox PARC (Dick Lyon).

STATUS
Generally used.

SEE ALSO
PLA description.

LAYOUT SCALE
8 λ per inch.

CIF ID
30

LOGIC DIAGRAM

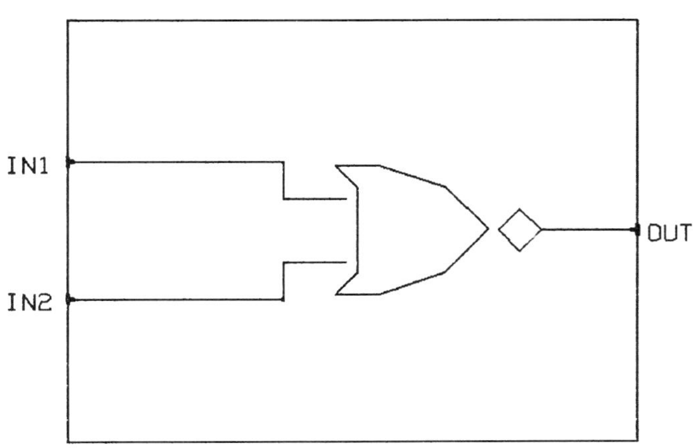

IN1
(5,53)
IN2
(13,53)
CLK
(0,51)
GND
(0,38)
VDD
(0,10)
OUT1
(3,0)
OUT2
(11,0)

NAME
PlaOut—output buffer

SYNOPSIS
PlaOut is the standard PLA output driver.

PROPERTIES
Size

16 x 53 λ.

Interface

IN1	input,	4:1 ratio	5 □ C_g	
IN2	input,	4:1 ratio	5.5 □ C_g	
OUT1	output,	gate	1 □ C_g	[1 / ¼] □ *R*
OUT2	output,	gate	1 □ C_g	[1 / ¼] □ *R*

DC current

0.4 mA (worst case).

DESCRIPTION
PlaOut inverts and buffers two separate output lines of a PLA.

Note: the clock line provides continuity in case, for example, a *PlaClockOut* is being used.

TESSELATION
Abuts horizontally and vertically, with no overlap.

ORIGIN
Xerox PARC (Dick Lyon).

STATUS
Generally used.

SEE ALSO
PLA description.

LAYOUT SCALE
8 λ per inch.

CIF ID
26

LOGIC DIAGRAM

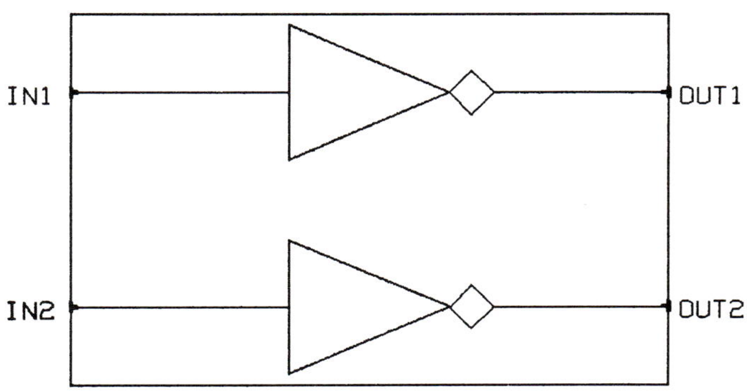

PlaCell - 12

PlaGround - 13

PlaPullup - 14

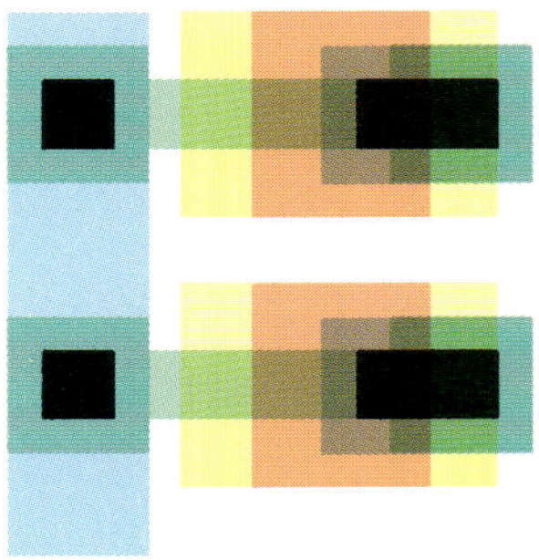

PlaConnect - 15

PlaProgTop - 21

PlaProgBottom - 22

PlaProgLeft - 23

PlaProgRight - 24

PlaConnectSpace - 16

PlaOrSpace - 17

PlaHoleWires - 28

PlaOutSpace - 29

PlaGroundSpace - 18

PlaPulllupSpace - 25

NAME

PlaPieces—Individual PLA Pieces

SYNOPSIS

Various cells for assembling PLAs.

PROPERTIES

Size

PlaCell:	16 λ x 16 λ
PlaGround:	16 λ x 8 λ
PlaPullup:	15 λ x 16 λ
PlaConnect:	13 λ x 16 λ
PlaConnectSpace:	13 λ x 8 λ
PlaOrSpace:	16 λ x 8 λ
PlaGroundSpace:	8 λ x 8 λ
PlaProgTop:	4 λ x 9 λ
PlaProgBottom:	4 λ x 9 λ
PlaProgLeft:	9 λ x 4 λ
PlaProgRight:	9 λ x 4 λ
PlaPullupSpace:	15 λ x 8 λ
PlaHoleWires:	13 λ x 45 λ
PlaOutSpace:	8 λ x 45 λ

DC current

PlaPullup:	0.2 mA (worst case).

DESCRIPTION

A whole host of cells are available for assembling PLA planes; they are mainly used by the automatic PLA generator. Designers should rarely have to deal with them, except for PlaHoleWires, which can be used to join the input and output drivers together.

ORIGIN

Xerox PARC (Dick Lyon).

STATUS

Generally used.

LAYOUT SCALE

10 λ per inch.

CIF ID

PlaCell:	12
PlaGround:	13
PlaPullup:	14
PlaConnect:	15
PlaConnectSpace:	16
PlaOrSpace:	17
PlaGroundSpace:	18
PlaProgTop:	21
PlaProgBottom:	22
PlaProgLeft:	23
PlaProgRight:	24
PlaPullupSpace:	25
PlaHoleWires:	28
PlaOutSpace:	29

OUT
(2,71)
OUTbar
(10,71)
GND
(0,68)
VDD
(0,49)
PHI2
(0,7.5)
PHI1
(0,3)
IN
(11,0)

NAME

PlaStaticIn—static clocked input driver

SYNOPSIS

PlaStaticIn is a static clocked input driver for PLAs.

PROPERTIES

Size

16 x 71 λ.

Interface

IN	input,	8:1 ratio	10.5 □ C_g	
$\phi 2$	input,	switch control	1.8 □ C_g	
$\phi 1$	input,	switch control	1.8 □ C_g	
OUT	output,	gate	0.5 □ C_g	[1 / ¼] □ *R*
$\overline{OUT}$	output,	gate	0.5 □ C_g	[1 / ¼] □ *R*

DC current

0.2 mA

DESCRIPTION

PlaStaticIn is a clocked buffer that consists of two inverters in series for producing both the true and complement versions of the input signal. Unlike the other PLA input drivers it holds its output while $\phi 2$ is true. This static storage can make testing easier. The cell can also be used as a one-bit register if $\phi 1$ is interpreted as *Load* and $\phi 2$ as $\overline{Load}$. Like all feedback circuits, though, it can be flipped if a large capacitance is suddenly switched onto its output. The outputs should run into gate logic to prevent this problem.

TESSELATION

Abuts horizontally and vertically, with no overlap.

ORIGIN

Stanford University (John Redford).

STATUS

Tested.

SEE ALSO

PlaIn, PlaShiftIn, Afterburner.

LAYOUT SCALE

8 λ per inch.

CIF ID

80

LOGIC DIAGRAM

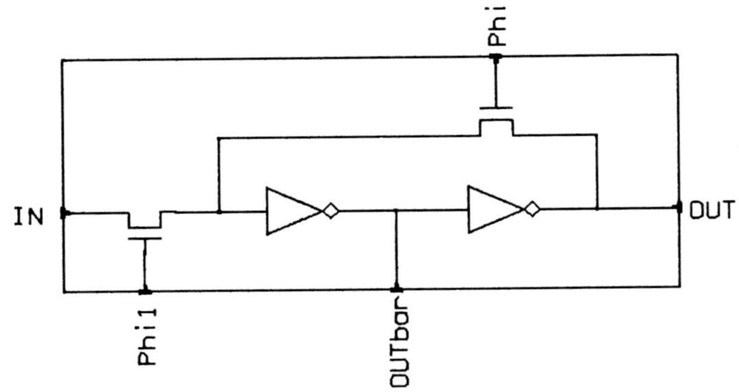

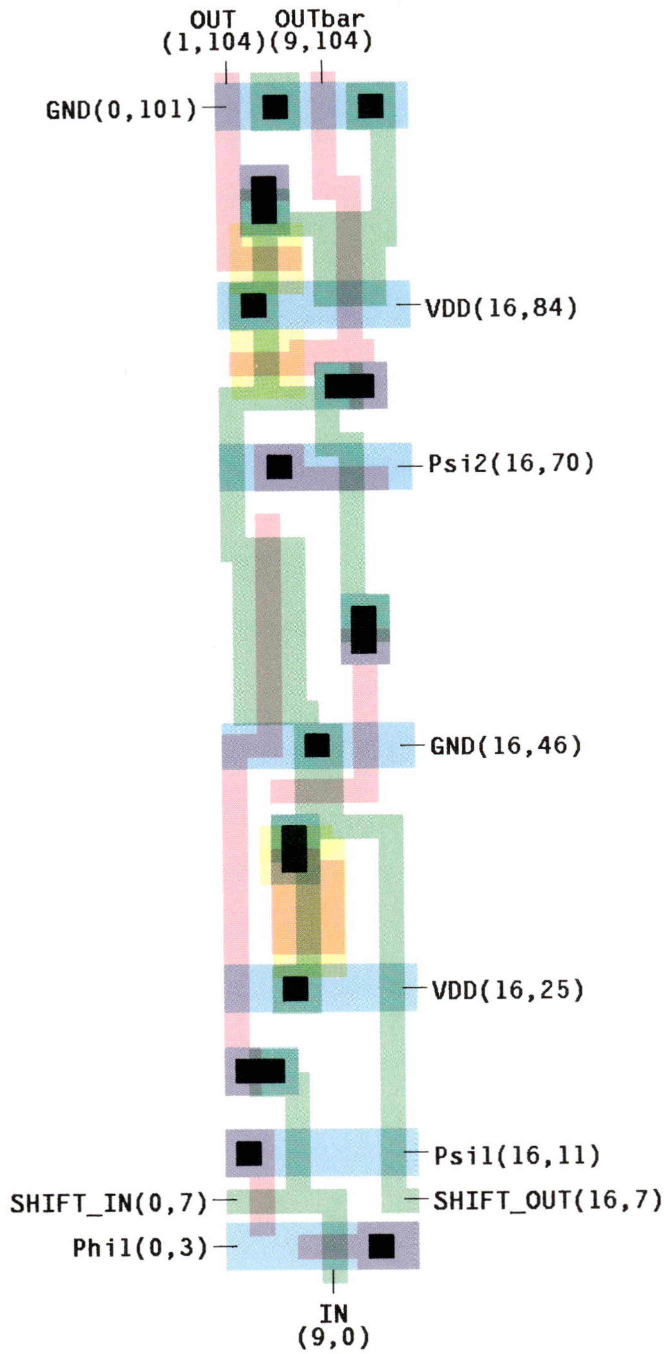
OUT
(1,104)
OUTbar
(9,104)
GND(0,101)
VDD(16,84)
Psi2(16,70)
GND(16,46)
VDD(16,25)
Psi1(16,11)
SHIFT_IN(0,7)
SHIFT_OUT(16,7)
Phi1(0,3)
IN
(9,0)

NAME

ScanClkIn—LSSD input driver

SYNOPSIS

ScanClkIn is a clocked PLA input driver that can convert to an LSSD shift register.

PROPERTIES

Size

16 x 104 λ.

Interface

$\phi 1$	input,	switch control	2.5 □ C_g	
$\psi 1$	input,	switch control	2.5 □ C_g	
$\psi 2$	input,	switch control	2.5 □ C_g	
IN	input,	switched	14.5 □ C_g	
Shift_in	input,	switched	14.5 □ C_g	
Shift_out	output,	gate	12 □ C_g	[4 / ½] □ *R*
OUT	output,	gate	4.5 □ C_g	[1 / ¼] □ *R*
$\overline{OUT}$	output,	gate	13 □ C_g	[1 / ⅛] □ *R*

DC current

0.2 mA.

DESCRIPTION

This cell behaves like a clocked PLA input driver under normal clocking conditions, and becomes a shift register when the LSSD clock lines $\psi 1$ and $\psi 2$ are used. Data moves from left to right.

TESSELATION

Abuts horizontally, with no overlap.

ORIGIN

Stanford University (Dave Noice).

STATUS

Tested.

SEE ALSO

Intro, ScanUnClkIn, ScanIn, ScanJoin, ScanClkOut, ScanOut.

LAYOUT SCALE

16 λ per inch.

CIF ID

510

LOGIC DIAGRAM

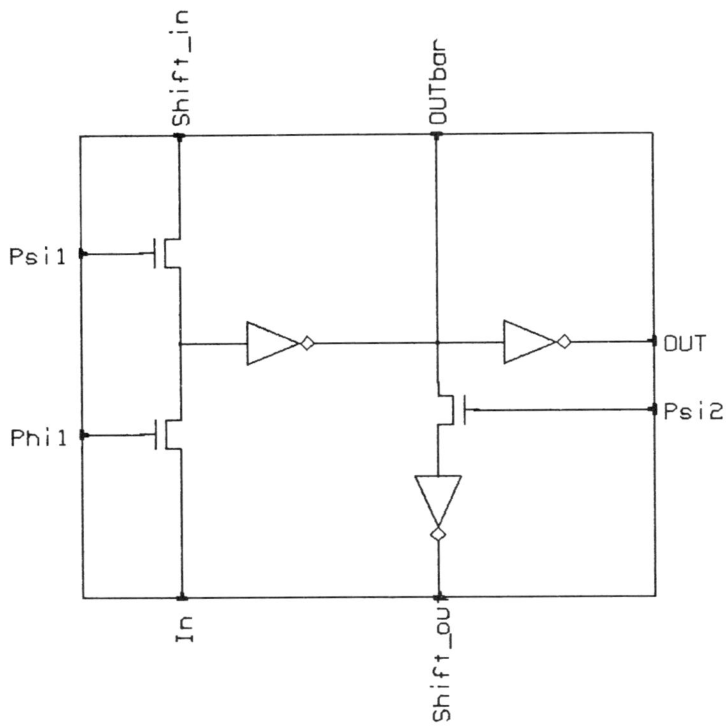

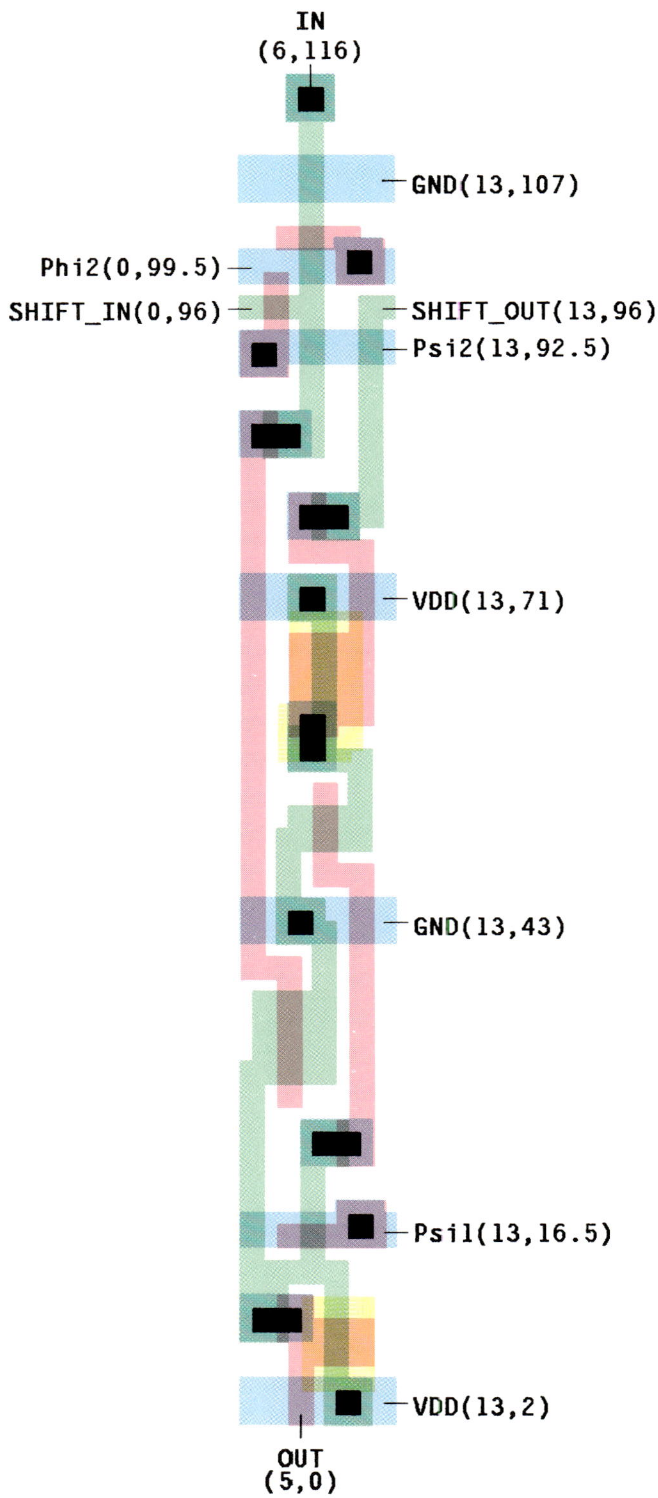
IN
(6,116)
GND(13,107)
Phi2(0,99.5)
SHIFT_IN(0,96)
SHIFT_OUT(13,96)
Psi2(13,92.5)
VDD(13,71)
GND(13,43)
Psi1(13,16.5)
VDD(13,2)
OUT
(5,0)

NAME

ScanClkOut—LSSD output buffer

SYNOPSIS

A PLA LSSD Output buffer.

PROPERTIES

Size

13 x 116 λ.

Interface

IN	input,	switched	12 □ C_g	
φ2	input,	switch control	2 □ C_g	
ψ1	input,	switch control	2 □ C_g	
ψ2	input,	switch control	2 □ C_g	
Shift_in	input,	switched	11 □ C_g	
Shift_out	output,	gate	12.5 □ C_g	[2 / ¼] □ *R*
OUT	output,	gate	6 □ C_g	[2 / ¼] □ *R*

DC Current

0.1 mA.

DESCRIPTION

Like ScanClkIn, this cell can be used to read and modify data on the output of a PLA. When the LSSD clocks ψ1 and ψ2 are used the cell behaves like a shift register. Data comes in SHIFT_IN and goes out SHIFT_OUT. Although the cell is 13 λ wide it can only handle one output line, unlike the regular PLA output buffers.

TESSELATION

Abuts horizontally, with no overlap.

ORIGIN

Stanford University (Dave Noice).

STATUS

Tested.

SEE ALSO

Intro, ScanOut, ScanClkIn, StScanClkIn, ScanJoin, StScanJoin.

LAYOUT SCALE

16 λ per inch.

CIF ID

518

LOGIC DIAGRAM

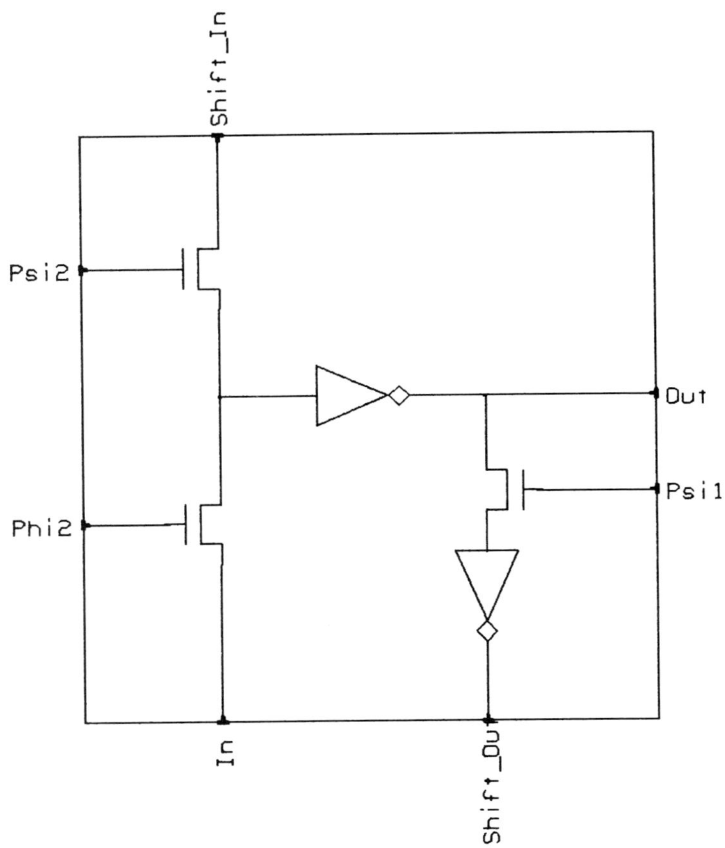

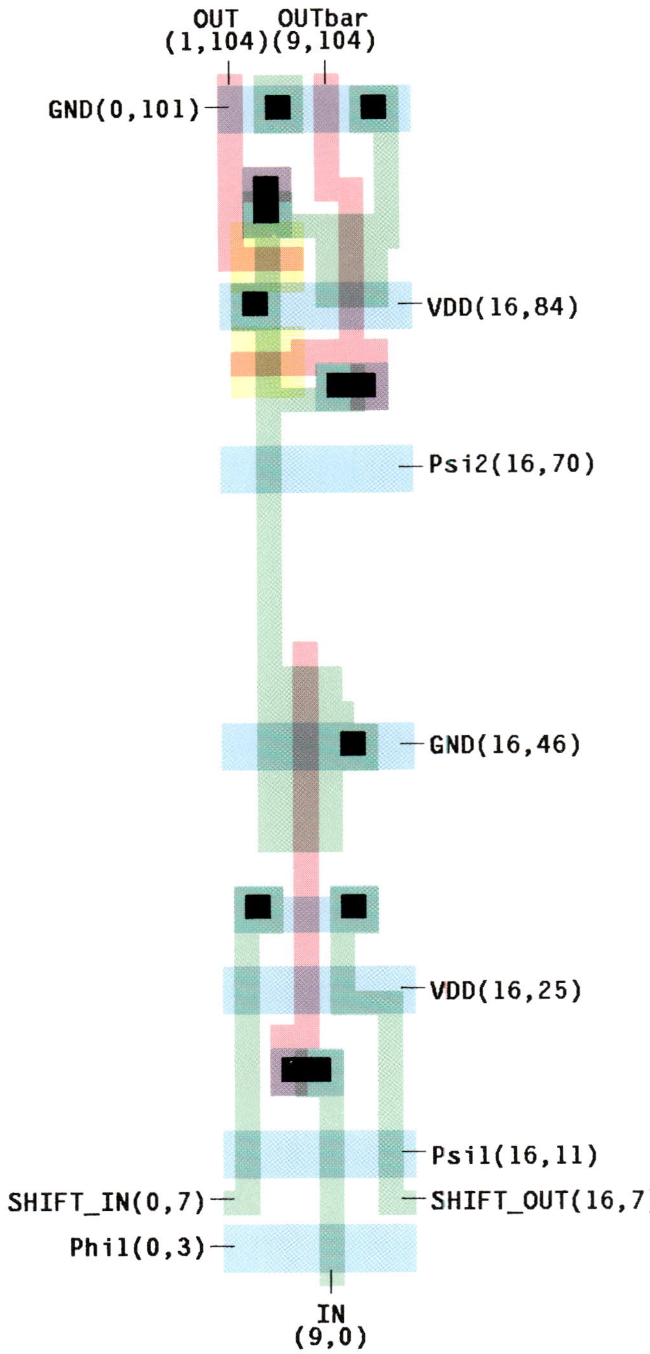
OUT
(1,104)
OUTbar
(9,104)
GND(0,101)
VDD(16,84)
Psi2(16,70)
GND(16,46)
VDD(16,25)
Psi1(16,11)
SHIFT_IN(0,7)
SHIFT_OUT(16,7)
Phi1(0,3)
IN
(9,0)

NAME

ScanIn—LSSD input driver dummy

SYNOPSIS

ScanIn is a normal PLA input driver that fits the pitch of the other LSSD input cells.

PROPERTIES

Size

16 x 104 λ.

Interface

$\phi 1$	input,	switch control	1.5 □ C_g	
$\psi 1$	input,	switch control	1.5 □ C_g	
$\psi 2$	input,	switch control	1.5 □ C_g	
IN	input,	switched	11 □ C_g	
Shift_in	input,	switched	10 □ C_g	
Shift_out	output,	gate	4 □ C_g	[4 / ½] □ *R*
OUT	output,	gate	4.5 □ C_g	[1 / ¼] □ *R*
$\overline{OUT}$	output,	gate	14 □ C_g	[1 / ⅛] □ *R*

DC current

0.2 mA.

DESCRIPTION

ScanIn is a normal PLA input driver that fits the pitch of the other LSSD input cells. It is used when parts of the PLA are used for combinational logic. In this case the inputs are not clocked so the input cannot be part of the LSSD chain.

TESSELATION

Abuts horizontally, with no overlap.

LOGIC DIAGRAM

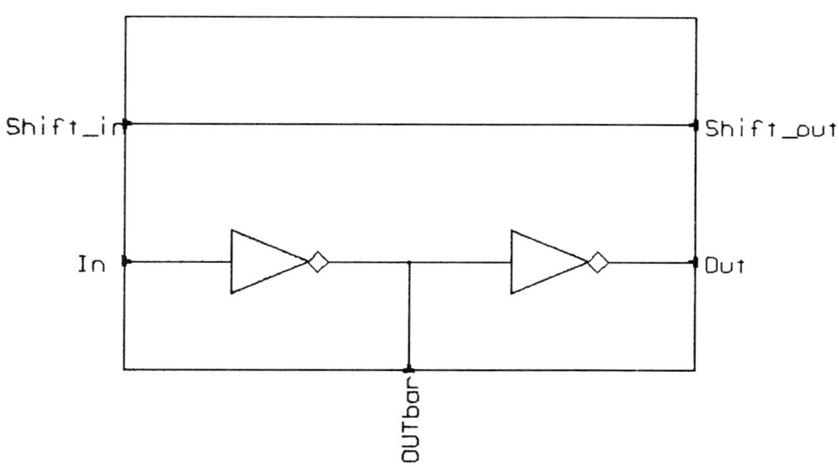

ORIGIN

Stanford University (Dave Noice).

STATUS

Tested.

SEE ALSO

ScanClkIn, ScanUnClkIn, ScanJoin, ScanClkOut, ScanClkIn.

LAYOUT SCALE

16 λ per inch.

CIF ID

511

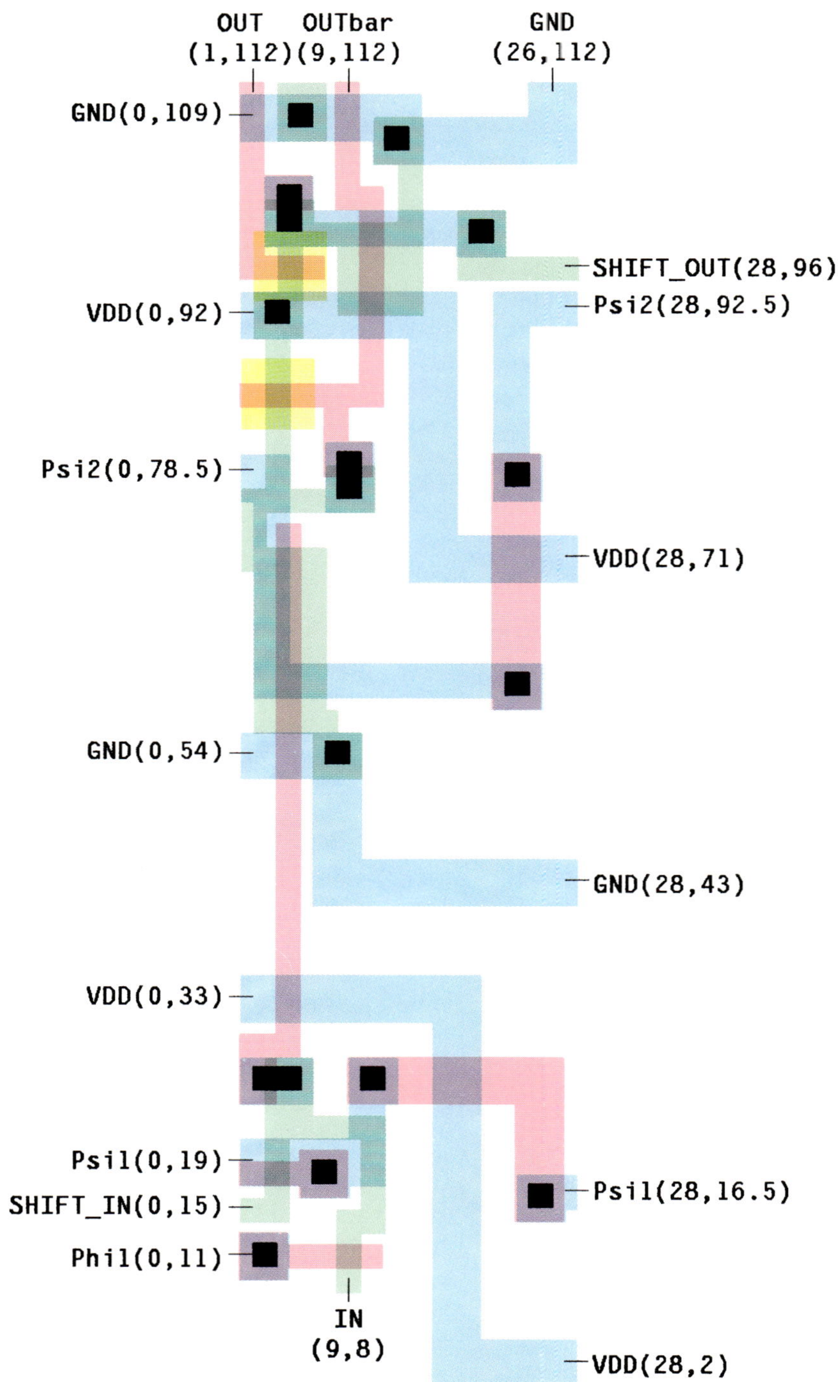
OUT
(1,112)
OUTbar
(9,112)
GND
(26,112)
GND(0,109)
SHIFT_OUT(28,96)
VDD(0,92)
Psi2(28,92.5)
Psi2(0,78.5)
VDD(28,71)
GND(0,54)
GND(28,43)
VDD(0,33)
Psi1(0,19)
Psi1(28,16.5)
SHIFT_IN(0,15)
Phi1(0,11)
IN
(9,8)
VDD(28,2)

NAME

ScanJoin—LSSD driver/buffer connect

SYNOPSIS

A LSSDin cell with connections added for linking it to a LSSDout cell.

PROPERTIES

Size

28 x 112 λ.

Interface

$\phi 1$	input,	switch control	2 □ C_g	
$\psi 1$	input,	switch control	5 □ C_g	
$\psi 2$	input,	switch control	5.5 □ C_g	
IN	input,	switched	15 □ C_g	
*Shift*_in	input,	switched	15 □ C_g	
*Shift*_out	output,	gate	7.5 □ C_g	[1 / ¼] □ *R*
$\underline{OUT}$	output,	gate	7.5 □ C_g	[1 / ¼] □ *R*
$\overline{OUT}$	output,	gate	12 □ C_g	[1 / ⅛] □ *R*

DC current

0.2 mA.

DESCRIPTION

ScanJoin has connections for all the clocks, power, ground, and data output from the LSSD shift register. It is only intended for LSSD schemes where the data shifts from the input side of the PLA to the output. Note that there is an extra one phase shift provided to match the *Shift_Out* output phase to the *Shift_in* of the LSSDout cell.

TESSELATION

Abuts horizontally, with no overlap.

ORIGIN

Stanford University (Dave Noice).

STATUS

Tested.

SEE ALSO

ScanClkIn, ScanClkOut, StScanJoin.

LAYOUT SCALE

16 λ per inch.

CIF ID

513

LOGIC DIAGRAM

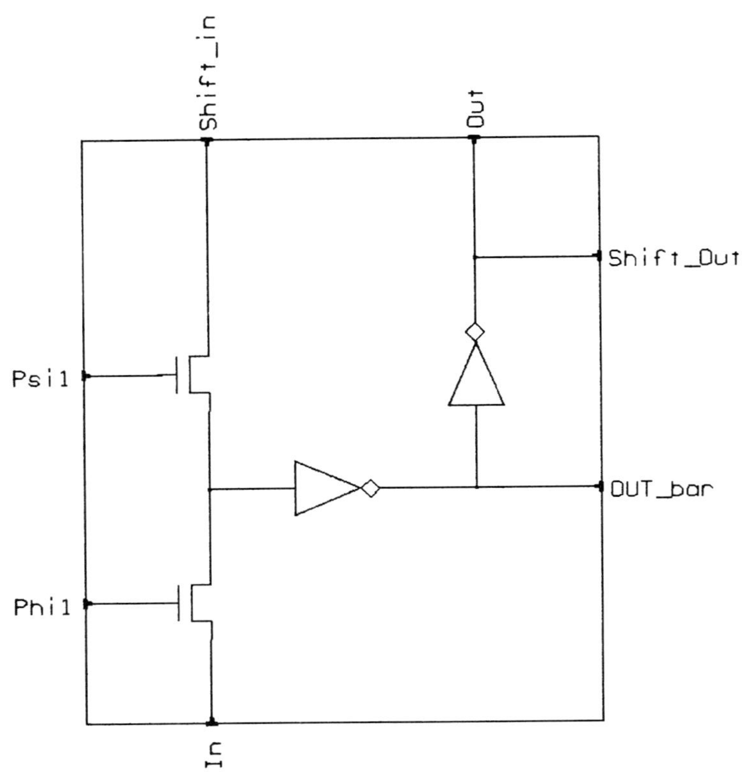

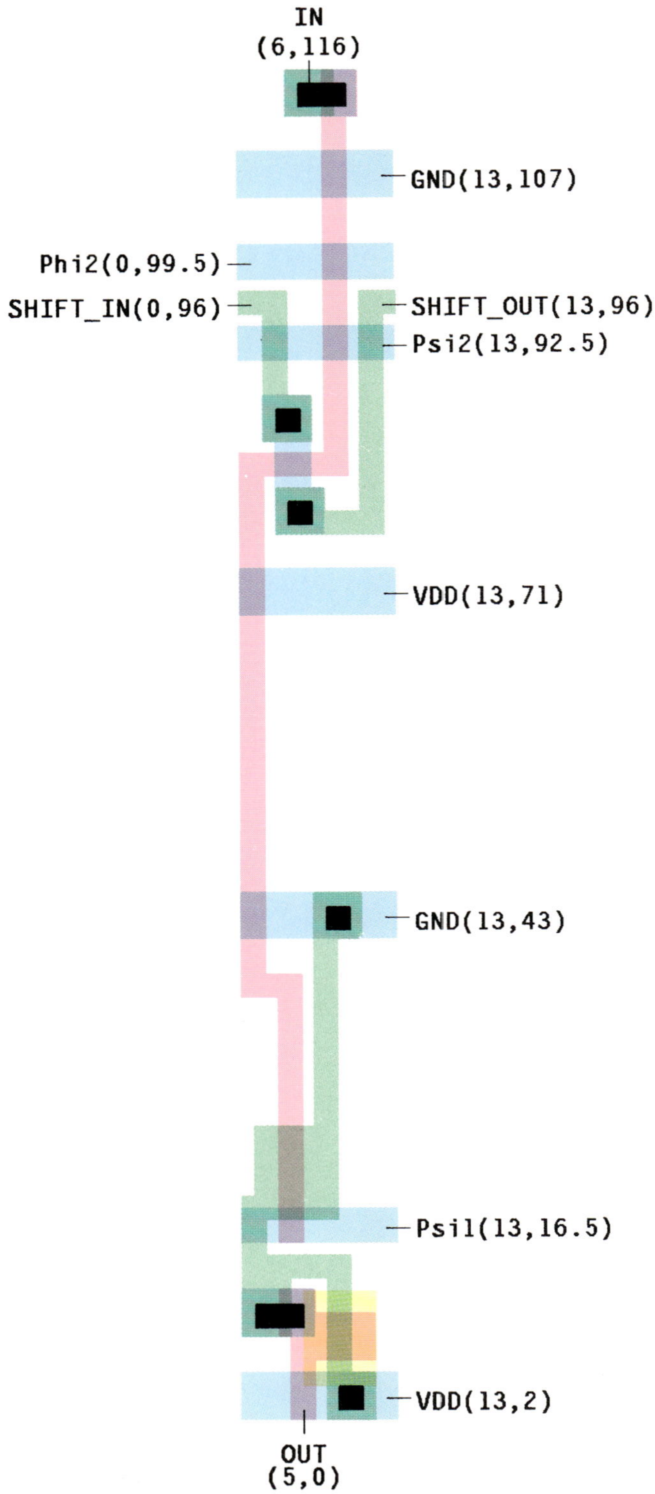
IN
(6,116)
GND(13,107)
Phi2(0,99.5)
SHIFT_IN(0,96)
SHIFT_OUT(13,96)
Psi2(13,92.5)
VDD(13,71)
GND(13,43)
Psi1(13,16.5)
VDD(13,2)
OUT
(5,0)

NAME

ScanOut—LSSD output buffer dummy

SYNOPSIS

A PLA Output buffer with connections to fit with LSSD output buffers.

PROPERTIES

Size

13 x 116 λ.

Interface

IN	input,	switched	9.5 □ C_g	
$\phi 2$	input,	switch control	1 □ C_g	
$\psi 2$	input,	switch control	1 □ C_g	
$\psi 1$	input,	switch control	1 □ C_g	
Shift_in	input,	switched	7 □ C_g	
Shift_out	output,	gate	7 □ C_g	[2 / ¼] □ *R*
OUT	output,	gate	6 □ C_g	[2 / ¼] □ *R*

DC Current

0.1 mA.

DESCRIPTION

This cell is a dummy cell that matches the *ScanClkOut* cell. It is useful for applications where some of the PLA outputs are for combinational logic—they are not clocked or useful in an LSSD chain.

TESSELATION

Abuts horizontally, with no overlap.

LOGIC DIAGRAM

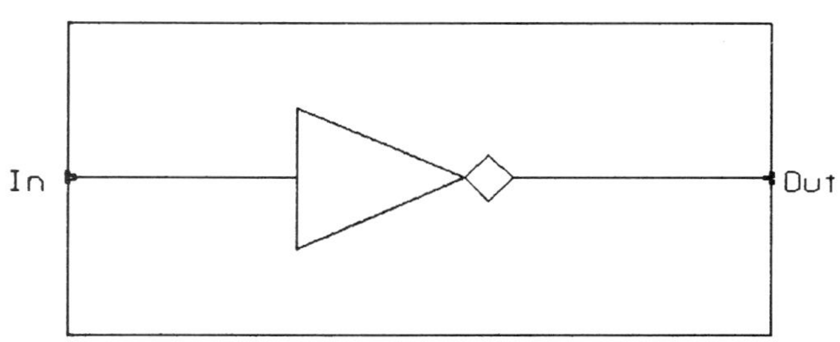

ORIGIN

Stanford University (Dave Noice).

STATUS

Tested.

SEE ALSO

Intro, ScanClkOut, ScanClkIn, StScanClkIn, ScanJoin, StScanJoin.

LAYOUT SCALE

16 λ per inch.

CIF ID

519

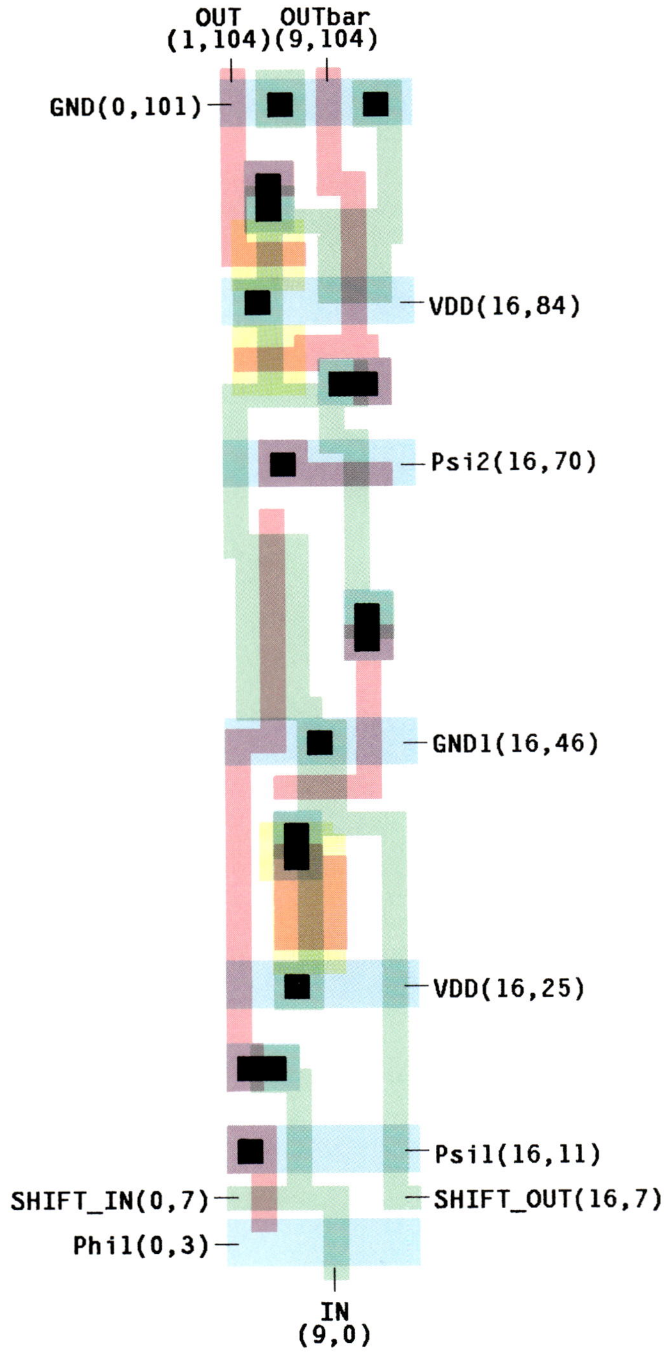
OUT
(1,104)
OUTbar
(9,104)
GND(0,101)
VDD(16,84)
Psi2(16,70)
GND1(16,46)
VDD(16,25)
Psi1(16,11)
SHIFT_IN(0,7)
SHIFT_OUT(16,7)
Phi1(0,3)
IN
(9,0)

NAME

ScanUnClkIn—LSSD input driver with unclocked input

SYNOPSIS

ScanUnClkIn is an unclocked LSSD input driver that can convert to an LSSD shift register.

PROPERTIES

Size

16 x 104 λ.

Interface

$\phi 1$	input,	switch control	2.5 □ C_g	
$\psi 1$	input,	switch control	2.5 □ C_g	
$\psi 2$	input,	switch control	2.5 □ C_g	
IN	input,	switched	14.5 □ C_g	
Shift_in	input,	switched	14.5 □ C_g	
Shift_out	output,	gate	12 □ C_g	[4 / ½] □ *R*
OUT	output,	gate	4.5 □ C_g	[1 / ¼] □ *R*
$\overline{OUT}$	output,	gate	13 □ C_g	[1 / ⅛] □ *R*

DC current

0.2 mA.

DESCRIPTION

ScanUnClkIn is identical to ScanClkIn except the input is not clocked by $\phi 1$. It is an unclocked LSSD input driver that can convert to an LSSD shift register. Any input to it *must* have been clocked since the input is a storage node. It is generally used when several inputs are multiplexed onto the storage node with qualified clocks. Data moves from left to right.

TESSELATION

Abuts horizontally, with no overlap.

ORIGIN

Stanford University (Dave Noice).

STATUS

Checked.

SEE ALSO

Intro, ScanClkIn, ScanIn, ScanJoin, ScanClkOut, ScanClkIn.

LAYOUT SCALE

16 λ per inch.

CIF ID

512

LOGIC DIAGRAM

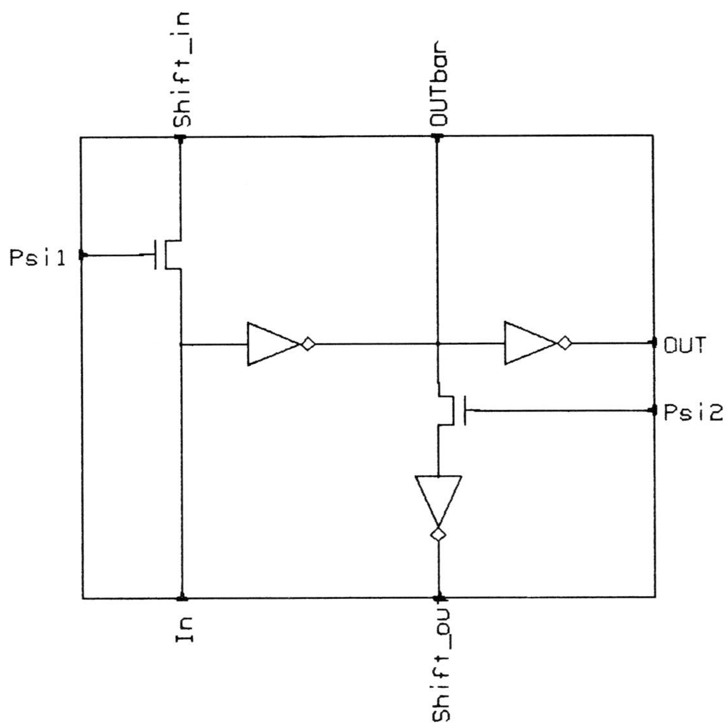

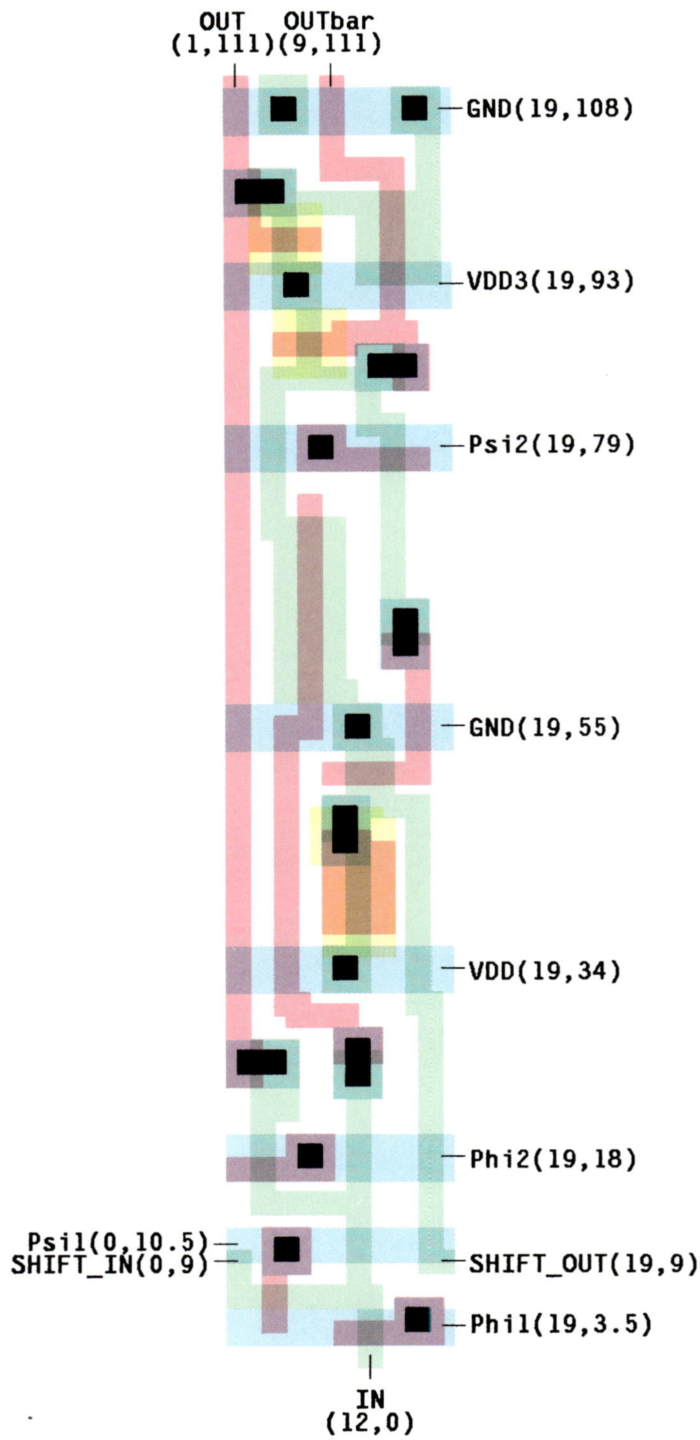
OUT
(1,111)
OUTbar
(9,111)
GND(19,108)
VDD3(19,93)
Psi2(19,79)
GND(19,55)
VDD(19,34)
Phi2(19,18)
Psi1(0,10.5)
SHIFT_IN(0,9)
SHIFT_OUT(19,9)
Phi1(19,3.5)
IN
(12,0)

NAME

StScanClkIn—static LSSD input driver

SYNOPSIS

StScanClkIn is a static, clocked PLA input driver that can convert to an LSSD shift register.

PROPERTIES

Size

19 x 111 λ.

Interface

$\phi 1$	input,	switch control	2.5 □ C_g	
$\phi 2$	input,	switch control	2.5 □ C_g	
$\psi 1$	input,	switch control	2.5 □ C_g	
$\psi 2$	input,	switch control	2.5 □ C_g	
IN	input,	switched	17 □ C_g	
Shift_In	input,	switched	17 □ C_g	
Shift_out	output,	gate	12.5 □ C_g	[4 / ½] □ *R*
OUT	output,	gate	10 □ C_g	[1 / ¼] □ *R*
$\overline{OUT}$	output,	gate	13 □ C_g	[1 / ⅛] □ *R*

DC current

0.2 mA.

DESCRIPTION

The only difference between this cell and the normal *ScanClkIn* cell is the feedback path used to restore the storage node every $\phi 2$. Note that the pitch of each cell is 19 λ wide. Thus, the PLA input plane pitch must be expanded from the normal 16 λ to 19 λ to fit properly.

TESSELATION

Abuts horizontally, with no overlap.

ORIGIN

Stanford University (Dave Noice).

STATUS

Checked.

SEE ALSO

Intro, ScanClkIn, StScanIn, StScanUnClkIn, StScanjoin, ScanClkOut, StScanClkIn.

LAYOUT SCALE

16 λ per inch.

CIF ID

514

LOGIC DIAGRAM

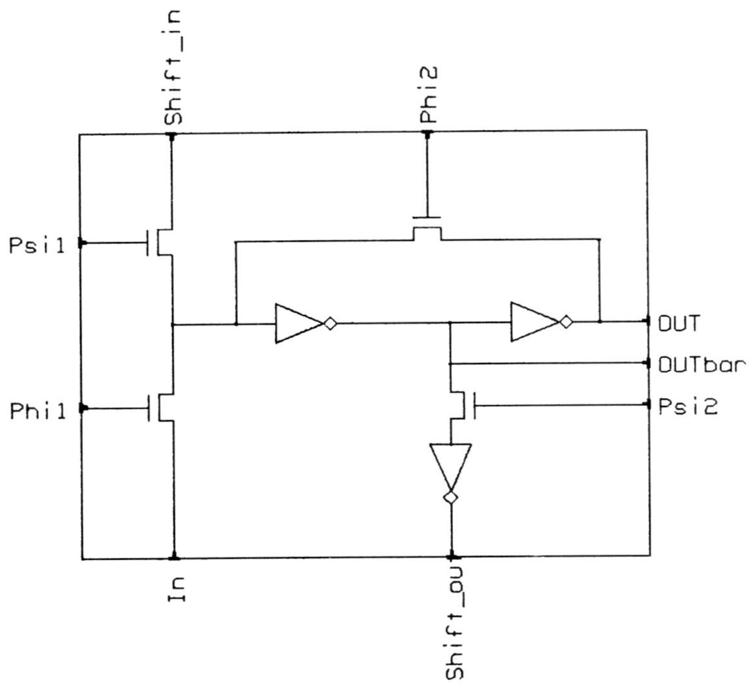

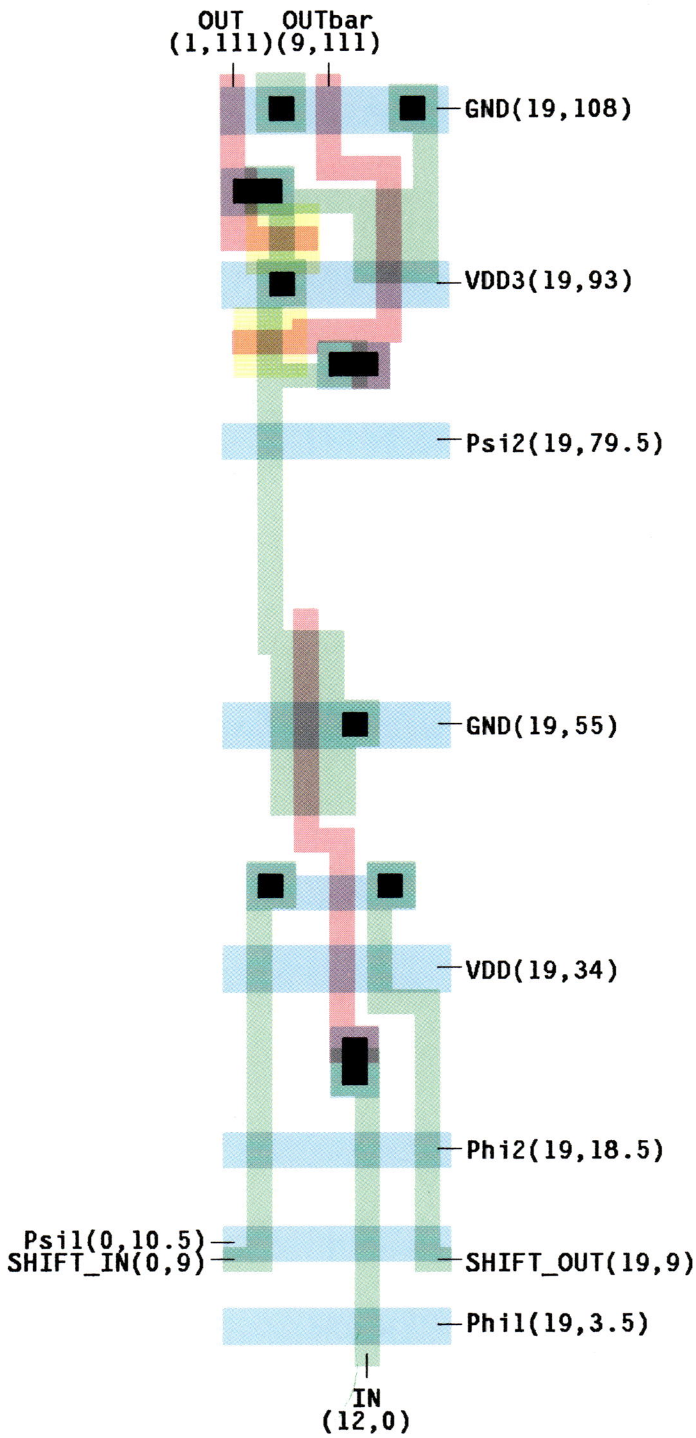
OUT
(1,111)
OUTbar
(9,111)
GND(19,108)
VDD3(19,93)
Psi2(19,79.5)
GND(19,55)
VDD(19,34)
Phi2(19,18.5)
Psi1(0,10.5)
SHIFT_IN(0,9)
SHIFT_OUT(19,9)
Phi1(19,3.5)
IN
(12,0)

NAME

StScanIn—static input driver dummy (unconnected to LSSD chain)

SYNOPSIS

StScanIn is a normal PLA input driver with connections to fit it in an LSSD chain.

PROPERTIES

Size

19 x 111 λ.

Interface

$\phi 1$	input,	switch control	1 □ C_g	
$\phi 2$	input,	switch control	1 □ C_g	
$\psi 1$	input,	switch control	1 □ C_g	
$\psi 2$	input,	switch control	1 □ C_g	
IN	input,	switched	13 □ C_g	
Shift_in	input,	switched	10 □ C_g	
Shift_out	output,	gate	10 □ C_g	[4 / ½] □ *R*
OUT	output,	gate	4.5 □ C_g	[1 / ¼] □ *R*
$\overline{OUT}$	output,	gate	13 □ C_g	[1 / ⅛] □ *R*

DC current

0.2 mA.

DESCRIPTION

StScanIn is a normal PLA input driver that fits the pitch of the other static LSSD input cells. It is used when parts of the PLA are used for combinational logic in which case the input cannot be clocked or part of the LSSD chain.

Note that the pitch of each cell is 19 λ wide. Thus, the PLA input plane pitch must be expanded from the normal 16 λ to 19 λ to fit properly.

TESSELATION

Abuts horizontally, with no overlap.

LOGIC DIAGRAM

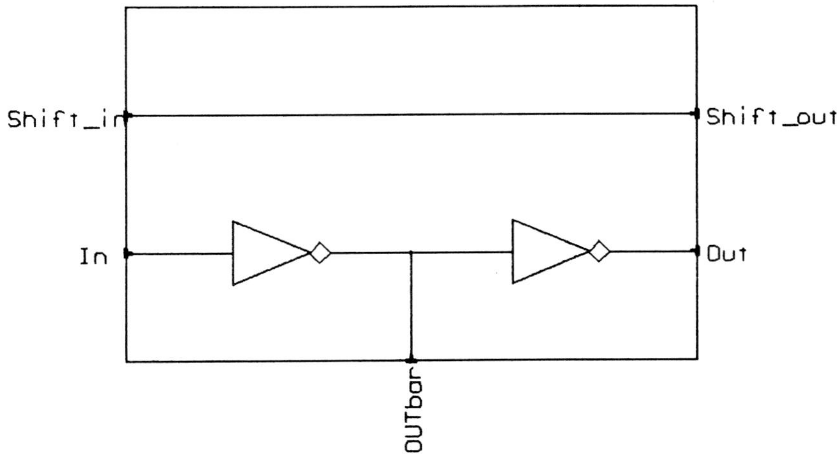

ORIGIN
Stanford University (Dave Noice).

STATUS
Checked.

SEE ALSO
Intro, StScanClkIn, StScanUnClkIn, StScanJoin, ScanClkOut.

LAYOUT SCALE
16 λ per inch.

CIF ID
515

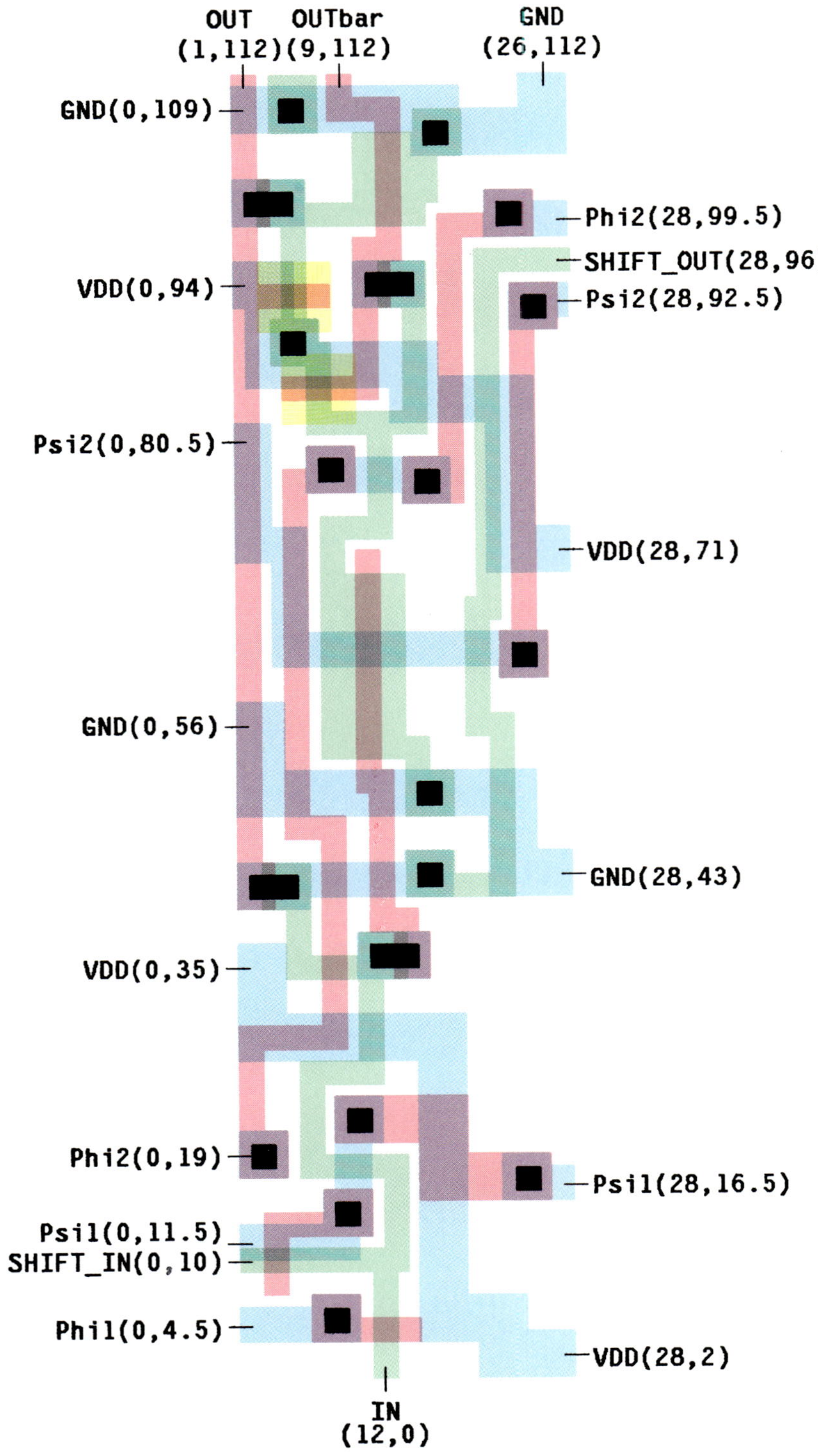
OUT
(1,112)
OUTbar
(9,112)
GND
(26,112)
GND(0,109)
Phi2(28,99.5)
SHIFT_OUT(28,96)
VDD(0,94)
Psi2(28,92.5)
Psi2(0,80.5)
VDD(28,71)
GND(0,56)
GND(28,43)
VDD(0,35)
Phi2(0,19)
Psi1(28,16.5)
Psi1(0,11.5)
SHIFT_IN(0,10)
Phi1(0,4.5)
VDD(28,2)
IN
(12,0)

NAME

StScanJoin—Static LSSD driver/buffer connect

SYNOPSIS

A Static LSSDin cell with connections added for linking it to a LSSDout cell.

PROPERTIES

Size

28 x 112 λ.

Interface

$\phi 1$	input,	switch control	2 □ C_g	
$\phi 2$	input,	switch control	7.5 □ C_g	
$\psi 1$	input,	switch control	5 □ C_g	
$\psi 2$	input,	switch control	4.5 □ C_g	
IN	input,	switched	17 □ C_g	
Shift_in	input,	switched	17 □ C_g	
Shift_out	output,	gate	19.5 □ C_g	[1 / ¼] □ *R*
OUT	output,	gate	19.5 □ C_g	[1 / ¼] □ *R*
$\overline{OUT}$	output,	gate	15 □ C_g	[1 / ⅛] □ *R*

DC current

0.2 mA.

DESCRIPTION

StScanJoin is the static version of *ScanJoin*. It has connections for all the clocks, power, ground, and the data output from the static LSSD shift register. It is only intended for LSSD schemes where the data shifts from the input side of the PLA to the output. Note that there is an extra one-phase shift provided to match the *Shift_Out* output phase to the *Shift_in* of the ScanClkOut cell.

TESSELATION

Abuts horizontally, with no overlap.

ORIGIN

Stanford University (Dave Noice).

STATUS

Checked.

SEE ALSO

Intro, StScanClkIn, ScanClkOut.

LAYOUT SCALE

16 λ per inch.

CIF ID

517

LOGIC DIAGRAM

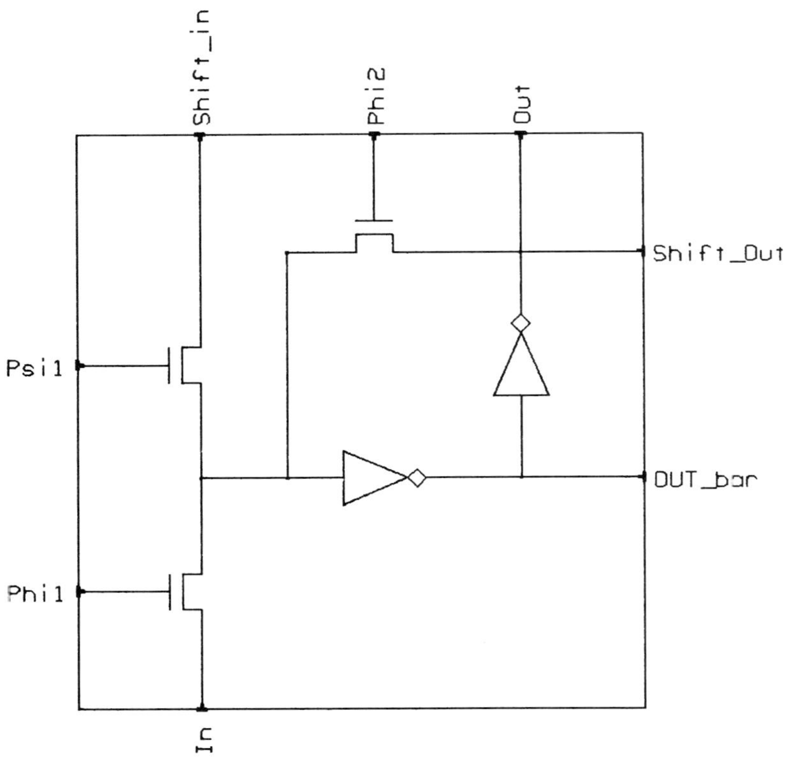

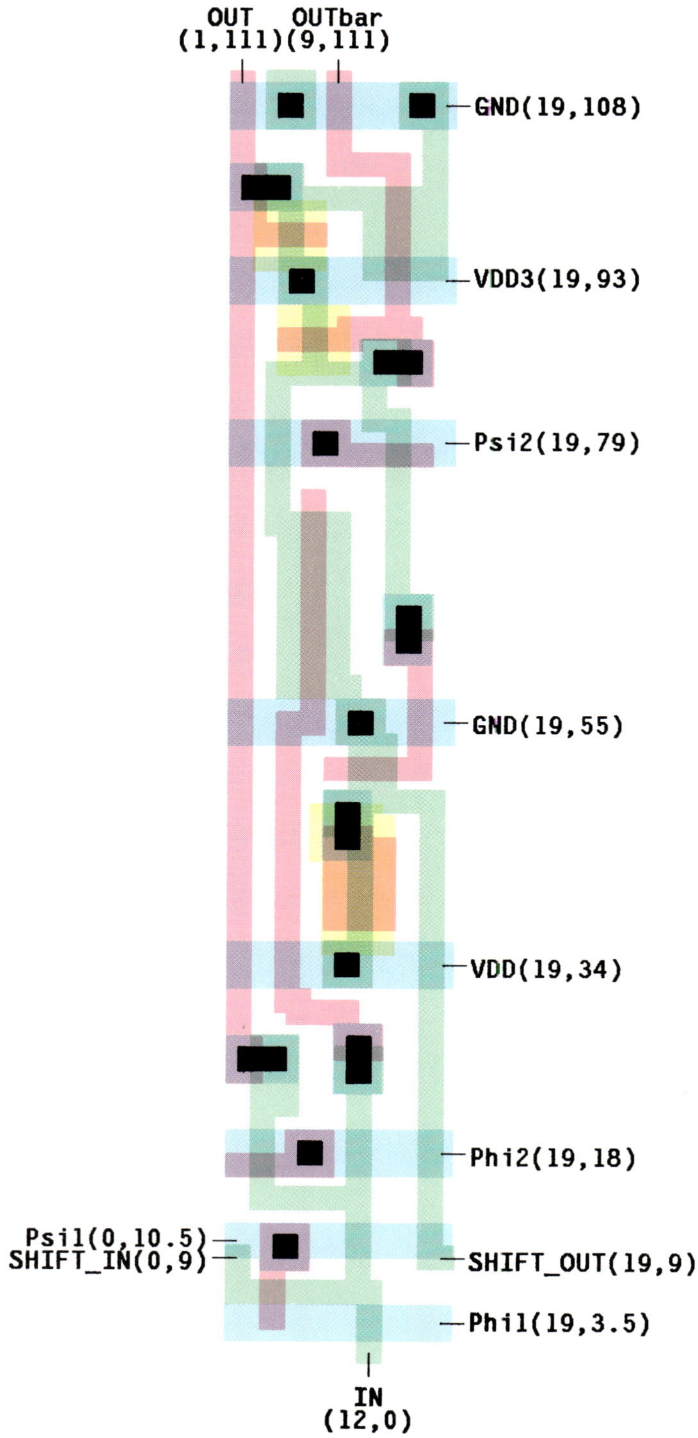
OUT
(1,111)
OUTbar
(9,111)
GND(19,108)
VDD3(19,93)
Psi2(19,79)
GND(19,55)
VDD(19,34)
Phi2(19,18)
Psi1(0,10.5)
SHIFT_IN(0,9)
SHIFT_OUT(19,9)
Phi1(19,3.5)
IN
(12,0)

NAME

StScanUnClkIn—static LSSD input driver with unclocked input

SYNOPSIS

StScanUnClkIn is a static, unclocked PLA input driver that can convert to an LSSD shift register.

PROPERTIES

Size

19 x 111 λ.

Interface

$\phi 1$	input,	switch control	2.5 □ C_g	
$\phi 2$	input,	switch control	2.5 □ C_g	
$\psi 1$	input,	switch control	2.5 □ C_g	
$\psi 2$	input,	switch control	2.5 □ C_g	
IN	input,	switched	16 □ C_g	
Shift_in	input,	switched	17 □ C_g	
Shift_out	output,	gate	12.5 □ C_g	[4 / ½] □ *R*
OUT	output,	gate	10 □ C_g	[1 / ¼] □ *R*
$\overline{OUT}$	output,	gate	13 □ C_g	[1 / ⅛] □ *R*

DC current

0.2 mA.

DESCRIPTION

StScanUnClkIn is a static, unclocked PLA input driver that can convert to an LSSD shift register. Any input to it *must* have been previously clocked with a pass transistor since the input is a storage node. It is generally used when several inputs are multiplexed onto the storage node with qualified clocks.

Note that the pitch of each cell is 19 λ wide. Thus, the PLA input plane pitch must be expanded from the normal 16 λ to 19 λ to fit properly.

TESSELATION

Abuts horizontally, with no overlap.

ORIGIN

Stanford University (Dave Noice).

STATUS

Checked.

SEE ALSO

Intro, StScanClkIn, StScanIn, StScanJoin, ScanClkOut, ScanClkIn.

LAYOUT SCALE

16 λ per inch.

CIF ID

516

LOGIC DIAGRAM

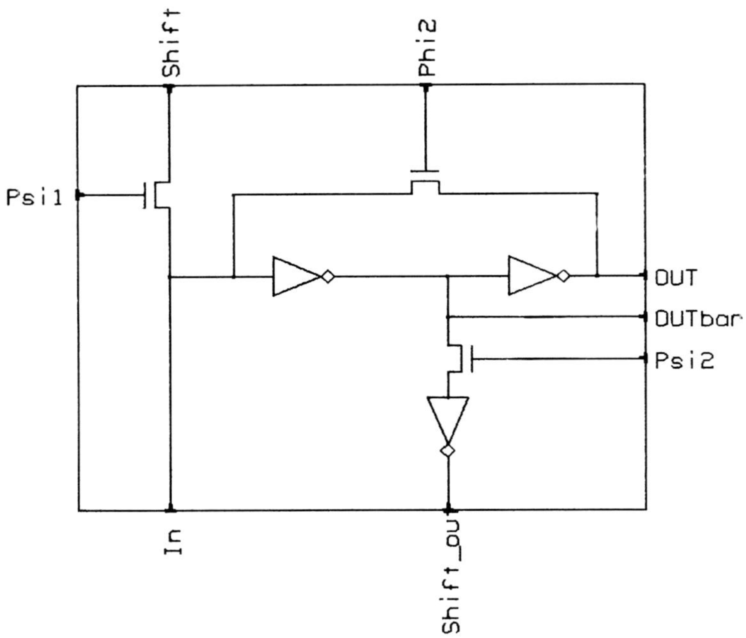

9 / Serial Arithmetic

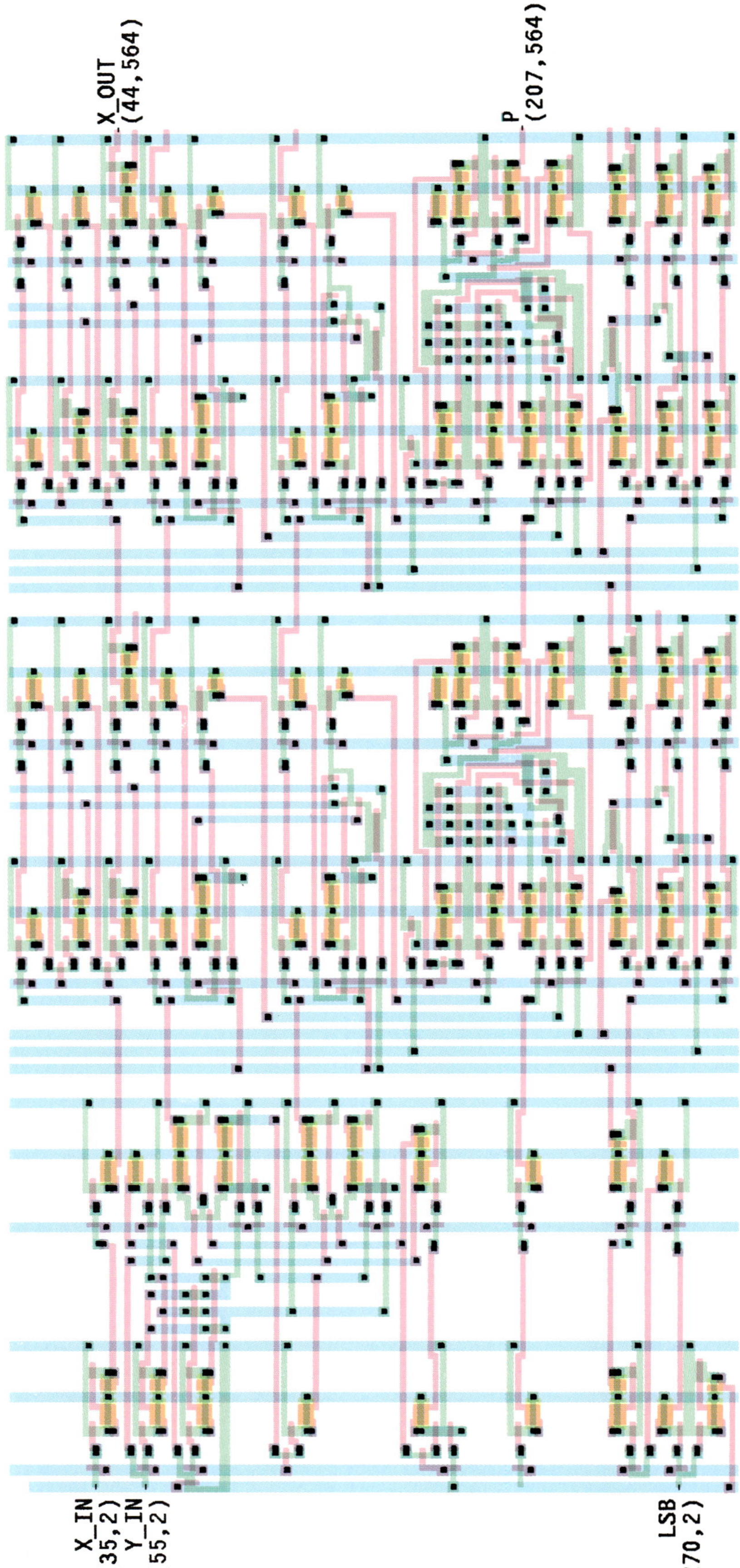
X_OUT
(44,564)
P
(207,564)
X_IN
(35,2)
Y_IN
(55,2)
LSB
(270,2)

NAME

intro—introduction to serial arithmetic

SYNOPSIS

Cells for constructing custom serial adders and multipliers.

DESCRIPTION

This section describes components for implementing a particular scheme of serial arithmetic processing using 2's complement representation. Numbers are coded least-significant-bit first on a data line, and an accompanying control line is used to mark the beginnings of numbers. Negative logic (low equals 1) is generally used. All cells are designed to fit into a vertical metal grid, supplying power and clocks and repeating at 200λ intervals:

Signal	*Center*
$\phi 1$	45λ
VDD	75λ
GND	96λ
$\phi 2$	145λ
VDD	175λ
GND	196λ

The multiplier cells use a five-level recoding scheme. A summary of it is given in the various cell sections, but a fuller description can be found in "Two's Complement Pipeline Multipliers" by R. F. Lyon, *IEEE Transactions on Communications,* April 1976.

Lyon's multiplier multiplies a $2N$-bit Y by a K-bit X, where N is the number of multiplier stages and $K > 2N$, to yield a K-bit product P:

$$P = floor\,(X * Y * 2^{-2(N-1)})$$

where X, Y and P are regarded as integers.

X and Y are input serially, least significant bits (LSBs) first. An active-high timing signal *LSB(0)* indicates the presence of the LSBs. Bits $2N+1$ through K on the Y line are ignored. The LSB of the product output appears $3N+1$ cycles after the LSBs of X and Y entered the inputs. The multiplier is fully pipelined so that a second pair of operands may enter before the first product is complete.

All input data is stable during $\phi 1$, and outputs change during $\phi 2$. Maximum clock rate is 12 MHz nominal, but in practice, a 18-bit multiplier seems to run at 3.3 MHz.

LAYOUT

An adder is fairly simple. Read the appropriate manual pages.

A multiplier consists of a recoder followed by N multiplier stages. Timing cells are attached underneath each recoder and multiplier stage. The next page shows a block diagram. Note that a cell with a patch is superimposed on the last timing delay cell.

If it is desired to multiply X by a pair of Ys to yield two products the double multiplier configuration may be used to save area. Also shown is a quadruple multiplier configuration computing a pair of products for each of two Xs. In double and quadruple configurations K must be the same for all products since common timing is used.

All cells are laid out on the standard 200λ grid distributing clocks, power, and ground and so are compatible with other cells in Lyon's signal processing family. These include adders, limiters, subtractors, *etc.*

Cells used in the multiplier are:

Recoder	DblRecoder	RecoderTime
Mult	DblMult	
TimingDelay	LastDelay	

ORIGIN

Xerox PARC (Dick Lyon); modified at Stanford University (Lyle Smith)

FURTHER EXPLANATION (Lyle Smith)

In general, the product of a 2N-bit and a K-bit 2's-complement number will be 2N+K−1 bits long. The serial multiplier outputs bits 2(N−1) through bits K+2N− (the top k bits of the product starting from the MSB-2 and going down; see figure below). The result is effectively the product scaled by 2**(−2(N−1)) with the digits to the right of the decimal point truncated. If the inputs are too big the output will overflow and the results will no longer be valid. In order to prevent overflows there has to be a range restriction on the inputs. The restriction can take many forms; and all that counts is that it works. Examples of input restrictions that can be used are 1) the top three bits of X or Y must be the same, or 2) the top two bits of X and Y must be the same.

If pps0 is always logic low (since we are dealing with negative logic this means +5 volts) then the lower 2N bits of the product are truncated to form the product, giving

$$P = floor\,(\,X*Y*2^{\,-2(N-1)})$$

Pps0 can also be used to round the final product or to add an offset to it. Pps0 is effectively added to the product before it is truncated, as shown in the figure below. The maximum value possible for the offset is (2**(K−2)) − 1 and the minimum offset is 0. Also the offset is limited to positive numbers (because the top 2N+1 bits of the offset are implicitly zeros). In summary,

$$P = floor\,[(X*Y+pps0*2^{-2})*2^{-2(N-1)}]$$

The most common use of pps0 is for rounding the product. This is done by having pps0 logic zero at all times except when the MSB of the 2N bit multiplier is being input. The resulting product is then

$$P = floor\,(X*Y*2^{-2(N-1)}+.5)$$

where we assuming that X and Y are being treated as integers.

FURTHER EXPLANATION DIAGRAMS

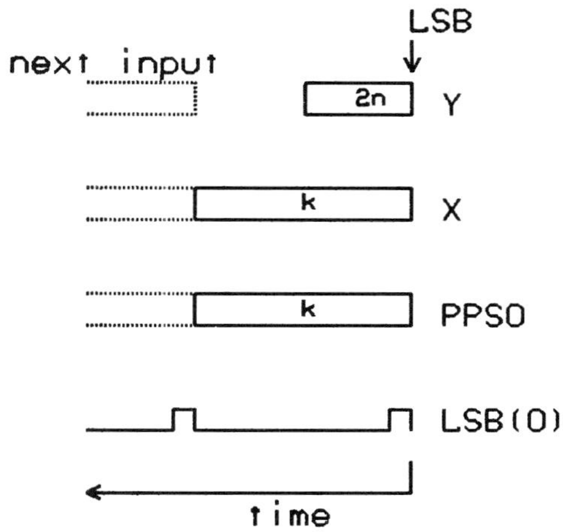

input format

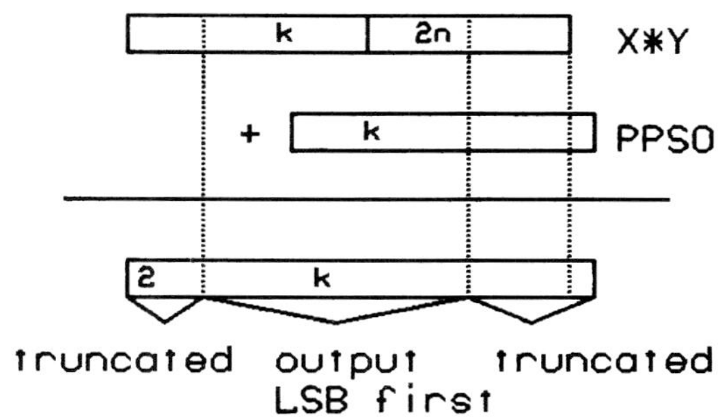

final output

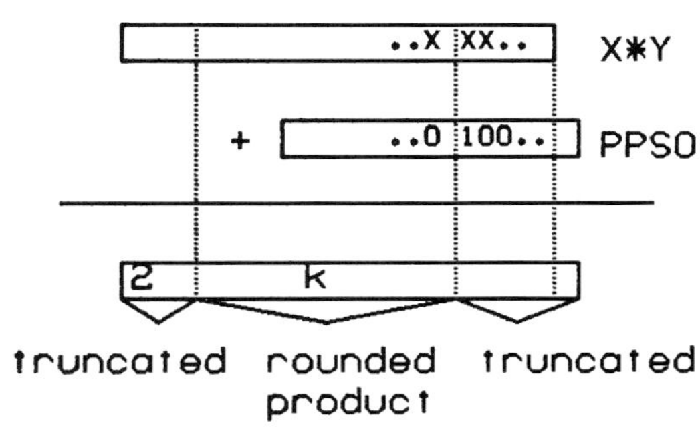

example using PPSO for rounding

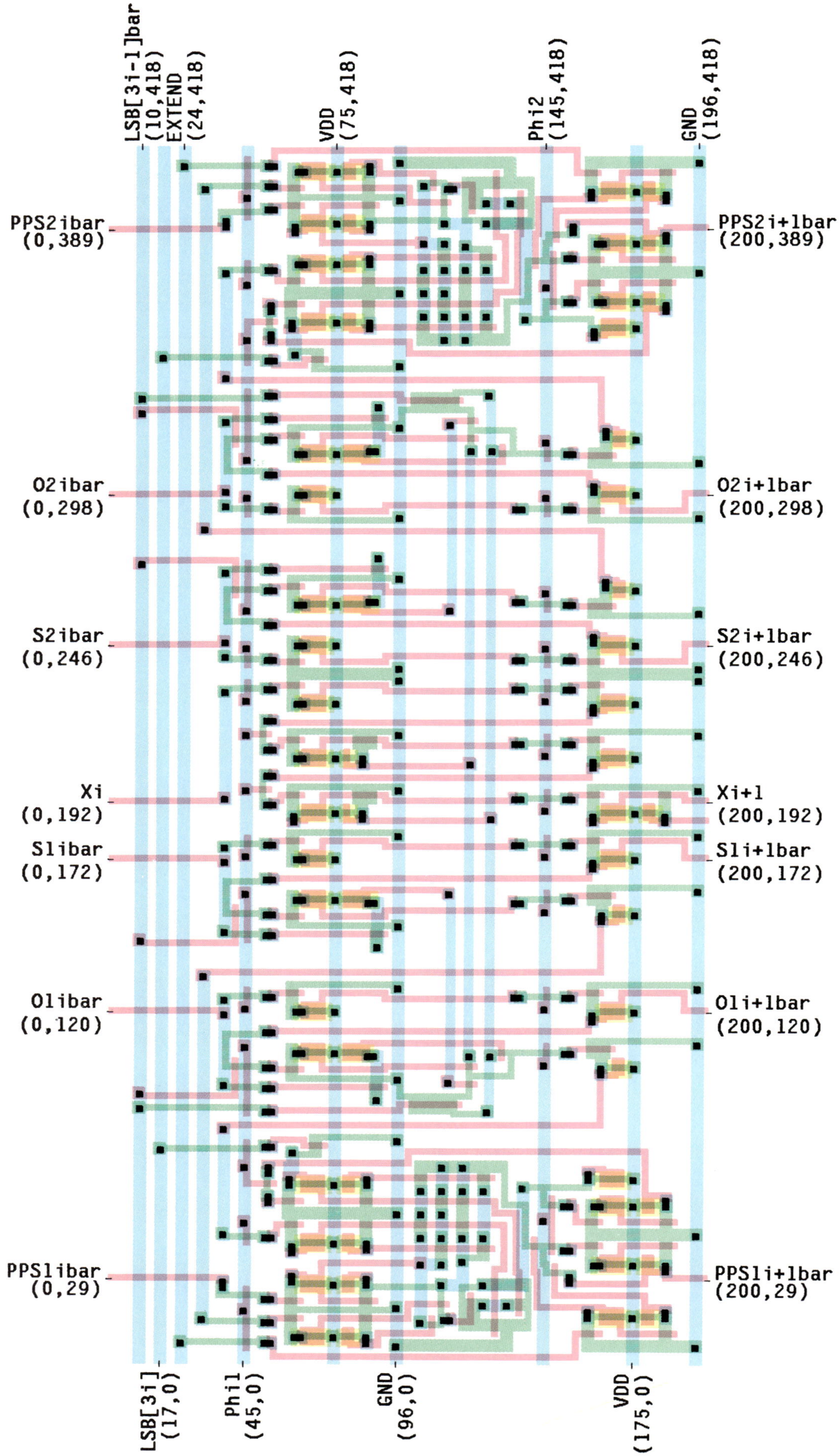
LSB[3i-1]bar
(10,418)
EXTEND
(24,418)
VDD
(75,418)
Phi2
(145,418)
GND
(196,418)
PPS2ibar
(0,389)
PPS2i+1bar
(200,389)
O2ibar
(0,298)
O2i+1bar
(200,298)
S2ibar
(0,246)
S2i+1bar
(200,246)
Xi
(0,192)
Xi+1
(200,192)
S1ibar
(0,172)
S1i+1bar
(200,172)
O1ibar
(0,120)
O1i+1bar
(200,120)
PPS1ibar
(0,29)
PPS1i+1bar
(200,29)
LSB[3i]
(17,0)
Phi1
(45,0)
GND
(96,0)
VDD
(175,0)

NAME

DblMult—dual serial multiplier

SYNOPSIS

Basic multiplier module for a dual serial multiplier.

PROPERTIES

Size

200 x 418 λ

Interface

Xi	input,		6.25 □ C_g	
$\overline{S1i}$	input,	switched	6.5 □ C_g	
$\overline{O1i}$	input,	switched	6.5 □ C_g	
$\overline{PPS1i}$	input,	switched	4 □ C_g	
$\overline{S2i}$	input,	switched	6.5 □ C_g	
$\overline{O2i}$	input,	switched	6.5 □ C_g	
$\overline{PPS2i}$	input,	switched	4 □ C_g	
$\overline{LSB(3i-1)}$	input,	switch control	57 □ C_g	
$\overline{LSB(3i)}$	input,	switched	39 □ C_g	
$EXTEND$	input,	switched	37.5 □ C_g	
$Xi+1$	output,	gate	10.5 □ C_g	[4 / ½] □ R
$\overline{PPS1i+1}$	output,	gate	8 □ C_g	[2 / ½] □ R
$\overline{S1i+1}$	output,	gate	18 □ C_g	[4 / ½] □ R
$\overline{O1i+1}$	output,	gate	18 □ C_g	[4 / ½] □ R
$\overline{PPS2i+1}$	output,	gate	8 □ C_g	[2 / ½] □ R
$\overline{S2i+1}$	output,	gate	18 □ C_g	(4 / ½] □ R
$\overline{O2i+1}$	output,	gate	18 □ C_g	[4 / ½] □ R

DC Current

3.5 mA including associated timing delay stage.

DESCRIPTION

The *DblMult, DblRecoder, TimingDelay, RecoderTime,* and *LastDelay* cells implement a dual serial two's complement pipelined multiplier. The multiplier consists of a recoder stage and N multiplier stages and will multiply a two 2N-bit Ys and a K-bit X (K>2N) to yield two K-bit products after 3N+1 clock cycles. A new X operand can be put in every K cycles.

Inputs Xi, $\overline{S1i}$ and $\overline{S2i}$, $\overline{O1i}$, $\overline{O2i}$, $\overline{PPS1i}$, and $\overline{PPS2i}$ are received from the previous multiplier stage or recoder. *Si* and *Oi* are recoded versions of the Y operands. $\overline{LSB(3i\text{-}1)}$, $\overline{LSB(3i)}$, and *EXTEND* are three timing signals received from the associated timing delay stage. Outputs $Xi+1$ and $\overline{PPSi+1}$ come directly from gates and so may enter 4:1 inputs. On the last stage $\overline{PPSi+1}$ is the product output. The cell's logic is the same as two *Mult* cells that share the same X operand circuitry. *LastDelay* instead of *TimingDelay* should be used for the timing delay on the last stage.

ORIGIN

Xerox PARC (Dick Lyon).

STATUS

Generally used.

LAYOUT SCALE

50 λ per inch.

CIF ID

105

SEE ALSO

Mult, DblRecoder, TimingDelay.

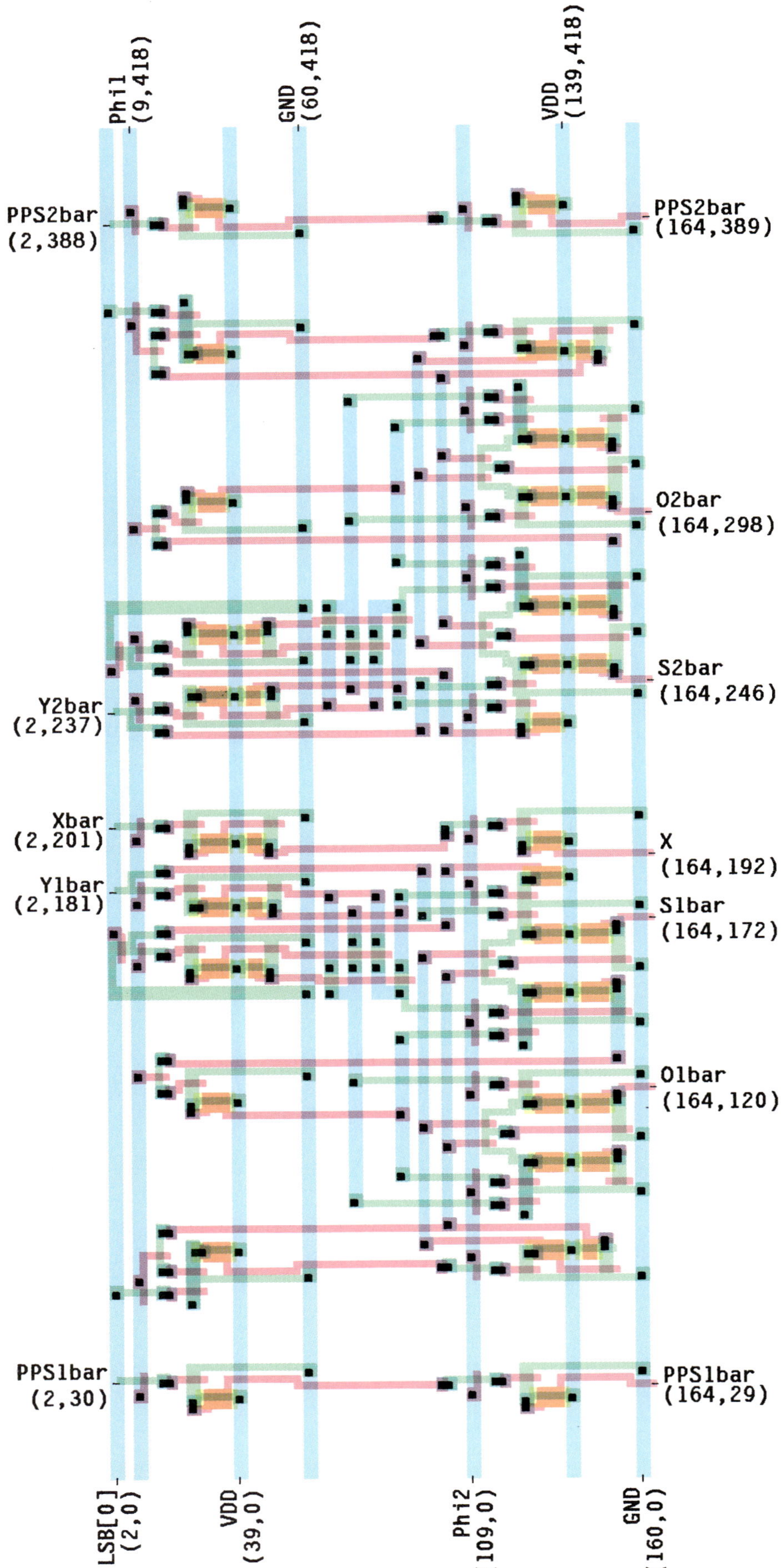
Phi1 (9,418)
GND (60,418)
VDD (139,418)
PPS2bar (2,388)
PPS2bar (164,389)
O2bar (164,298)
S2bar (164,246)
Y2bar (2,237)
Xbar (2,201)
X (164,192)
Y1bar (2,181)
S1bar (164,172)
O1bar (164,120)
PPS1bar (2,30)
PPS1bar (164,29)
LSB[0] (2,0)
VDD (39,0)
Phi2 (109,0)
GND (160,0)

NAME

DblRecoder—first stage of dual serial multiplier

SYNOPSIS

Y-operand recoder for dual serial two's complement multiplier.

PROPERTIES

Size

164 x 418 λ

Interface

X	output,	gate	4 □ C_g	[4 / ½] □ R
$\overline{S1}$	output,	gate	4 □ C_g	[4 / ½] □ R
$\overline{S2}$	output,	gate	4 □ C_g	[4 / ½] □ R
$\overline{O1}$	output,	gate	4 □ C_g	[4 / ½] □ R
$\overline{O2}$	output,	gate	4 □ C_g	[4 / ½] □ R
$\overline{PPS1}$	output,	gate	4 □ C_g	[4 / ½] □ R
$\overline{PPS2}$	output,	gate	4 □ C_g	[4 / ½] □ R
$LSB(O)$	input,	switch control	11 □ C_g	
$\overline{X}$	input,	switched	3 □ C_g	
$\overline{Y1}$	input,	switched	3 □ C_g	
$\overline{Y2}$	input,	switched	3 □ C_g	
$\overline{PPS1}$	input,	switched	3 □ C_g	
$\overline{PPS2}$	input,	switched	3 □ C_g	

DC Current

1.3 mA

DESCRIPTION

This is the first stage of the dual multiplier. It recodes two Y operands into two-bit digits for use in the later stages. It consists of two single recoders sharing the same X operand, so see the documentation for *Recoder* for the details of the logic and recoding scheme. The *Subtract* and *Zero* signals are multiplexed onto one wire, *S*, with *Zero* coming out on even and *Subtract* on odd clock cycles. The same is true for the *One* and *Two* signals. The partial product input, $\overline{PPS}$, is usually tied to logic zero (physical VDD), but can be used for rounding by adding in a one (physical ground) with a significance of ½ LSB. The X operand is passed through unchanged. The *LSB(0)* input comes from the associated cell *RecoderTime*.

ORIGIN

Xerox PARC (Dick Lyon).

STATUS

Generally used.

LAYOUT SCALE

50 λ per inch.

CIF ID

108

SEE ALSO

RecoderTime, Mult, Recoder.

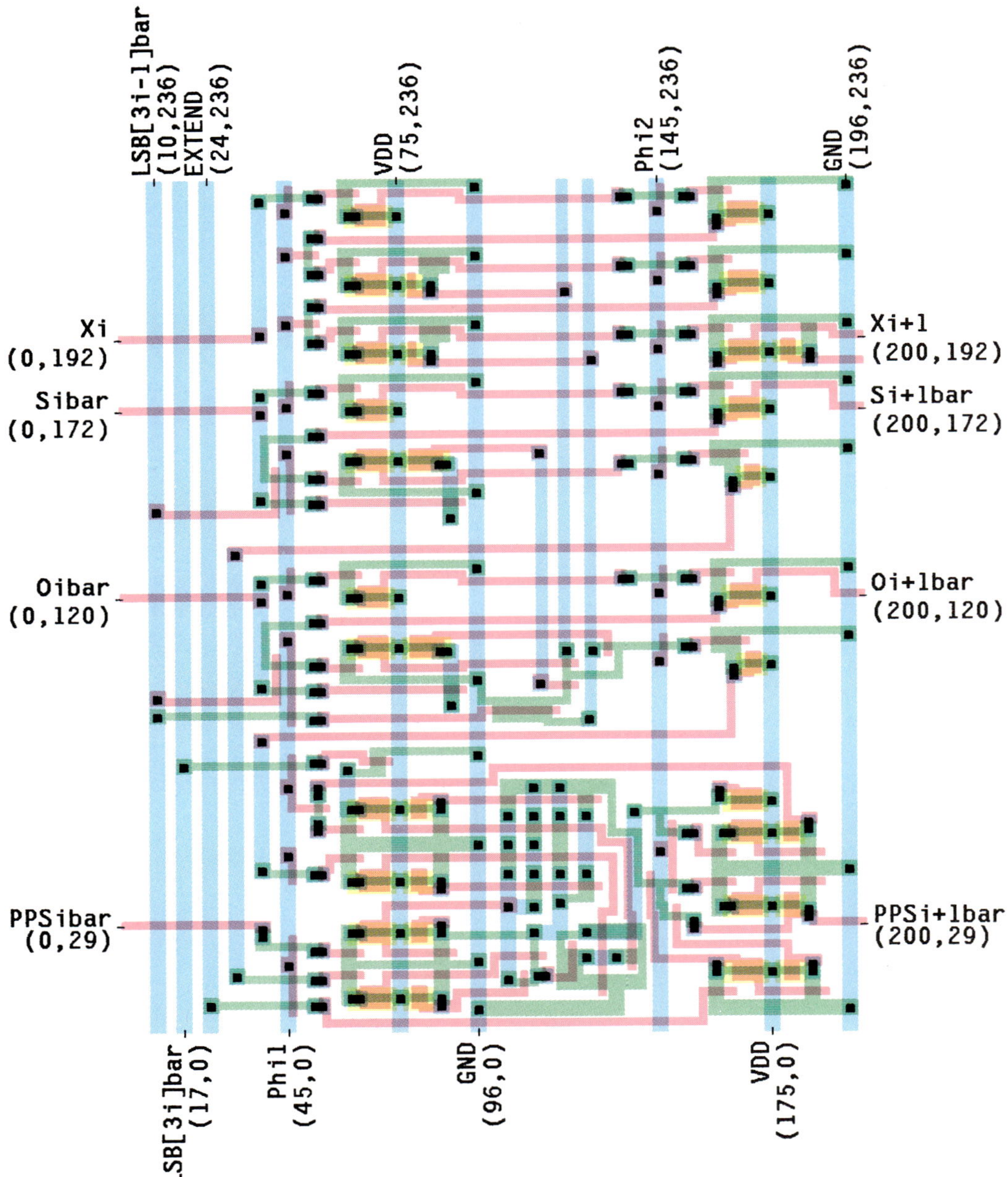

LSB[3i-1]bar (10,236)
EXTEND (24,236)
VDD (75,236)
Phi2 (145,236)
GND (196,236)
Xi (0,192)
Sibar (0,172)
Oibar (0,120)
PPSibar (0,29)
Xi+1 (200,192)
Si+1bar (200,172)
Oi+1bar (200,120)
PPSi+1bar (200,29)
LSB[3i]bar (17,0)
Phi1 (45,0)
GND (96,0)
VDD (175,0)

NAME

Mult—serial multiplier

SYNOPSIS

Basic multiplier module for serial two's complement multiplier.

PROPERTIES

Size

200 x 236 λ

Interface

X(3i-2)	input,	switched	6.25 □ C_g	
$\overline{Si}$	input,	switched	6.5 □ C_g	
$\overline{Oi}$	input,	switched	6.5 □ C_g	
$\overline{PPSi}$	input,	switched	4 □ C_g	
$\overline{LSB(3i-1)}$	input,	switch control	30.5 □ C_g	
$\overline{LSB(3i)}$	input,	switched	21.5 □ C_g	
EXTEND	input,	switched	20.75 □ C_g	
X(3i+1)	output,	gate	5 □ C_g	[4 / ½] □ *R*
$\overline{PPSi+1}$	output,	gate	8 □ C_g	[2 / ½] □ *R*
$\overline{Si+1}$	output,	gate	18 □ C_g	[4 / ½] □ *R*
$\overline{Oi+1}$	output,	gate	18 □ C_g	[4 / ½] □ *R*

DC Current

2.3 mA including associated *TimingDelay* stage.

DESCRIPTION

The *Mult, Recoder, TimingDelay, RecoderTime* and *LastDelay* cells implement a serial two's complement pipelined multiplier. The multiplier consists of a recoder stage and N multiplier stages and will multiply a 2N-bit Y and a K-bit X (K > 2N) to yield a K-bit product after 3N+1 clock cycles. A new X operand can be put in every K cycles.

Inputs X(3i-2), $\overline{Si}$, $\overline{Zi}$, $\overline{Oi}$, $\overline{Ti}$, and $\overline{PPSi}$ are received from the previous multiplier stage or recoder. *Si* and *Oi* are recoded versions of the Y operand. *LSB(3i-1), LSB(3i)*, and *EXTEND* are three timing signals received from the associated timing delay stage. Outputs *X(3j-2)* and $\overline{PPSi+1}$ come directly from gates and so may enter 4:1 inputs. On the last stage $\overline{PPSi+1}$ is the product output.

LastDelay instead of *TimingDelay* should be used for the timing delay on the last stage. The diagram on the next page shows the cell's logic.

ORIGIN

Xerox PARC (Dick Lyon).

STATUS

Generally used.

LAYOUT SCALE

50 λ per inch.

CIF ID

104

SEE ALSO

Lyon's multiplier introduction, documentation on other cells listed above. Double and quadruple versions are available which compute 2 or 4 products; see DblMult and DblRecoder cell documentation.

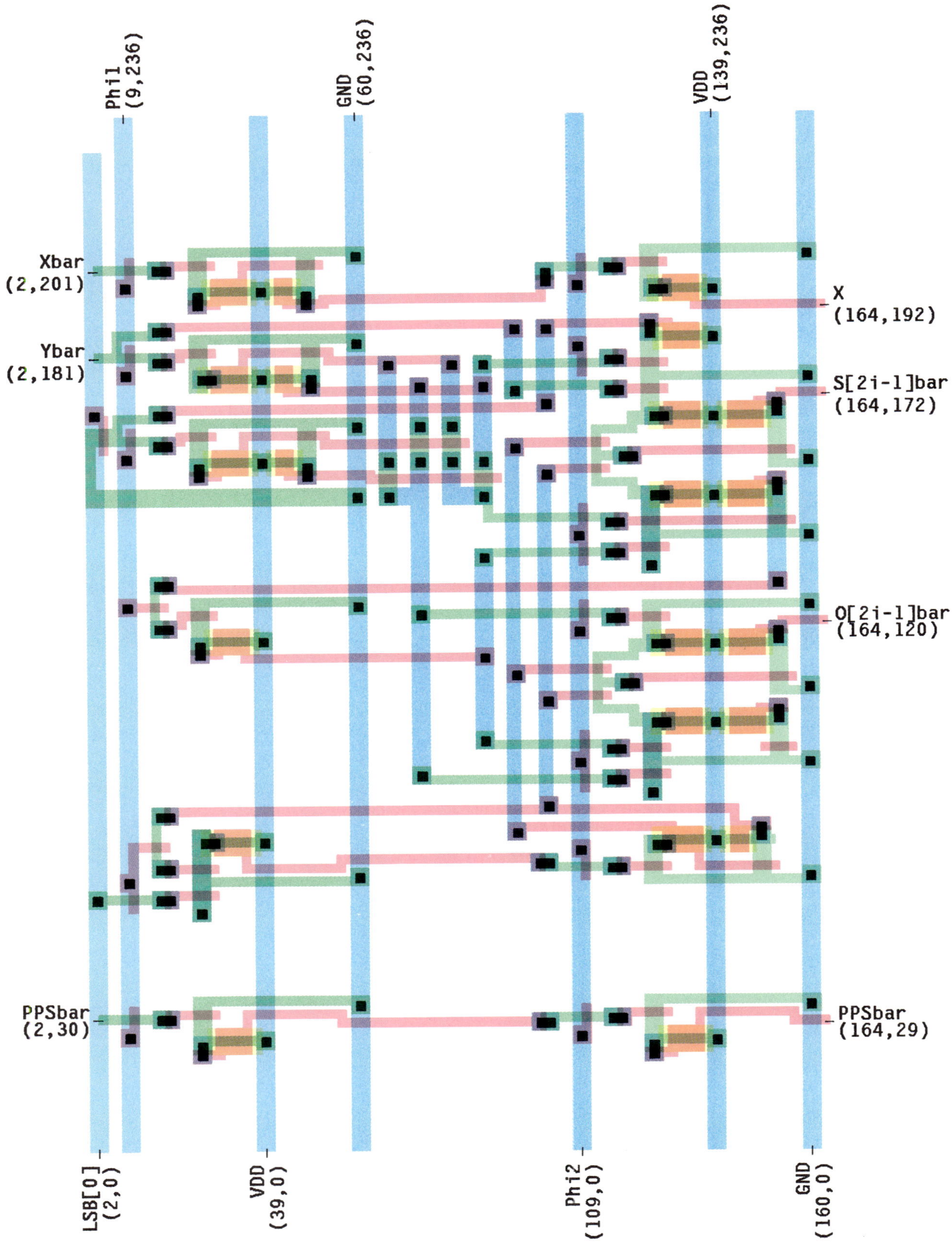
Phi1
(9,236)
GND
(60,236)
VDD
(139,236)
Xbar
(2,201)
X
(164,192)
Ybar
(2,181)
S[2i-1]bar
(164,172)
O[2i-1]bar
(164,120)
PPSbar
(2,30)
PPSbar
(164,29)
LSB[0]
(2,0)
VDD
(39,0)
Phi2
(109,0)
GND
(160,0)

NAME

Recoder—first stage of serial multiplier

SYNOPSIS

Y-operand for recoder serial two's complement multiplier.

PROPERTIES

Size

164 x 236 λ

Interface

$\overline{X}$	output	gate	4 □ C_g	[4 / ½] □ *R*
$\overline{S[2i-1]}$	output,	gate	4 □ C_g	[4 / ½] □ *R*
$\overline{O[2i-1]}$	output,	gate	4 □ C_g	[4 / ½] □ *R*
$\overline{PPS}$	output,	gate	4 □ C_g	[4 / ½] □ *R*
$\overline{LSB(O)}$	input,		11 □ C_g	
$\overline{X}$	input,		3 □ C_g	
$\overline{Y}$	input,		3 □ C_g	
$\overline{PPS}$	input,		3 □ C_g	

DC Current

0.7 mA

DESCRIPTION

This is the first stage of the multiplier. It recodes the Y operand into a two-bit digit for use in the later stages. The block diagram on the next page shows the cell's logic and the recoding scheme. The *Subtract* and *Zero* signals are multiplexed onto one wire, *S(2i-1)*, with *Zero* coming out on even and *Subtract* on odd clock cycles. The same is true for the *One* and *Two* signals. The partial product input, $\overline{PPS}$, is usually tied to logic zero (physical VDD), but can be used for rounding by adding in a one (physical ground) with a significance of ½ LSB. The X operand is passed through unchanged. The *LSB(0)* signal comes from the associated *RecoderTime* cell.

ORIGIN

Xerox PARC (Dick Lyon).

STATUS

Generally used.

LAYOUT SCALE

32 λ per inch.

CIF ID

107

SEE ALSO

Mult, RecoderTime. double and quadruple versions are available which compute 2 or 4 products; see DblMult and DblRecoder cell documentation.

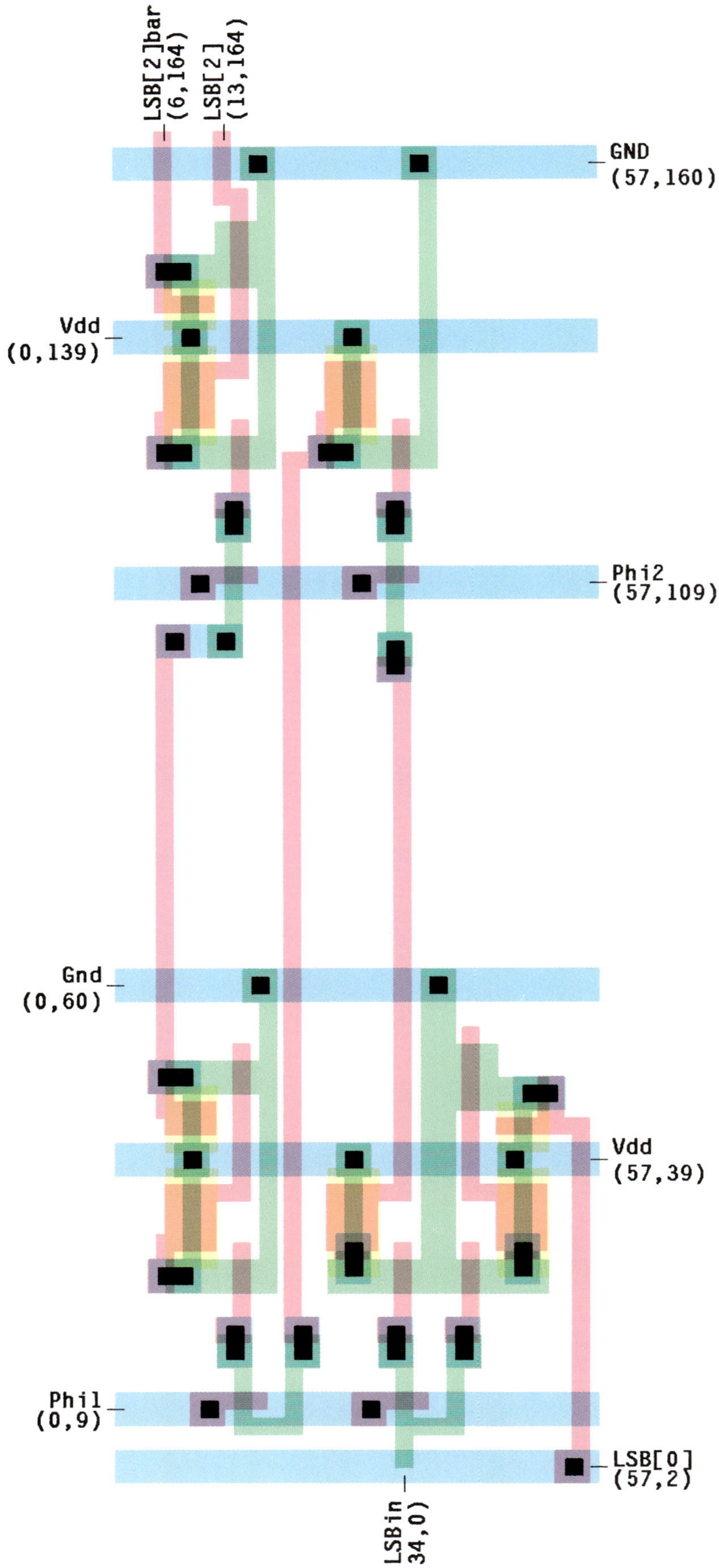

LSB[2]bar (6,164)
LSB[2] (13,164)
GND (57,160)
Vdd (0,139)
Phi2 (57,109)
Gnd (0,60)
Vdd (57,39)
Phi1 (0,9)
LSB[0] (57,2)
LSBin (34,0)

NAME

RecoderTime—timing for serial multiplier

SYNOPSIS

Timing for first stage of the serial multiplier.

PROPERTIES

Size

57 x 164 λ

Interface

LSBin	input,	switch control	7 □ C_g	
LSB(0)	output,	gate	2 □ C_g	[1 / ¼] □ *R*
$\overline{LSB(2)}$	output,	gate	2 □ C_g	[1 / ¼] □ *R*
LSB(2)	output,	gate	6 □ C_g	[4 / ½] □ *R*

DC Current

0.4 mA

DESCRIPTION

RecoderTime handles the control signal timing for the first stage of Lyon's serial pipelined multiplier. *LSB(0)* is sent to the associated recoder stage, either *Recoder* or *DblRecoder*. *LSB(2)* and $\overline{LSB(2)}$ are sent to the following *TimingDelay* cell. The output is the same as the input delayed by two clock cycles.

ORIGIN

Xerox PARC (Dick Lyon).

STATUS

Generally used.

LAYOUT SCALE

20 λ per inch.

CIF ID

109

SEE ALSO

Recoder, DblRecoder.

LOGIC DIAGRAM

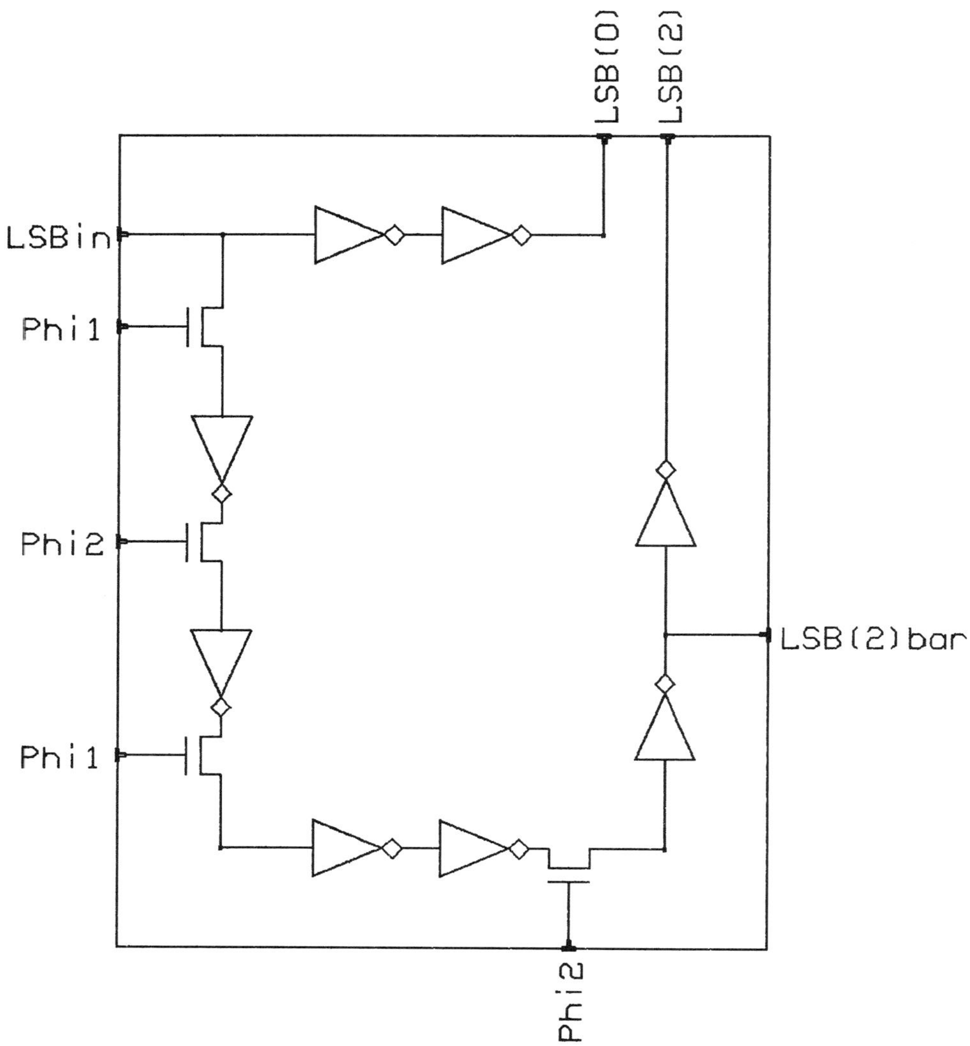

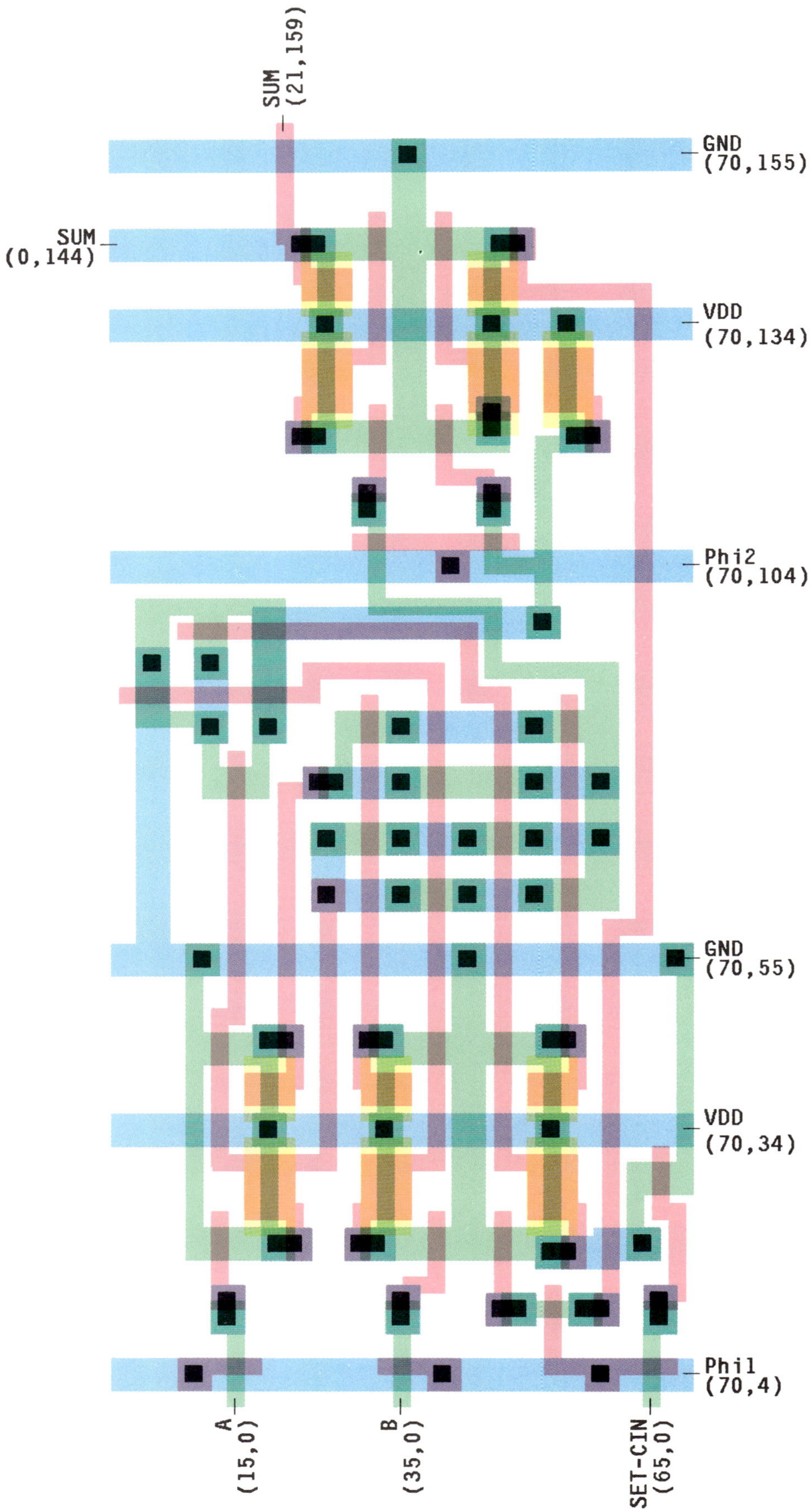
SUM
(21,159)
GND
(70,155)
SUM
(0,144)
VDD
(70,134)
Phi2
(70,104)
GND
(70,55)
VDD
(70,34)
Phi1
(70,4)
A
(15,0)
B
(35,0)
SET-CIN
(65,0)

NAME

S_Add—serial adder

SYNOPSIS

Serial adder.

PROPERTIES

Size

78 x 159λ

Interface

A	input,	switched	4 □ C_g
B	input,	switched	4 □ C_g
SET_CIN	input,	switched	4 □ C_g
$\phi 1$	input,	switch control	10 □ C_g
$\phi 2$	input,	switch control	8 □ C_g
Sum	output,	gate	6.75 □ C_g [2 / ½] □ *R*

DC Current

0.5 mA

DESCRIPTION

A and B are serial input operands that come in LSB first. SET_CIN actually clears the carry in; it should be high during the LSB time and low otherwise. Like the other multiplier cells this uses negative logic (low equals 1).

TESSELATION

Abuts horizontally with no overlap.

ORIGIN

Xerox PARC (Dick Lyon).

STATUS

Generally used.

LAYOUT SCALE

20 λ per inch.

CIF ID

129

SEE ALSO

S_AddSub.

LOGIC DIAGRAM

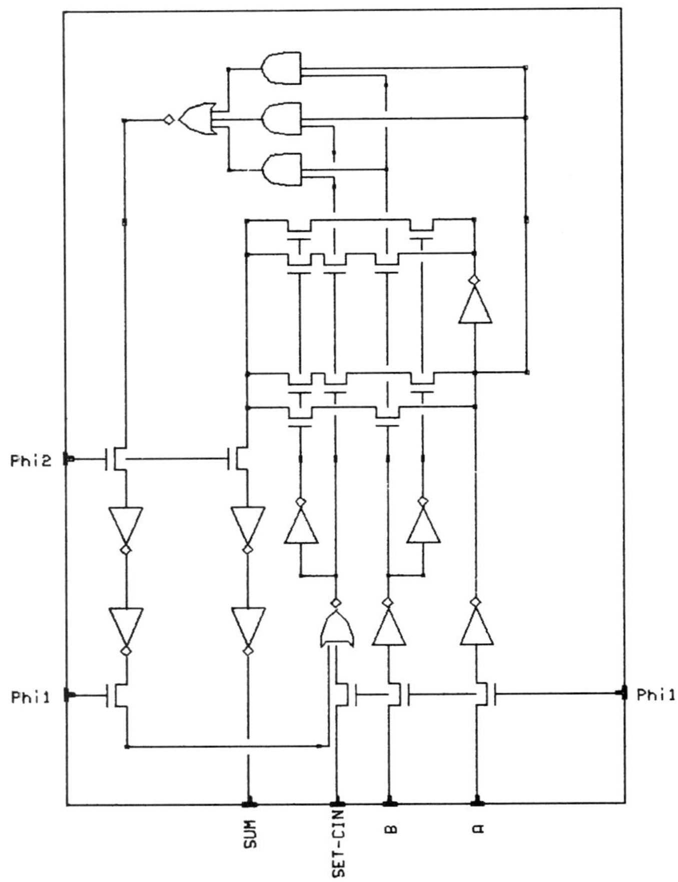

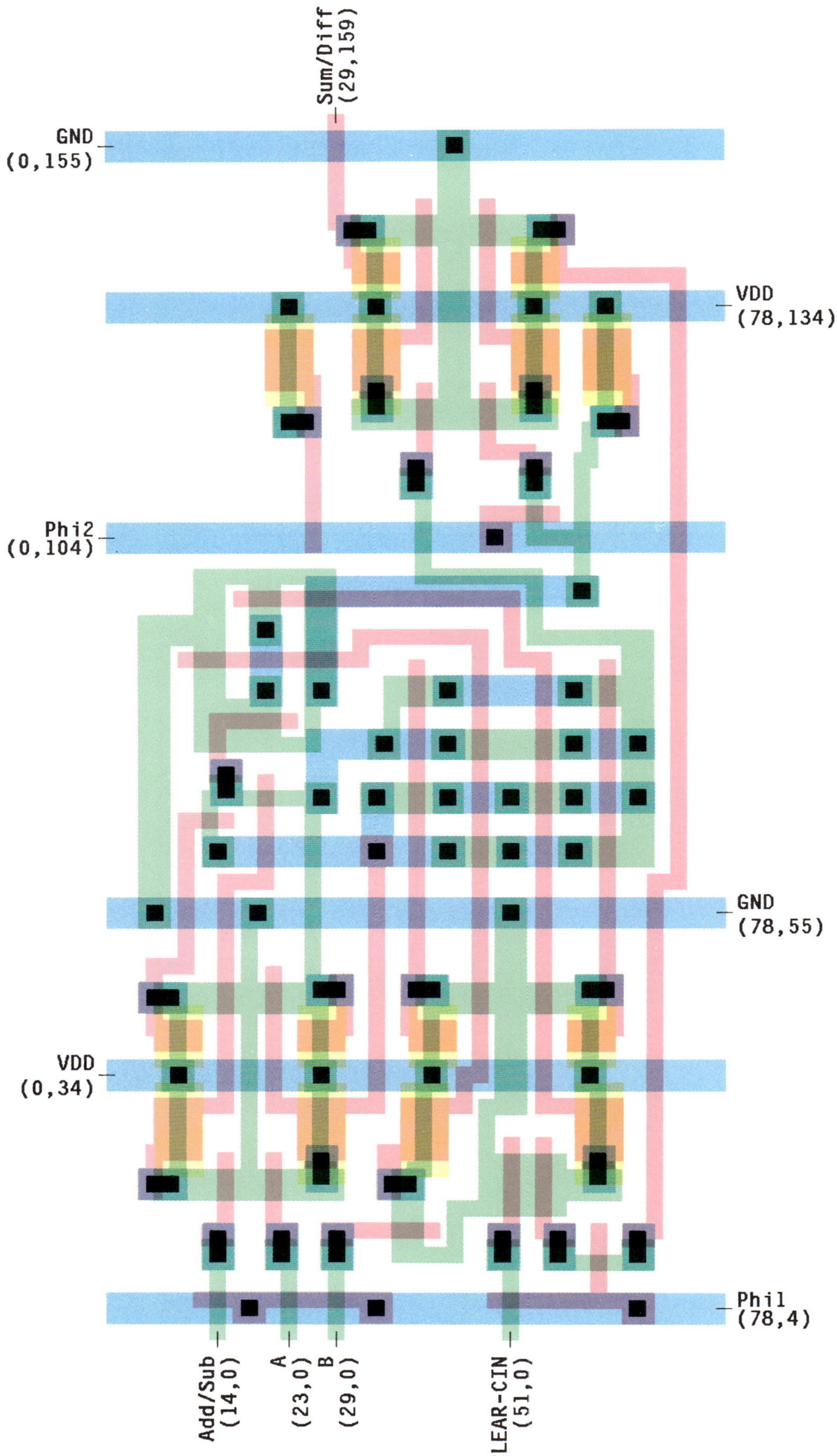
Sum/Diff
(29,159)
GND
(0,155)
VDD
(78,134)
Phi2
(0,104)
GND
(78,55)
VDD
(0,34)
Phi1
(78,4)
Add/Sub
(14,0)
A
(23,0)
B
(29,0)
CLEAR-CIN
(51,0)

NAME

S_AddSub—serial adder/subtractor

SYNOPSIS

Serial adder/subtractor.

PROPERTIES

Size

78 x 159λ

Interface

A	input,	switched	4 □ C_g	
B	input,	switched	4 □ C_g	
CLEAR_CIN	input,	switched	4 □ C_g	
Add/$\overline{Sub}$	input,	switched	4 □ C_g	
$\phi 1$	input,	switch control	11.5 □ C_g	
$\phi 2$	input,	switch control	7.25 □ C_g	
Sum	output,	gate	5.25 □ C_g	[2 / ½] □ *R*

DC Current

0.5 mA

DESCRIPTION

A and B are serial input operands that come in LSB first. Each pair of bits are added (or subtracted) together with the carry (or borrow) from the previous pair of bits. The operands can be of any length. The output starts appearing on the next clock cycle after the LSB of the input. *CLEAR_CIN* clears the carry in; it should be high during the LSB time and low otherwise. The output will be $A + B$ if *Add/Sub* is high and $A - B$ if *Add/Sub* is low. This cell will fit on the same power and ground pitch as the multiplier cells, but unlike them uses positive logic (high equals 1). The negative logic version is called S-AddSubNeg. It is used in the multiplier cells.

TESSELATION

Abuts horizontally with an optional 8λ overlap.

ORIGIN

Xerox PARC (Dick Lyon), modified at Stanford University (Lyle Smith).

STATUS

Tested.

LAYOUT SCALE

20 λ per inch.

CIF ID

141

SEE ALSO

S_Add.

LOGIC DIAGRAM

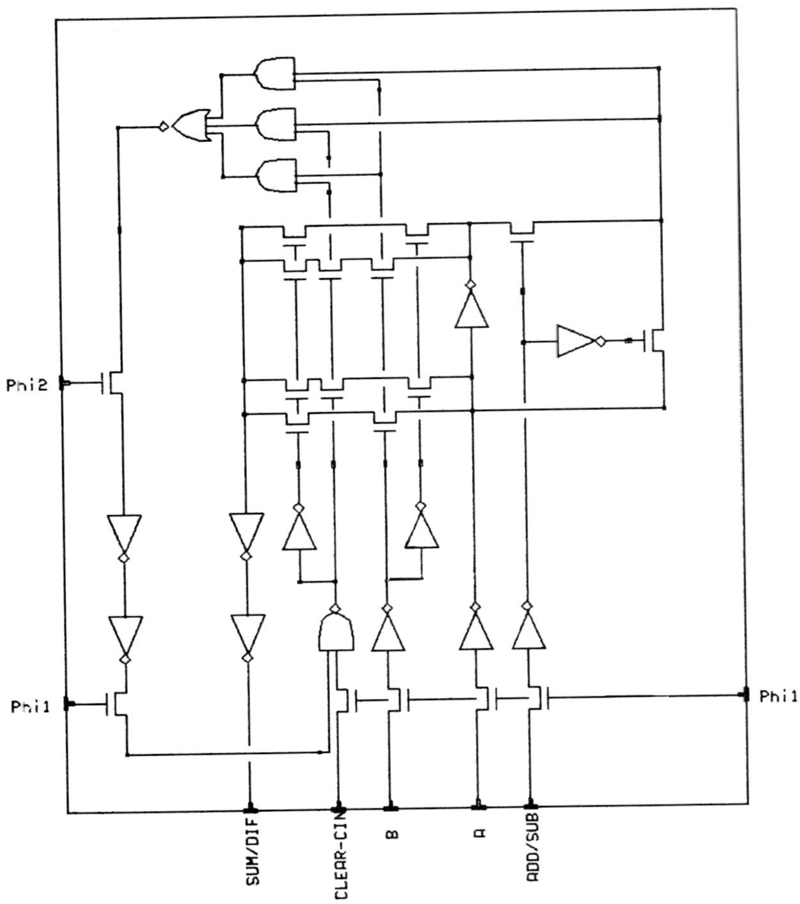

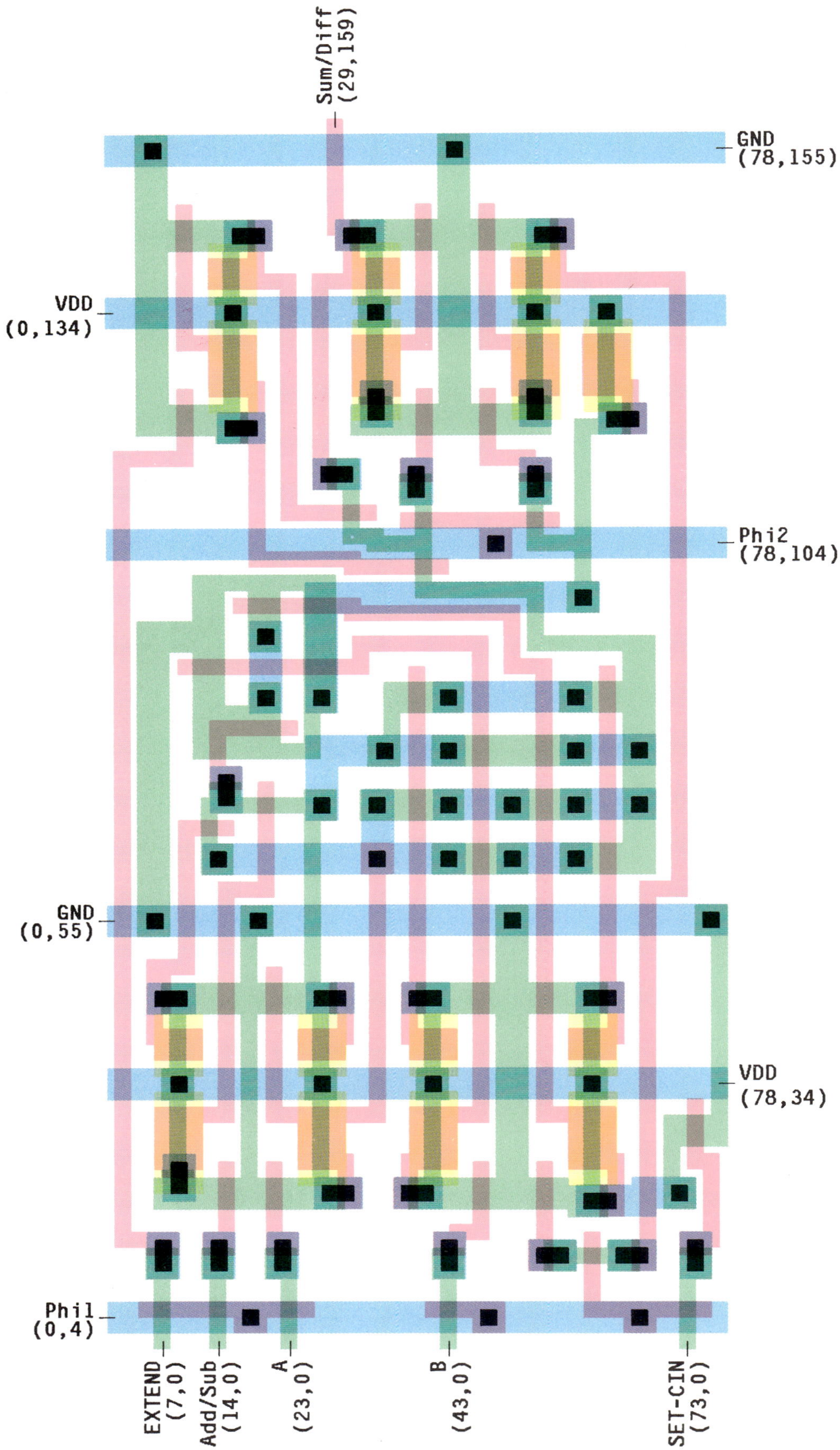
Sum/Diff
(29,159)
GND
(78,155)
VDD
(0,134)
Phi2
(78,104)
GND
(0,55)
VDD
(78,34)
Phi1
(0,4)
EXTEND
(7,0)
Add/Sub
(14,0)
A
(23,0)
B
(43,0)
SET-CIN
(73,0)

NAME

S_AddSubNeg—serial adder/subtractor with negative logic

SYNOPSIS

Serial adder/subtractor with negative logic.

PROPERTIES

Size

78 x 159λ

Interface

A	input,	switched	4 □ C_g
B	input,	switched	4 □ C_g
SET_CIN	input,	switched	4 □ C_g
Add/$\overline{Sub}$	input,	switched	4 □ C_g
EXTEND	input,	switched	4 □ C_g
$\phi 1$	input	switch control	12.5 □ C_g
$\phi 2$	input,	switch control	8.75 □ C_g
Sum/Diff	output,	gate	8 □ C_g [2 / ½] □ *R*

DC Current

0.5 mA

DESCRIPTION

This is the negative logic version of serial adder/subtractor. A and B are serial input operands that come in LSB first. Each pair of bits are added (or subtracted) together with the carry (or borrow) from the previous pair of bits. The operands can be of any length. The output starts appearing on the next clock cycle after the LSB of the input. *SET_CIN* clears the carry in; it should be high during the LSB time and low otherwise. The output will be $A + B$ if *Add/Sub* is high and A − B if *Add/Sub* is low. This cell will fit on the same power and ground pitch as the multiplier cells. The *Extend* signal is used for sign extension; it should be low during addition and subtraction but high for sign extension.

ORIGIN

Xerox PARC (Dick Lyon).

STATUS

Tested.

LAYOUT SCALE

20 λ per inch.

CIF ID

110

SEE ALSO

S_Add, S_AddSub.

LOGIC DIAGRAM

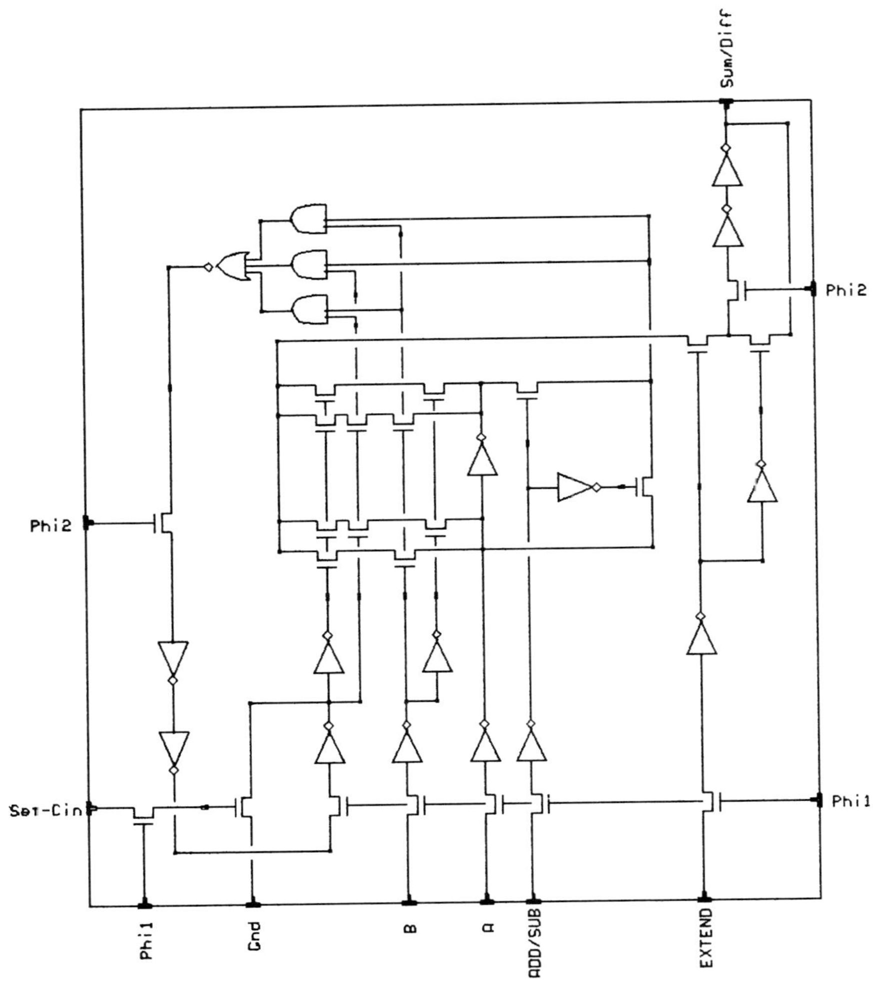

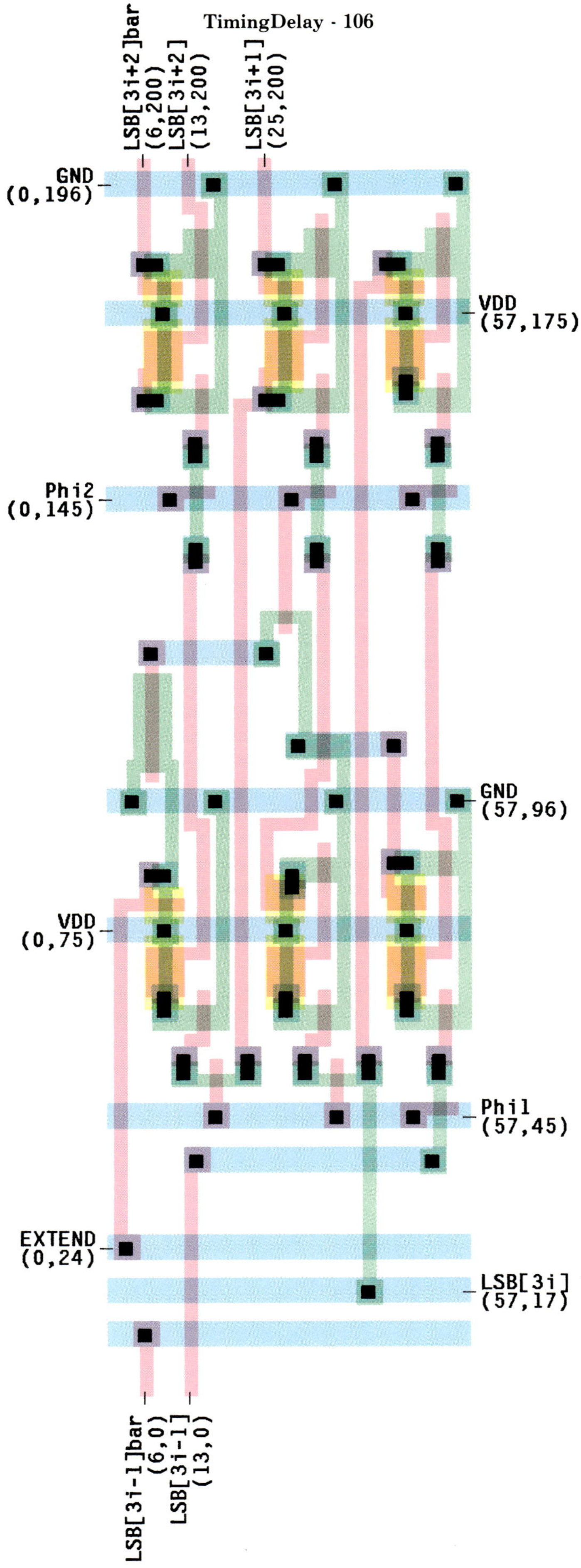
LSB[3i+2]bar (6,200)
LSB[3i+2] (13,200)
LSB[3i+1] (25,200)
GND (0,196)
VDD (57,175)
Phi2 (0,145)
GND (57,96)
VDD (0,75)
Phi1 (57,45)
EXTEND (0,24)
LSB[3i] (57,17)
LSB[3i-1]bar (6,0)
LSB[3i-1] (13,0)

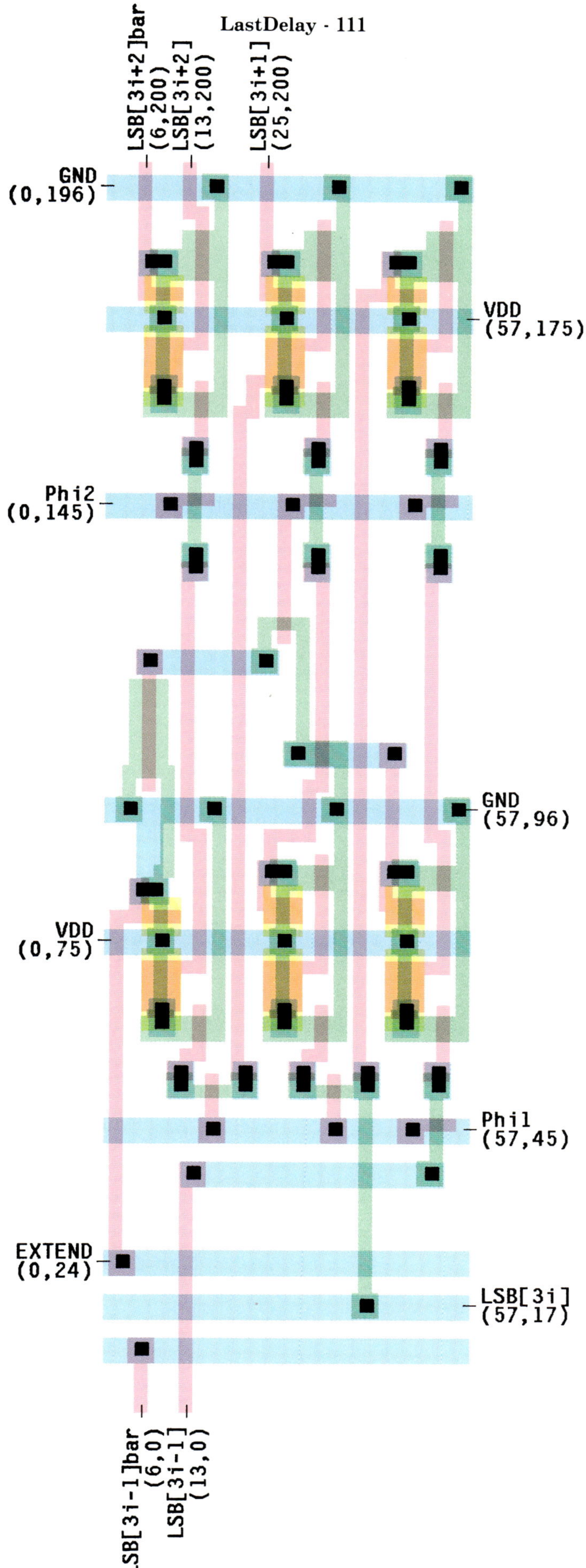
LSB[3i+2]bar (6,200)
LSB[3i+2] (13,200)
LSB[3i+1] (25,200)
GND (0,196)
VDD (57,175)
Phi2 (0,145)
GND (57,96)
VDD (0,75)
Phi1 (57,45)
EXTEND (0,24)
LSB[3i] (57,17)
LSB[3i-1]bar (6,0)
LSB[3i-1] (13,0)

NAME

TimingDelay, LastDelay—timing for serial multiplier

SYNOPSIS

Timing for basic serial multiplier stage.

PROPERTIES

Size

57 x 200 λ

Interface

Signal	Direction	Type	Capacitance	Resistance
LSB(3i−1)	input,	switched	6 □ C_g	
$\overline{LSB(3i-1)}$	input,	switch control	4.75 □ C_g	
LSB(3i+2)	output,	gate	12 □ C_g	[4 / ½] □ *R*
$\overline{LSB(3i+2)}$	output,	gate	4.5 □ C_g	[1 / ¼] □ *R*
LSB(3i+1)	output,	gate	4.5 □ C_g	[1 / ¼] □ *R*
LSB(3i)	output,	gate	20 □ C_g	[1 / ¼] □ *R*
EXTEND	output,	gate	14 □ C_g	[1 / ¼] □ *R*

DC Current

2.3 mA including associated multiplier stage.

DESCRIPTION

These cells are used for the control signal timing for *Mult* and *DblMult*. The output is the same as the input delayed by three clock cycles. The *Extend* signal is used for sign extension of the partial product. A diagram of the cell's logic may be seen in the documentation for *Mult*. *LastDelay* is used on the last stage of the multiplier and *TimingDelay* is used everywhere else.

ORIGIN

Xerox PARC (Dick Lyon).

STATUS

Generally used.

LAYOUT SCALE

25 λ per inch.

CIF ID

TimingDelay - 106
LastDelay - 111

SEE ALSO

Mult, DblMult.

LOGIC DIAGRAM (*TimingDelay*)

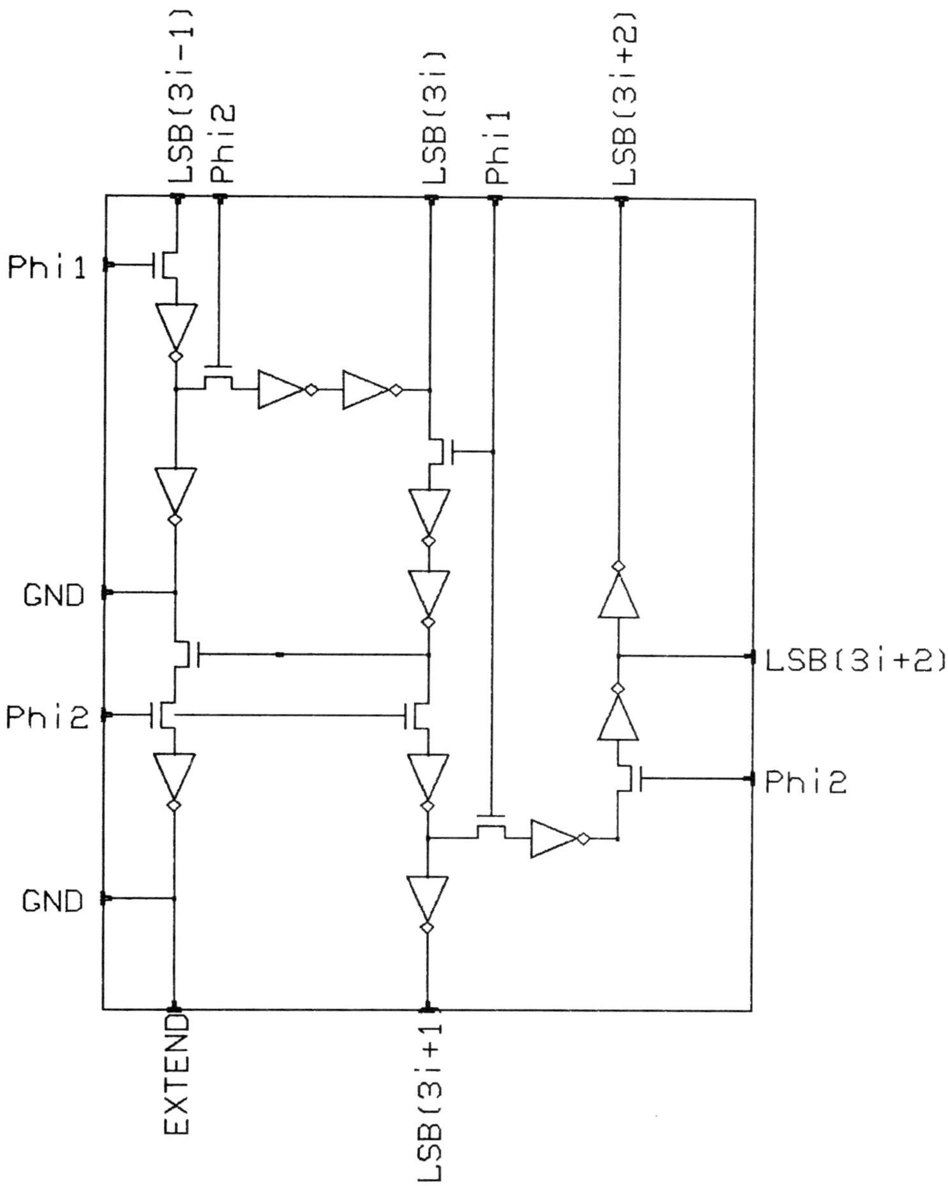

LOGIC DIAGRAM (*LastDelay*)

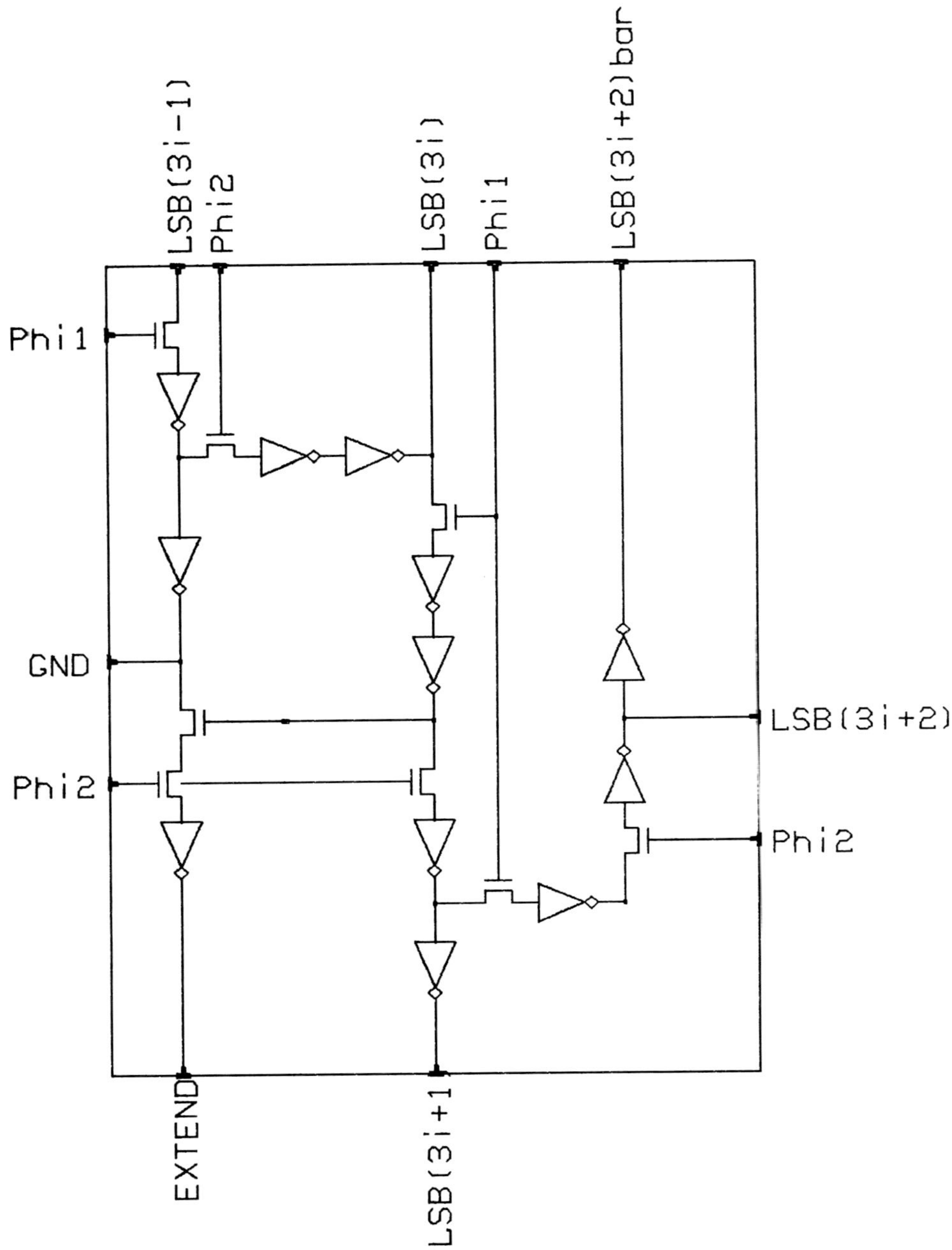

APPENDIX

This appendix contains the CIF descriptions of the library cells. Each section appears in the order it did in the library and individual cells are labeled by name. These CIF descriptions are available in machine-readable form from Addison-Wesley.

Adder_M

DS 170 125/1; L NP; B 4 422 64,211; B 8 4 30,2; B 36 4 30,6; B 8 6 40,23; B 8 6 20,23; B 4 22 38,37; B 4 22 22,37; B 8 8 54,38; B 8 8 6,38; B 4 6 56,45; B 4 6 4,45; B 14 4 51,50; B 14 4 9,50; B 22 4 47,54; B 22 4 13,54; B 14 4 51,70; B 4 6 56,75; B 14 4 9,70; B 4 6 4,75; B 8 8 54,82; B 8 8 6,82; B 4 16 6,94; B 4 46 38,79; B 4 16 54,94; B 4 46 22,79; B 10 4 9,104; B 6 4 23,104; B 4 108 24,160; B 6 4 37,104; B 4 52 36,132; B 6 4 53,104; B 4 64 52,138; B 4 108 12,160; B 6 4 35,160; B 4 16 34,170; B 14 4 47,172; B 6 4 35,180; B 4 32 36,198; B 8 6 30,221; B 18 4 35,226; B 4 10 28,233; B 6 8 51,234; B 4 40 52,258; B 10 4 9,242; B 4 34 6,261; B 14 8 19,258; B 4 4 24,264; B 14 8 39,258; B 4 4 34,264; B 6 8 23,270; B 6 8 35,270; B 4 24 22,286; B 4 8 34,278; B 12 4 10,280; B 12 4 48,280; B 8 4 32,284; B 4 28 30,300; B 18 4 15,300; B 10 4 53,310; B 8 4 54,306; B 4 26 50,325; B 8 6 40,313; B 12 4 26,316; B 4 104 22,370; B 8 8 8,326; B 4 10 10,335; B 14 10 43,347; B 4 60 48,382; B 12 10 36,369; B 14 10 9,369; B 8 4 6,384; B 4 36 8,404; B 4 8 4,378; B 6 8 47,416; B 6 10 11,235; L ND; B 4 14 42,7; B 4 14 18,7; B 8 8 40,18; B 8 8 20,18; B 12 4 38,30; B 28 2 30,33; B 12 4 22,30; B 8 30 30,59; B 8 2 48,45; B 44 10 30,39; B 8 2 12,45; B 4 8 50,70; B 4 12 50,52; B 8 8 50,62; B 4 8 10,70; B 4 12 10,52; B 8 8 10,62; B 44 12 30,80; B 28 4 30,88; B 8 10 30,95; B 8 2 4,115; B 8 2 4,109; B 60 4 30,112; B 4 8 58,226; B 4 62 58,187; B 4 10 58,147; B 4 10 58,133; B 4 10 58,119; B 60 4 30,126; B 8 2 4,129; B 8 2 4,123; B 60 4 30,140; B 8 2 4,143; B 8 2 4,137; B 60 4 30,154; B 8 2 4,157; B 8 2 4,151; B 46 4 25,166; B 4 8 46,172; B 4 8 4,226; B 4 8 4,214; B 4 10 4,201; B 4 24 4,180; B 8 8 44,180; B 46 4 25,194; B 8 2 44,197; B 8 2 44,191; B 46 4 25,208; B 8 2 44,211; B 8 2 44,205; B 22 4 13,220; B 24 4 48,220; B 4 8 22,226; B 4 8 38,226; B 8 8 6,234; B 8 8 56,234; B 20 4 30,232; B 4 10 38,239; B 8 8 20,248; B 8 8 40,248; B 4 12 18,258; B 4 12 40,258; B 8 2 16,265; B 20 2 48,265; B 12 12 52,280; B 10 2 17,267; B 22 8 47,270; B 22 6 11,271; B 18 4 9,288; B 12 12 6,280; B 8 8 48,290; B 34 8 27,308; B 8 14 14,297; B 4 20 16,322; B 14 12 51,324; B 4 42 56,355; B 18 4 49,332; B 4 2 42,335; B 16 4 10,334; B 4 18 4,345; B 8 8 40,340; B 4 10 42,349; B 8 4 40,356; B 12 4 8,356; B 8 4 10,360; B 10 4 39,360; B 4 10 36,367; B 4 10 10,367; B 8 4 38,374; B 20 4 18,378; B 12 8 22,384; B 8 4 12,374; B 24 4 46,378; B 16 8 50,384; B 18 4 25,390; B 8 4 30,394; B 8 4 56,394; B 20 4 50,390; B 4 10 42,397; B 4 26 54,409; B 44 4 22,404; B 4 4 2,420; B 4 8 2,410; B 46 4 23,416; B 8 2 42,419; B 8 2 42,413; L NM; B 8 2 30,1; B 68 6 34,5; B 8 12 40,20; B 8 12 20,20; B 12 8 50,38; B 12 8 10,38; B 68 8 34,62; B 12 8 50,82; B 12 8 10,82; B 68 8 34,96; B 68 6 34,111; B 8 2 4,115; B 68 6 34,125; B 8 2 4,129; B 68 6 34,139; B 8 2 4,143; B 68 6 34,153; B 8 2 4,157; B 8 2 44,183; B 68 6 34,179; B 8 2 44,197; B 68 6 34,193; B 8 2 44,211; B 68 6 34,207; B 68 6 34,221; B 8 2 30,225; B 12 8 54,234; B 12 8 8,234; B 68 8 34,248; B 12 8 20,270; B 12 8 38,270; B 68 8 34,290; B 8 4 40,314; B 32 8 52,308; B 18 8 9,308; B 68 6 34,325; B 8 2 8,329; B 8 12 40,342; B 68 8 34,358; B 8 12 38,374; B 8 12 12,374; B 68 8 34,392; B 12 8 44,416; L NI; B 12 32 50,60; B 12 32 10,60; B 12 12 12,154; B 34 12 23,166; B 12 12 12,208; B 12 16 18,258; B 12 16 40,258; B 12 20 42,348; B 12 20 36,368; B 12 18 10,369; B 12 12 8,404; B 24 14 30,113; B 12 14 12,127; B 12 14 36,127; B 12 26 52,147; B 12 14 24,141; B 12 14 24,195; B 14 12 21,416; L NC; B 4 4 30,4; B 4 8 40,20; B 4 8 20,20; B 8 4 50,38; B 8 4 10,38; B 4 4 50,62; B 4 4 10,62; B 8 4 50,82; B 8 4 10,82; B 4 4 30,96; B 4 4 4,112; B 4 4 4,126; B 4 4 4,140; B 4 4 4,154; B 4 4 44,180; B 4 4 44,194; B 4 4 44,208; B 4 4 30,222; B 8 4 54,234; B 8 4 8,234; B 4 4 20,248; B 4 4 40,248; B 8 4 20,270; B 8 4 38,270; B 4 4 14,290; B 4 4 48,290; B 4 8 40,310; B 4 4 54,308; B 4 4 14,308; B 4 4 8,326; B 4 8 40,342; B 4 4 10,358; B 4 4 40,358; B 4 8 12,374; B 4 8 38,374; B 4 4 56,392; B 4 4 30,392; B 8 4 44,416; DF;

AddRestore

DS 171 125/1; L NP; B 4 422 64,211; B 8 4 30,2; B 36 4 30,6; B 8 6 40,23; B 8 6 20,23; B 4 22 38,37; B 4 22 22,37; B 8 8 54,38; B 8 8 6,38; B 4 6 56,45; B 4 6 4,45; B 14 4 51,50; B 14 4 9,50; B 22 4 47,54; B 22 4 13,54; B 14 4 51,70; B 4 6 56,75; B 14 4 9,70; B 4 6 4,75; B 8 8 54,82; B 8 8 6,82; B 4 16 6,94; B 4 46 38,79; B 4 16 54,94; B 4 46 22,79; B 10 4 9,104; B 6 4 23,104; B 4 108 24,160; B 6 4 37,104; B 4 52 36,132; B 6 4 53,104; B 4 64 52,138; B 4 108 12,160; B 6 4 35,160; B 4 16 34,170; B 14 4 47,172; B 6 4 35,180; B 4 32 36,198; B 8 6 30,221; B 18 4 35,226; B 4 10 28,233; B 6 8 51,234; B 4 40 52,258; B 10 4 9,242; B 4 34 6,261; B 14 8 19,258; B 4 4 24,264; B 14 8 39,258; B 4 4 34,264; B 6 8 23,270; B 6 8 35,270; B 4 24 22,286; B 4 8 34,278; B 12 4 10,280; B 12 4 48,280; B 8 4 32,284; B 4 28 30,300; B 18 4 15,300; B 10 4 53,310; B 8 4 54,306; B 4 26 50,325; B 8 6 40,313; B 12 4 26,316; B 4 8 38,320; B 4 104 22,370; B 8 8 8,326; B 12 4 34,326; B 4 36 30,346; B 4 10 10,335; B 14 10 43,347; B 4 60 48,382; B 14 4 35,366; B 12 6 36,371; B 14 10 9,369; B 8 4 6,384; B 4 36 8,404; B 4 8 4,378; B 6 8 47,416; B 6 10 11,235; L ND; B 4 14 42,7; B 4 14 18,7; B 8 8 40,18; B 8 8 20,18; B 12 4 38,30; B 28 2 30,33; B 12 4 22,30; B 8 30 30,59; B 8 2 48,45; B 44 10 30,39; B 8 2 12,45; B 4 8 50,70; B 4 12 50,52; B 8 8 50,62; B 4 8 10,70; B 4 12 10,52; B 8 8 10,62; B 44 12 30,80; B 28 4 30,88; B 8 10 30,95; B 8 2 4,115; B 8 2 4,109; B 60 4 30,112; B 4 8 58,226; B 4 62 58,187; B 4 10 58,147; B 4 10

58,133; B 4 10 58,119; B 60 4 30,126; B 8 2 4,129; B 8 2 4,123; B 60 4 30,140; B 8 2 4,143; B 8 2 4,137; B 60 4 30,154; B 8 2 4,157; B 8 2 4,151; B 46 4 25,166; B 4 8 46,172; B 4 8 4,226; B 4 8 4,214; B 4 10 4,201; B 4 24 4,180; B 8 8 44,180; B 46 4 25,194; B 8 2 44,197; B 8 2 44,191; B 46 4 25,208; B 8 2 44,211; B 8 2 44,205; B 22 4 13,220; B 24 4 48,220; B 4 8 22,226; B 4 8 38,226; B 8 8 6,234; B 8 8 56,234; B 20 4 30,232; B 4 10 38,239; B 8 8 20,248; B 8 8 40,248; B 4 12 18,258; B 4 12 40,258; B 8 2 16,265; B 20 2 48,265; B 12 12 52,280; B 10 2 17,267; B 22 8 47,270; B 22 6 11,271; B 18 4 9,288; B 12 12 6,280; B 8 8 48,290; B 34 8 27,308; B 8 14 14,297; B 4 20 16,322; B 14 12 51,324; B 4 42 56,355; B 18 4 49,332; B 4 2 42,335; B 16 4 10,334; B 4 18 4,345; B 8 8 40,340; B 4 10 42,349; B 8 4 40,356; B 12 4 8,356; B 8 4 10,360; B 10 4 39,360; B 4 10 36,367; B 4 10 10,367; B 8 4 38,374; B 20 4 18,378; B 12 8 22,384; B 8 4 12,374; B 24 4 46,378; B 16 8 50,384; B 18 4 25,390; B 8 4 30,394; B 8 4 56,394; B 20 4 50,390; B 4 10 42,397; B 4 26 54,409; B 44 4 22,404; B 4 4 2,420; B 4 8 2,410; B 46 4 23,416; B 8 2 42,419; B 8 2 42,413; L NM; B 8 2 30,1; B 68 6 34,5; B 8 12 40,20; B 8 12 20,20; B 12 8 50,38; B 12 8 10,38; B 68 8 34,62; B 12 8 50,82; B 12 8 10,82; B 68 8 34,96; B 68 6 34,111; B 8 2 4,115; B 68 6 34,125; B 8 2 4,129; B 68 6 34,139; B 8 2 4,143; B 68 6 34,153; B 8 2 4,157; B 8 2 44,183; B 68 6 34,179; B 8 2 44,197; B 68 6 34,193; B 8 2 44,211; B 68 6 34,207; B 68 6 34,221; B 8 2 30,225; B 12 8 54,234; B 12 8 8,234; B 68 8 34,248; B 12 8 20,270; B 12 8 38,270; B 68 8 34,290; B 8 12 40,310; B 18 8 59,308; B 18 8 9,308; B 68 6 34,325; B 8 2 8,329; B 8 12 40,342; B 68 8 34,358; B 8 12 38,374; B 8 12 12,374; B 68 8 34,392; B 12 8 44,416; L NI; B 12 32 50,60; B 12 32 10,60; B 12 12 12,154; B 34 12 23,166; B 12 12 12,208; B 12 16 18,258; B 12 16 40,258; B 12 20 42,348; B 12 20 36,368; B 12 18 10,369; B 12 12 8,404; B 24 14 30,113; B 12 14 12,127; B 12 14 36,127; B 12 26 52,147; B 12 14 24,141; B 12 14 24,195; B 14 12 21,416; L NC; B 4 4 30,4; B 4 8 40,20; B 4 8 20,20; B 8 4 50,38; B 8 4 10,38; B 4 4 50,62; B 4 4 10,62; B 8 4 50,82; B 8 4 10,82; B 4 4 30,96; B 4 4 4,112; B 4 4 4,126; B 4 4 4,140; B 4 4 4,154; B 4 4 44,180; B 4 4 44,194; B 4 4 44,208; B 4 4 30,222; B 8 4 54,234; B 8 4 8,234; B 4 4 20,248; B 4 4 40,248; B 8 4 20,270; B 8 4 38,270; B 4 4 14,290; B 4 4 48,290; B 4 8 40,310; B 4 4 54,308; B 4 4 14,308; B 4 4 8,326; B 4 8 40,342; B 4 4 10,358; B 4 4 40,358; B 4 8 12,374; B 4 8 38,374; B 4 4 56,392; B 4 4 30,392; B 8 4 44,416; DF;

AddOut

DS 172 125/1; L NP; B 40 4 20,64; B 10 6 9,11; B 4 38 12,33; B 14 4 23,60; B 4 10 28,53; L ND; B 12 4 6,2; B 8 4 4,54; B 4 4 6,50; B 8 6 8,7; B 16 6 12,45; B 14 26 11,29; B 4 4 18,50; B 8 4 20,58; B 24 4 28,54; B 8 4 36,50; L NM; B 8 12 8,8; B 8 12 20,58; B 8 68 36,34; L NI; B 12 12 28,54; L NC; B 4 8 8,8; B 4 8 20,58; B 4 4 36,52; DF;

AddClkOut

DS 173 125/1; L NP; B 8 8 10,16; B 4 10 10,7; B 56 4 28,64; B 10 6 25,11; B 4 38 28,33; B 14 4 39,60; B 4 10 44,53; L ND; B 6 4 3,2; B 4 2 4,5; B 24 4 12,54; B 4 4 22,50; B 16 6 28,45; B 14 26 27,29; B 4 4 34,50; B 8 4 36,58; B 24 4 44,54; B 8 4 52,50; B 14 6 21,3; B 26 4 15,8; L NM; B 8 8 10,16; B 6 48 11,44; B 6 12 11,6; B 8 12 24,8; B 8 12 36,58; B 8 68 52,34; L NI; B 12 12 44,54; L NC; B 4 4 10,16; B 4 8 24,8; B 4 8 36,58; B 4 4 52,52; DF;

StaticAdder

DS 175 125/1; L NM; B 68 6 34,5; B 68 8 34,62; B 68 8 34,96; B 68 6 34,111; B 68 6 34,125; B 68 6 34,139; B 68 6 34,153; B 68 6 34,179; B 68 6 34,193; B 68 6 34,207; B 68 6 34,221; B 68 8 34,248; B 68 8 34,290; B 10 8 5,308; B 68 8 34,392; B 68 8 34,358; B 8 8 4,112; B 8 8 4,126; B 8 8 4,140; B 8 8 4,154; B 12 8 8,234; B 12 8 10,38; B 12 8 10,82; B 8 8 10,62; B 8 8 10,358; B 8 12 10,338; B 8 8 12,290; B 8 8 12,308; B 8 12 12,374; B 12 8 20,270; B 8 8 20,248; B 8 12 20,20; B 8 8 30,222; B 8 8 30,4; B 8 8 30,96; B 8 8 30,392; B 12 8 38,270; B 8 12 38,374; B 8 8 40,248; B 8 8 40,358; B 8 12 40,20; B 8 12 40,342; B 12 8 44,416; B 8 8 44,180; B 8 8 44,194; B 8 8 44,208; B 8 8 48,290; B 12 8 50,38; B 12 8 50,82; B 8 8 50,62; B 12 8 54,234; B 18 8 59,308; B 8 8 54,308; B 8 8 56,392; L ND; B 4 10 38,367; B 4 18 2,413; B 4 4 2,404; B 38 4 19,416; B 12 22 6,279; B 8 8 4,112; B 8 8 4,126; B 8 8 4,140; B 8 8 4,154; B 4 68 4,200; B 4 4 4,166; B 40 4 22,404; B 8 8 6,234; B 42 4 25,166; B 34 4 23,194; B 34 4 23,208; B 16 4 14,220; B 8 8 10,62; B 8 8 10,358; B 8 8 10,336; B 8 4 12,44; B 4 34 10,59; B 4 4 10,76; B 48 4 32,112; B 48 4 32,126; B 48 4 32,140; B 48 4 32,154; B 4 12 10,368; B 4 42 10,333; B 8 10 12,299; B 8 8 12,290; B 8 8 12,308; B 8 8 12,38; B 8 8 12,82; B 8 8 12,376; B 6 4 13,76; B 4 10 14,269; B 8 8 18,270; B 4 14 18,7; B 4 14 18,259; B 12 16 22,36; B 28 16 30,82; B 12 16 22,384; B 8 8 20,308; B 8 8 20,248; B 8 8 20,18; B 4 12 22,226; B 4 4 22,220; B 4 4 22,232; B 14 4 29,232; B 24 16 36,304; B 8 60 30,62; B 8 8 30,96; B 8 8 30,392; B 12 16 38,36; B 8 8 38,376; B 4 24 38,232; B 4 4 38,220; B 8 8 40,248; B 8 8 40,358; B 8 8 40,18; B 8 8 40,270; B 8 8 40,340; B 18 4 47,220; B 4 14 40,259; B 8 8 42,416; B 4 14 42,7; B 4 16 42,396; B 4 4 42,404; B 4 6 42,333; B 4 10 42,349; B 8 8 44,180; B 8 8 44,194; B 8 8 44,208; B 4 10 44,269; B 14 16 49,384; B 8 4 48,44; B 6 4 47,76; B 4 10 46,171; B 4 4 46,166; B 4 8 46,294; B 16 16 52,326; B 8 8 48,290; B 8 8 48,38; B 8 8 48,82; B 12 22 52,275; B 8 8 50,62; B 4 34 50,59; B 4 4 50,76; B 4 26 54,409; B 8 8 56,392; B 8 8 56,234; B 4 124 58,172; B 4 56 58,362; L NP; B 4 14 4,49; B 14 4 9,70; B 4 18 4,77; B 4 10 4,379; B 4 4 4,384; B 8 8 6,38; B 14 10 9,369; B 4 20 6,96; B 4 4 6,384; B 4 38 6,261; B 4 4 6,242; B 4 4 6,280; B 16 4 12,300; B 12 12 10,346; B 6 8 7,38; B 6 8 7,82; B 4 38 8,403; B 4 4 8,384; B 6 4 9,242; B 10 4 11,280; B 10 8 11,52; B 8 6 10,341; B 6 4 11,104; B 6 8 11,234; B 8 6 12,371; B 4 108 12,160; B 4 12 12,236; B 4 4 12,242; B 14 4 19,6; B 10 8 17,258; B 4 4 18,54; B 8 6 20,23; B 4 12 20,294; B 4 4 20,288; B 4 4 20,300; B 4 24 22,36; B 4 56 22,80; B 4 106 22,369; B 4 4 22,316; B 4 14 22,281; B 2 4 21,288; B 4 4 22,288; B 6 8 23,270; B 4 112 24,158; B 8 4 26,316; B 4 12 24,260; B 4 14 28,231; B 8 8 30,222; B 8 8 30,4; B 4 20 30,336; B 4 4 30,326; B 4 4 30,346; B 4 32 30,300; B 4 4 30,284; B 4 4 30,316; B 8 4 34,326; B 14 4 37,346; B 14 4 37,226; B 4 4 32,284; B 12 10 36,369; B 4 24 34,170; B 4

30 34,269; B 4 4 34,284; B 6 8 35,270; B 14 4 41,6; B 4 60 36,132; B 4 36 36,196; B 8 6 38,371; B 4 24 38,36; B 4 56 38,80; B 4 34 38,309; B 4 4 38,326; B 10 8 41,258; B 14 10 43,347; B 8 6 40,23; B 8 6 40,345; B 4 4 42,54; B 10 4 45,172; B 10 4 47,280; B 10 4 49,70; B 10 8 49,52; B 6 8 47,416; B 4 68 48,386; B 6 8 51,234; B 4 72 52,138; B 4 42 52,259; B 4 4 52,280; B 4 30 52,323; B 8 8 54,38; B 8 8 54,308; B 6 8 53,38; B 6 8 53,82; B 4 20 54,96; B 4 14 56,49; B 4 18 56,77; B 4 422 64,211; L NC; B 4 4 4,112; B 4 4 4,126; B 4 4 4,140; B 4 4 4,154; B 8 4 8,234; B 8 4 10,38; B 8 4 10,82; B 4 4 10,62; B 4 4 10,358; B 4 8 10,338; B 4 4 12,290; B 4 4 12,308; B 4 8 12,374; B 8 4 20,270; B 4 4 20,248; B 4 8 20,20; B 4 4 30,222; B 4 4 30,4; B 4 4 30,96; B 4 4 30,392; B 8 4 38,270; B 4 8 38,374; B 4 4 40,248; B 4 4 40,358; B 4 8 40,20; B 4 8 40,342; B 8 4 44,416; B 4 4 44,180; B 4 4 44,194; B 4 4 44,208; B 4 4 48,290; B 8 4 50,38; B 8 4 50,82; B 4 4 50,62; B 8 4 54,234; B 4 4 54,308; B 4 4 56,392; L NI; B 12 12 8,404; B 12 32 10,60; B 12 18 10,369; B 12 22 10,345; B 12 12 12,126; B 12 24 12,160; B 12 12 12,208; B 12 16 18,258; B 12 12 22,416; B 24 12 30,112; B 12 12 24,140; B 22 12 29,166; B 12 12 24,194; B 12 12 36,126; B 12 20 36,370; B 12 16 40,258; B 12 20 42,348; B 12 32 50,60; B 12 12 52,140; B 12 12 52,154; DF;

ALU

DS 302 125/1; L NP; B 6 4 5,116; B 4 390 4,313; B 4 114 6,57; B 8 8 14,314; B 14 10 17,369; B 4 54 12,401; B 20 4 22,122; B 8 6 16,49; B 4 16 14,112; B 4 36 14,70; B 18 4 21,338; B 26 4 25,354; B 6 12 15,98; B 12 4 16,430; B 6 8 17,272; B 4 2 16,309; B 4 12 16,298; B 16 4 20,90; B 26 4 29,266; B 8 8 22,20; B 8 4 22,2; B 4 76 20,470; B 14 4 21,306; B 4 14 24,31; B 6 10 23,63; B 16 4 28,70; B 18 4 27,6; B 14 4 33,314; B 4 20 28,326; B 4 64 30,388; B 4 122 30,185; B 6 4 31,422; B 4 84 32,466; B 12 4 40,338; B 4 12 36,346; B 16 8 32,288; B 6 10 37,279; B 10 4 41,122; B 6 12 39,36; B 4 22 38,53; B 20 4 30,74; B 4 44 38,98; B 22 4 47,364; B 14 10 43,379; B 10 4 41,414; B 4 28 38,398; B 4 20 38,302; B 4 92 44,462; B 4 12 44,330; B 4 122 44,185; B 6 4 45,322; B 8 4 48,304; B 4 14 46,313; B 26 4 59,266; B 6 10 51,279; B 4 10 50,297; B 6 12 53,36; B 16 8 56,288; B 4 22 54,53; B 6 4 55,122; B 4 44 54,98; B 4 80 56,468; B 4 4 56,418; B 4 42 56,387; B 4 122 56,185; B 14 8 61,412; B 6 8 57,424; B 14 34 69,379; B 14 4 73,122; B 4 14 68,31; B 8 8 70,20; B 18 4 65,6; B 8 4 70,2; B 4 122 68,185; B 6 10 69,63; B 20 4 62,74; B 16 4 64,70; B 4 16 70,342; B 6 8 71,272; B 8 8 74,314; B 10 4 73,332; B 8 4 74,328; B 4 18 72,301; B 8 6 76,49; B 4 36 74,414; B 4 36 78,70; B 6 12 77,98; B 16 4 72,90; B 4 16 78,112; B 4 508 86,254; L ND; B 8 8 12,272; B 4 20 10,286; B 4 86 10,225; B 4 12 10,172; B 4 12 10,156; B 4 12 10,140; B 4 4 12,506; B 4 12 12,494; B 4 12 12,478; B 4 12 12,462; B 8 8 16,44; B 8 4 20,98; B 4 6 18,37; B 8 40 20,346; B 4 32 18,382; B 8 8 20,402; B 8 4 22,196; B 8 4 22,212; B 8 4 22,228; B 8 4 22,244; B 4 10 22,91; B 4 24 22,314; B 16 12 30,266; B 8 4 28,64; B 16 8 26,82; B 8 6 30,257; B 26 4 21,298; B 8 2 30,295; B 4 12 30,72; B 4 30 30,15; B 4 12 30,288; B 22 4 27,32; B 8 4 34,36; B 10 2 29,61; B 8 10 32,277; B 8 34 44,359; B 8 8 46,114; B 4 22 44,387; B 8 8 46,402; B 28 10 46,49; B 8 34 46,77; B 28 6 46,107; B 28 6 46,97; B 24 8 44,330; B 16 12 58,266; B 8 4 58,36; B 8 6 58,257; B 8 2 58,301; B 8 2 58,295; B 4 12 58,288; B 8 10 56,277; B 32 4 46,56; B 44 2 46,59; B 10 2 63,61; B 8 4 64,64; B 4 12 62,72; B 4 30 62,15; B 58 4 39,454; B 8 2 64,457; B 8 2 64,451; B 58 4 39,470; B 8 2 64,473; B 8 2 64,467; B 58 4 39,486; B 8 2 64,489; B 8 2 64,483; B 58 4 39,502; B 8 2 64,505; B 8 2 64,499; B 4 12 62,412; B 4 10 64,351; B 10 8 65,402; B 6 2 63,419; B 26 4 67,298; B 24 8 70,424; B 16 8 66,82; B 12 4 68,358; B 8 4 70,362; B 4 34 68,381; B 4 10 70,91; B 60 4 46,102; B 8 4 72,98; B 22 4 65,32; B 4 6 74,37; B 8 8 76,44; B 8 8 76,272; B 20 4 72,344; B 8 4 78,348; B 8 8 78,438; B 74 4 45,132; B 8 4 78,128; B 74 4 45,148; B 8 4 78,144; B 74 4 45,164; B 8 4 78,160; B 74 4 45,180; B 8 4 78,176; B 62 4 49,192; B 62 4 49,208; B 62 4 49,224; B 62 4 49,240; B 4 20 78,286; B 4 26 78,255; B 4 12 78,232; B 4 12 78,216; B 4 12 78,200; B 4 6 80,431; B 14 2 27,301; L NM; B 90 8 45,4; B 90 8 45,454; B 90 8 45,470; B 90 8 45,486; B 90 8 45,502; B 90 8 45,314; B 40 8 20,330; B 90 8 45,346; B 90 8 45,402; B 90 8 45,438; B 90 8 45,258; B 90 8 45,298; B 90 8 45,130; B 90 8 45,146; B 90 8 45,162; B 90 8 45,178; B 90 8 45,194; B 90 8 45,210; B 90 8 45,226; B 90 8 45,242; B 12 8 14,272; B 12 8 18,100; B 8 12 16,46; B 8 12 20,364; B 12 8 26,62; B 12 8 34,278; B 12 8 36,34; B 8 12 44,374; B 90 8 45,114; B 90 8 45,20; B 90 8 45,82; B 12 8 54,278; B 12 8 56,34; B 12 8 60,424; B 12 8 66,62; B 8 12 70,362; B 12 8 74,100; B 42 8 69,330; B 12 8 74,272; B 8 12 76,46; L NI; B 12 18 18,369; B 12 12 20,486; B 12 12 20,502; B 12 12 22,90; B 12 16 30,72; B 12 16 30,288; B 12 12 30,208; B 12 12 30,132; B 12 12 30,148; B 12 12 32,454; B 24 12 38,470; B 12 12 44,502; B 12 18 44,379; B 12 12 44,224; B 12 12 44,164; B 12 12 56,454; B 12 12 56,486; B 12 12 56,208; B 24 12 50,240; B 12 12 56,148; B 24 12 50,180; B 12 16 58,288; B 12 16 62,72; B 12 16 62,412; B 12 42 68,379; B 12 12 68,192; B 12 12 68,224; B 12 12 68,132; B 12 12 68,164; B 12 12 70,90; B 14 12 31,192; L NC; B 8 4 14,272; B 4 4 14,314; B 8 4 18,100; B 4 8 16,46; B 4 4 20,330; B 4 4 20,346; B 4 8 20,364; B 4 4 20,402; B 4 4 22,82; B 4 4 22,20; B 4 4 22,4; B 4 4 22,194; B 4 4 22,210; B 4 4 22,226; B 4 4 22,242; B 8 4 26,62; B 4 4 30,82; B 4 4 30,258; B 4 4 30,298; B 8 4 34,278; B 8 4 36,34; B 4 4 36,330; B 4 4 44,346; B 4 8 44,374; B 4 4 46,114; B 4 4 46,402; B 4 4 52,330; B 8 4 54,278; B 8 4 56,34; B 8 4 60,424; B 4 4 58,258; B 4 4 58,298; B 4 4 62,82; B 8 4 66,62; B 4 4 64,454; B 4 4 64,470; B 4 4 64,486; B 4 4 64,502; B 4 4 66,402; B 4 4 70,82; B 4 4 70,20; B 4 4 70,4; B 4 8 70,362; B 8 4 74,100; B 8 4 74,272; B 4 4 74,314; B 4 4 74,330; B 4 8 76,46; B 4 4 78,346; B 4 4 78,438; B 4 4 78,130; B 4 4 78,146; B 4 4 78,162; B 4 4 78,178; DF;

ALURestore

DS 303 125/1; L NP; B 6 4 5,116; B 4 390 4,313; B 4 114 6,57; B 8 8 14,314; B 14 10 17,369; B 4 54 12,401; B 20 4 22,122; B 8 6 16,49; B 4 16 14,112; B 4 36 14,70; B 18 4 21,338; B 26 4 25,354; B 6 12 15,98; B 12 4 16,430; B 6 8 17,272; B 4 2 16,309; B 4 12 16,298; B 16 4 20,90; B 26 4 29,266; B 8 8 22,20; B 8 4 22,2; B 4 76 20,470; B 14 4 21,306; B 4 14 24,31; B 6 10 23,63; B 16 4 28,70; B 18 4 27,6; B 14 4 33,314; B 4 20 28,326; B 4 64 30,388; B 4 122 30,185; B 6 4 31,422; B 4 84 32,466; B 12 4 40,338; B 4 12 36,346; B 16 8 32,288; B 6 10 37,279; B 10 4 41,122; B 6 12 39,36; B 4 22 38,53; B

20 4 30,74; B 4 44 38,98; B 22 4 47,364; B 14 10 43,379; B 10 4 41,414; B 4 28 38,398; B 4 20 38,302; B 4 92 44,462; B 4 12 44,330; B 4 122 44,185; B 6 4 45,322; B 8 4 48,304; B 4 14 46,313; B 26 4 59,266; B 6 10 51,279; B 4 10 50,297; B 6 12 53,36; B 16 8 56,288; B 4 22 54,53; B 4 44 54,98; B 4 28 56,348; B 6 4 55,122; B 4 80 56,468; B 4 4 56,418; B 4 42 56,387; B 4 122 56,185; B 6 8 57,330; B 14 8 61,412; B 6 8 57,424; B 14 34 69,379; B 14 4 73,122; B 4 14 68,31; B 8 8 70,20; B 18 4 65,6; B 8 4 70,2; B 4 122 68,185; B 6 10 69,63; B 20 4 62,74; B 16 4 64,70; B 4 16 70,342; B 6 8 71,272; B 8 8 74,314; B 10 4 73,332; B 8 4 74,328; B 4 18 72,301; B 8 6 76,49; B 4 36 74,414; B 4 36 78,70; B 6 12 77,98; B 16 4 72,90; B 4 16 78,112; B 4 508 86,254; L ND; B 8 8 12,272; B 4 20 10,286; B 4 86 10,225; B 4 12 10,172; B 4 12 10,156; B 4 12 10,140; B 4 4 12,506; B 4 12 12,494; B 4 12 12,478; B 4 12 12,462; B 8 8 16,44; B 8 4 20,98; B 4 6 18,37; B 8 40 20,346; B 4 32 18,382; B 8 8 20,402; B 8 4 22,196; B 8 4 22,212; B 8 4 22,228; B 8 4 22,244; B 4 10 22,91; B 4 24 22,314; B 16 12 30,266; B 8 4 28,64; B 16 8 26,82; B 8 6 30,257; B 26 4 21,298; B 8 2 30,295; B 4 12 30,72; B 4 30 30,15; B 4 12 30,288; B 22 4 27,32; B 8 4 34,36; B 10 2 29,61; B 8 10 32,277; B 8 34 44,359; B 8 8 46,114; B 4 22 44,387; B 8 8 46,402; B 28 10 46,49; B 8 34 46,77; B 28 6 46,107; B 28 6 46,97; B 24 8 44,330; B 16 12 58,266; B 8 4 58,36; B 8 6 58,257; B 8 2 58,301; B 8 2 58,295; B 4 12 58,288; B 8 10 56,277; B 32 4 46,56; B 44 2 46,59; B 10 2 63,61; B 8 4 64,64; B 4 12 62,72; B 4 30 62,15; B 58 4 39,454; B 8 2 64,457; B 8 2 64,451; B 58 4 39,470; B 8 2 64,473; B 8 2 64,467; B 58 4 39,486; B 8 2 64,489; B 8 2 64,483; B 58 4 39,502; B 8 2 64,505; B 8 2 64,499; B 4 12 62,412; B 4 10 64,351; B 10 8 65,402; B 6 2 63,419; B 26 4 67,298; B 24 8 70,424; B 16 8 66,82; B 12 4 68,358; B 8 4 70,362; B 4 34 68,381; B 4 10 70,91; B 60 4 46,102; B 8 4 72,98; B 22 4 65,32; B 4 6 74,37; B 8 8 76,44; B 8 8 76,272; B 20 4 72,344; B 8 4 78,348; B 8 8 78,438; B 74 4 45,132; B 8 4 78,128; B 74 4 45,148; B 8 4 78,144; B 74 4 45,164; B 8 4 78,160; B 74 4 45,180; B 8 4 78,176; B 62 4 49,192; B 62 4 49,208; B 62 4 49,224; B 62 4 49,240; B 4 20 78,286; B 4 26 78,255; B 4 12 78,232; B 4 12 78,216; B 4 12 78,200; B 4 6 80,431; B 14 2 27,301; L NM; B 90 8 45,4; B 90 8 45,454; B 90 8 45,470; B 90 8 45,486; B 90 8 45,502; B 90 8 45,314; B 40 8 20,330; B 90 8 45,346; B 90 8 45,402; B 90 8 45,438; B 90 8 45,258; B 90 8 45,298; B 90 8 45,130; B 90 8 45,146; B 90 8 45,162; B 90 8 45,178; B 90 8 45,194; B 90 8 45,210; B 90 8 45,226; B 90 8 45,242; B 12 8 14,272; B 12 8 18,100; B 8 12 16,46; B 8 12 20,364; B 12 8 26,62; B 12 8 34,278; B 12 8 36,34; B 8 12 44,374; B 90 8 45,114; B 90 8 45,20; B 90 8 45,82; B 12 8 54,330; B 12 8 54,278; B 12 8 56,34; B 12 8 60,424; B 12 8 66,62; B 8 12 70,362; B 12 8 74,100; B 22 8 79,330; B 12 8 74,272; B 8 12 76,46; L NI; B 12 18 18,369; B 12 12 20,486; B 12 12 20,502; B 12 12 22,90; B 12 16 30,72; B 12 16 30,288; B 12 12 30,208; B 12 12 30,132; B 12 12 30,148; B 12 12 32,454; B 24 12 38,470; B 12 12 44,502; B 12 18 44,379; B 12 12 44,224; B 12 12 44,164; B 12 12 56,454; B 12 12 56,486; B 12 12 56,208; B 24 12 50,240; B 12 12 56,148; B 24 12 50,180; B 12 16 58,288; B 12 16 62,72; B 12 16 62,412; B 12 42 68,379; B 12 12 68,192; B 12 12 68,224; B 12 12 68,132; B 12 12 68,164; B 12 12 70,90; B 14 12 31,192; L NC; B 8 4 14,272; B 4 4 14,314; B 8 4 18,100; B 4 8 16,46; B 4 4 20,330; B 4 4 20,346; B 4 8 20,364; B 4 4 20,402; B 4 4 22,82; B 4 4 22,20; B 4 4 22,4; B 4 4 22,194; B 4 4 22,210; B 4 4 22,226; B 4 4 22,242; B 8 4 26,62; B 4 4 30,82; B 4 4 30,258; B 4 4 30,298; B 8 4 34,278; B 8 4 36,34; B 4 4 36,330; B 4 4 44,346; B 4 8 44,374; B 4 4 46,114; B 4 4 46,402; B 8 4 54,330; B 8 4 54,278; B 8 4 56,34; B 8 4 60,424; B 4 4 58,258; B 4 4 58,298; B 4 4 62,82; B 8 4 66,62; B 4 4 64,454; B 4 4 64,470; B 4 4 64,486; B 4 4 64,502; B 4 4 66,402; B 4 4 70,82; B 4 4 70,20; B 4 4 70,4; B 4 8 70,362; B 8 4 74,100; B 8 4 74,272; B 4 4 74,314; B 4 4 74,330; B 4 8 76,46; B 4 4 78,346; B 4 4 78,438; B 4 4 78,130; B 4 4 78,146; B 4 4 78,162; B 4 4 78,178; DF;

ALUBufPair

DS 310 125/1; L NP; B 6 8 7,94; B 10 4 9,106; B 4 6 6,101; B 8 8 14,18; B 4 14 12,7; B 4 28 12,122; B 4 22 24,113; B 6 8 25,4; B 4 28 26,22; B 6 4 25,100; B 4 28 38,22; B 6 4 39,100; B 4 22 40,113; B 6 8 39,4; B 8 8 50,18; B 10 4 55,106; B 4 14 52,7; B 4 28 52,122; B 6 8 57,94; B 4 6 58,101; B 6 8 7,32; B 8 8 24,46; B 8 8 40,46; B 6 8 57,32; B 14 4 11,84; B 14 4 11,64; B 12 4 22,74; B 16 4 12,68; B 4 2 18,71; B 4 18 26,59; B 12 4 42,74; B 4 18 38,59; B 4 2 46,71; B 14 4 53,84; B 16 4 52,68; B 14 4 53,64; B 4 26 6,49; B 4 26 58,49; B 4 4 6,88; B 4 4 58,88; B 4 22 26,87; B 4 22 38,87; L ND; B 8 6 4,135; B 8 2 4,127; B 4 4 6,2; B 20 4 10,130; B 8 2 12,97; B 8 8 16,148; B 20 4 14,6; B 8 4 20,2; B 4 12 18,138; B 4 8 18,124; B 4 24 32,44; B 28 12 32,110; B 24 14 32,21; B 8 4 44,2; B 8 8 48,148; B 32 4 32,118; B 4 12 46,138; B 4 8 46,124; B 20 4 50,6; B 8 2 52,97; B 20 4 54,130; B 8 6 60,135; B 8 2 60,127; B 4 4 58,2; B 10 4 13,34; B 4 20 16,46; B 10 4 51,34; B 48 4 32,30; B 4 20 48,46; B 8 8 10,76; B 4 12 12,66; B 4 14 32,71; B 24 4 32,62; B 8 8 54,76; B 4 12 52,66; B 44 4 32,58; B 4 10 12,85; B 4 10 52,85; B 12 2 50,95; B 12 2 14,95; B 48 4 32,92; B 24 12 32,84; B 4 10 32,99; L NM; B 12 8 10,94; B 8 8 16,148; B 12 8 22,4; B 64 16 32,112; B 64 12 32,132; B 64 8 32,18; B 12 8 42,4; B 8 8 48,148; B 12 8 54,94; B 12 8 10,32; B 64 8 32,46; B 12 8 54,32; B 64 8 32,76; L NI; B 12 12 12,130; B 12 12 52,130; B 12 16 12,66; B 12 16 52,66; B 12 12 12,84; B 12 12 52,84; L NC; B 4 8 4,132; B 8 4 10,94; B 4 4 14,18; B 4 4 16,148; B 8 4 22,4; B 4 8 32,110; B 8 4 42,4; B 4 4 48,148; B 4 4 50,18; B 8 4 54,94; B 4 8 60,132; B 8 4 10,32; B 4 4 24,46; B 4 4 40,46; B 8 4 54,32; B 4 4 10,76; B 4 4 54,76; DF;

AddBufPair

DS 311 125/1; L NP; B 6 8 7,94; B 10 4 9,106; B 4 6 6,101; B 8 8 14,18; B 4 14 12,7; B 4 28 12,122; B 4 22 24,113; B 6 8 25,4; B 4 28 26,22; B 6 4 25,100; B 4 28 38,22; B 6 4 39,100; B 4 22 40,113; B 6 8 39,4; B 8 8 50,18; B 10 4 55,106; B 4 14 52,7; B 4 28 52,122; B 6 8 57,94; B 4 6 58,101; B 8 8 24,46; B 8 8 40,46; B 6 8 7,32; B 6 8 57,32; B 14 4 11,64; B 12 4 22,74; B 16 4 12,68; B 4 2 18,71; B 4 18 26,59; B 12 4 42,74; B 4 18 38,59; B 4 2 46,71; B 16 4 52,68; B 14 4 53,64; B 4 26 6,49; B 4 26 58,49; B 14 4 53,84; B 14 4 11,84; B 4 4 6,88; B 4 4 58,88; B 4 22 26,87; B 4 22 38,87; L ND; B 8 6 4,135; B 8 2 4,127; B 4 4 6,2; B 20 4 10,130; B 8 2 12,97; B 8 8 16,148; B 20 4 14,6; B 8 4 20,2; B 4 12 18,138; B 4 8 18,124; B 4 24 32,44; B 28 12 32,110; B 24 14 32,21; B 8 4 44,2; B 8 8 44,148; B 32 4 32,118; B 4

12 46,138; B 4 8 46,124; B 20 4 50,6; B 8 2 52,97; B 20 4 54,130; B 8 6 60,135; B 8 2 60,127; B 4 4
58,2; B 10 4 13,34; B 4 20 16,46; B 10 4 51,34; B 48 4 32,30; B 4 20 48,46; B 8 8 10,76; B 4 12 12,66;
B 4 14 32,71; B 24 4 32,62; B 8 8 54,76; B 4 12 52,66; B 44 4 32,58; B 4 10 12,85; B 4 10 52,85; B 12
2 50,95; B 12 2 14,95; B 48 4 32,92; B 24 12 32,84; B 4 10 32,99; L NM; B 12 8 10,94; B 8 8 16,148; B
12 8 22,4; B 12 8 42,4; B 8 8 44,148; B 6 4 17,154; B 6 4 45,154; B 12 8 54,94; B 64 8 32,18; B 64 16
32,112; B 64 12 32,132; B 64 8 32,46; B 12 8 10,32; B 12 8 54,32; B 64 8 32,76; L NI; B 12 12 12,130;
B 12 12 52,130; B 12 16 12,66; B 12 16 52,66; B 12 12 12,84; B 12 12 52,84; L NC; B 4 8 4,132; B 8 4
10,94; B 4 4 14,18; B 4 4 16,148; B 8 4 22,4; B 4 8 32,110; B 8 4 42,4; B 4 4 44,148; B 4 4 50,18; B 8 4
54,94; B 4 8 60,132; B 4 4 24,46; B 4 4 40,46; B 8 4 10,32; B 8 4 54,32; B 4 4 10,76; B 4 4 54,76; DF;

InvSB4

DS 234 125/1; L NP; B 4 20 10,34; B 12 4 14,22; B 4 16 10,8; B 6 4 11,46; B 4 34 12,65; B 4 2 18,19; B
14 4 25,32; B 4 2 30,35; B 16 4 24,16; B 14 4 25,12; B 4 2 30,9; B 6 8 31,40; B 6 4 29,46; B 4 34
28,65; B 6 8 31,4; L ND; B 8 16 4,58; B 8 6 4,81; B 8 2 4,73; B 14 10 9,31; B 4 8 4,46; B 14 4 9,10; B 4
14 4,19; B 22 4 11,76; B 4 6 20,81; B 4 8 20,70; B 8 8 26,24; B 4 8 24,32; B 4 12 24,14; B 8 12 36,78;
B 22 4 29,64; B 18 12 31,56; B 28 6 16,39; B 8 2 26,43; B 28 4 16,6; B 8 4 26,2; L NM; B 40 8 20,24; B
40 12 20,56; B 40 12 20,78; B 12 8 28,40; B 12 8 28,4; L NI; B 12 12 12,76; B 12 12 24,32; B 12 16
24,14; L NC; B 4 8 4,56; B 4 8 4,78; B 4 4 26,24; B 4 8 36,56; B 4 8 36,78; B 8 4 28,40; B 8 4 28,4; DF;

NonInvSB4

DS 236 125/1; L NP; B 4 20 10,34; B 12 4 14,22; B 4 16 10,8; B 6 4 11,46; B 4 34 12,65; B 4 2 18,19; B
4 34 28,65; B 14 4 25,32; B 4 2 30,35; B 16 4 24,16; B 14 4 25,12; B 4 2 30,9; B 6 8 31,4; B 6 8 31,40;
B 6 4 29,46; L ND; B 8 12 4,78; B 14 10 9,31; B 4 8 4,46; B 14 4 9,10; B 4 14 4,19; B 18 12 9,56; B 22 4
11,64; B 4 6 20,81; B 4 8 20,70; B 8 8 26,24; B 4 8 24,32; B 4 12 24,14; B 22 4 29,76; B 8 6 36,81; B 8
2 36,73; B 8 16 36,58; B 28 4 16,6; B 8 4 26,2; B 28 6 16,39; B 8 2 26,43; L NM; B 40 8 20,24; B 40 12
20,56; B 40 12 20,78; B 12 8 28,4; B 12 8 28,40; L NI; B 12 12 24,32; B 12 16 24,14; B 12 12 28,76; L
NC; B 4 8 4,56; B 4 8 4,78; B 4 4 26,24; B 4 8 36,56; B 4 8 36,78; B 8 4 28,4; B 8 4 28,40; DF;

InvSB8

DS 251 125/1; L NP; B 4 20 10,42; B 12 4 14,30; B 4 24 10,12; B 6 4 11,54; B 4 34 12,73; B 4 2 18,27;
B 14 4 25,40; B 4 2 30,43; B 16 4 24,24; B 14 4 25,20; B 4 2 30,17; B 6 8 31,48; B 6 8 31,12; B 6 4
29,54; B 4 34 28,73; L ND; B 8 16 4,66; B 8 6 4,89; B 8 2 4,81; B 14 10 9,39; B 4 8 4,54; B 14 4 9,18; B
4 14 4,27; B 12 8 10,8; B 22 4 11,84; B 4 6 20,89; B 4 8 20,78; B 8 8 26,32; B 4 8 24,40; B 4 12 24,22;
B 8 12 36,86; B 28 6 16,47; B 8 2 26,51; B 28 4 16,14; B 8 4 26,10; B 18 12 31,64; B 22 4 29,72; L NM;
B 40 8 20,32; B 40 12 20,64; B 40 12 20,86; B 12 8 28,48; B 12 8 28,12; L NI; B 12 12 12,84; B 12 12
24,40; B 12 16 24,22; L NC; B 4 8 4,64; B 4 8 4,86; B 4 4 26,32; B 4 8 36,64; B 4 8 36,86; B 8 4 28,48;
B 8 4 28,12; DF;

NonInvSB8

DS 253 125/1; L NP; B 4 20 10,42; B 12 4 14,30; B 4 24 10,12; B 6 4 11,54; B 4 34 12,73; B 4 2 18,27;
B 4 34 28,73; B 14 4 25,40; B 4 2 30,43; B 16 4 24,24; B 14 4 25,20; B 4 2 30,17; B 6 8 31,48; B 6 4
29,54; B 6 8 31,12; L ND; B 8 12 4,86; B 14 10 9,39; B 4 8 4,54; B 14 4 9,18; B 4 14 4,27; B 12 8 10,8; B
18 12 9,64; B 22 4 11,72; B 4 6 20,89; B 4 8 20,78; B 8 8 26,32; B 4 8 24,40; B 4 12 24,22; B 22 4
29,84; B 8 6 36,89; B 8 2 36,81; B 8 16 36,66; B 28 6 16,47; B 8 2 26,51; B 28 4 16,14; B 8 4 26,10; L
NM; B 40 8 20,32; B 40 12 20,64; B 40 12 20,86; B 12 8 28,48; B 12 8 28,12; L NI; B 12 12 24,40; B 12
16 24,22; B 12 12 28,84; L NC; B 4 8 4,64; B 4 8 4,86; B 4 4 26,32; B 4 8 36,64; B 4 8 36,86; B 8 4
28,48; B 8 4 28,12; DF;

Vsclk

DS 520 125/1; L ND; B 8 16 4,16; B 8 4 22,154; B 8 8 66,114; B 26 4 17,132; B 4 22 2,141; B 4 4
38,112; B 4 4 62,112; B 4 4 34,74; B 4 4 34,90; B 4 4 30,48; B 4 4 30,60; B 2 4 3,154; B 8 4 4,200; B 4
8 8,46; B 4 8 16,46; B 4 12 24,114; B 4 12 32,114; B 10 4 5,200; B 4 4 32,90; B 4 4 24,90; B 8 4
52,200; B 8 8 64,138; B 4 4 26,6; B 4 4 38,6; B 4 12 46,132; B 4 12 54,132; B 2 4 21,200; B 22 4
19,168; B 8 8 62,174; B 14 4 15,118; B 8 8 4,92; B 4 4 46,172; B 4 4 58,172; B 8 8 58,8; B 14 4
37,132; B 4 4 28,132; B 4 8 8,8; B 4 8 16,8; B 2 4 59,172; B 4 4 58,136; B 8 8 4,12; B 2 4 33,114; B 2 4
63,112; B 8 8 20,6; B 4 4 28,132; B 8 8 64,88; B 14 4 15,90; B 6 4 21,22; B 2 4 7,24; B 8 8 4,120; B 2 4
9,74; B 10 4 53,56; B 4 8 32,166; B 4 8 40,166; B 10 4 13,62; B 12 4 50,18; B 8 8 4,170; B 4 4 58,86; B
4 4 38,86; B 4 4 12,200; B 4 4 20,200; B 4 4 20,74; B 4 4 12,74; B 4 4 26,200; B 4 4 46,200; B 8 8
4,76; B 8 8 4,64; B 4 8 28,60; B 4 8 20,60; B 8 8 60,20; B 8 8 62,58; B 4 4 46,56; B 4 4 34,56; B 8 8
60,202; B 8 8 4,26; B 4 4 42,18; B 4 4 30,18; B 4 8 16,22; B 4 8 8,22; B 8 8 4,202; B 2 4 41,168; B 2 4
17,8; B 4 4 4,154; B 4 4 16,154; B 2 4 7,10; B 4 4 2,154; B 4 4 2,132; B 14 4 47,6; B 8 8 30,152; B 2 4
17,48; B 2 4 7,48; B 4 12 30,52; B 4 4 8,182; B 4 4 16,182; B 4 4 30,60; B 4 4 30,60; B 8 8 28,180; B
16 4 24,48; B 4 16 34,80; B 4 4 34,90; B 2 4 35,90; B 8 8 4,184; B 14 4 29,74; B 2 4 29,60; B 8 8 4,50;
B 2 4 7,182; B 6 4 21,182; B 4 4 16,200; B 4 8 12,8; B 8 4 10,154; B 4 8 12,46; B 4 8 24,60; B 8 4 32,6;
B 4 4 12,182; B 16 4 48,86; B 4 12 28,114; B 8 4 36,18; B 4 8 36,166; B 4 8 12,22; B 8 4 40,56; B 4 4
16,74; B 16 4 36,200; B 20 4 50,112; B 4 4 28,90; B 8 4 52,172; B 4 12 50,132; B 8 8 24,198; B 8 8
28,20; B 8 8 32,58; B 8 8 36,88; B 8 8 30,130; B 8 8 36,112; B 8 8 44,170; L NI; B 12 16 32,8; B 16 12
52,172; B 10 12 35,58; B 10 12 39,112; B 28 12 50,112; B 24 12 36,200; B 10 12 39,88; B 10 12
47,170; B 16 12 40,56; B 16 12 32,6; B 24 12 48,86; B 10 12 31,20; B 10 12 27,198; B 16 12 36,18; L
NP; B 4 4 40,82; B 8 4 32,190; B 8 8 18,212; B 8 12 10,154; B 4 16 12,22; B 4 2 12,15; B 4 4 36,194; B
8 12 36,18; B 4 4 30,194; B 4 4 8,162; B 4 12 12,182; B 4 2 12,37; B 14 4 7,38; B 4 4 12,188; B 4 4

36,188; B 6 8 33,20; B 2 4 43,80; B 4 12 40,72; B 16 4 34,68; B 4 2 24,67; B 4 4 12,30; B 8 12 40,56; B 4 24 38,40; B 6 8 29,198; B 26 4 27,30; B 4 4 38,30; B 4 6 36,189; B 24 4 26,188; B 4 2 12,187; B 6 8 37,58; B 4 4 24,68; B 10 4 5,162; B 4 4 40,68; B 4 4 40,80; B 4 4 12,38; B 4 16 12,46; B 4 2 16,207; B 4 16 24,60; B 8 8 18,82; B 4 18 36,183; B 6 4 35,194; B 4 2 30,193; B 8 12 32,6; B 32 4 36,142; B 4 2 50,141; B 4 26 50,153; B 16 12 36,200; B 4 16 12,8; B 4 12 16,74; B 4 2 28,123; B 4 12 16,200; B 16 12 48,86; B 14 4 37,98; B 4 10 42,101; B 4 2 16,79; B 4 2 12,29; B 4 2 38,51; B 4 2 42,79; B 4 16 36,166; B 6 8 41,88; B 8 8 16,140; B 4 4 42,80; B 4 4 50,142; B 8 6 30,125; B 6 8 49,170; B 8 12 52,172; B 4 20 50,132; B 4 4 28,98; B 4 12 28,90; B 4 4 42,98; B 4 20 28,114; B 6 8 41,112; B 20 12 50,112; L NM; B 8 220 66,110; B 6 20 3,106; B 6 2 19,1; B 6 50 17,111; B 8 8 20,6; B 12 8 38,112; B 8 8 4,76; B 6 4 19,218; B 8 8 18,82; B 8 8 4,120; B 8 8 18,212; B 8 8 58,8; B 8 8 64,138; B 8 8 30,152; B 6 6 3,19; B 6 8 3,4; B 8 8 62,174; B 8 8 4,12; B 8 8 4,202; B 8 8 4,170; B 12 8 46,170; B 6 28 63,156; B 8 8 64,88; B 8 8 4,92; B 8 8 16,140; B 8 12 30,128; B 6 16 65,126; B 8 8 66,114; B 6 4 59,2; B 8 8 60,202; B 6 4 59,14; B 8 8 4,26; B 6 14 59,213; B 6 4 3,70; B 6 30 61,39; B 6 18 65,101; B 6 22 63,73; B 12 8 38,88; B 6 8 3,84; B 12 8 30,20; B 8 8 28,180; B 6 20 61,188; B 8 8 4,64; B 12 8 26,198; B 6 14 3,213; B 8 8 4,184; B 6 42 3,145; B 8 8 60,20; B 8 8 62,58; B 6 20 29,166; B 8 8 4,50; B 12 8 34,58; B 6 6 3,177; B 6 10 3,193; B 6 16 3,38; B 6 6 3,57; L NC; B 8 4 38,88; B 4 4 28,180; B 4 4 4,120; B 8 4 30,20; B 4 4 4,170; B 4 4 16,140; B 4 4 64,88; B 4 4 4,12; B 4 8 30,128; B 4 4 60,20; B 8 4 26,198; B 8 4 46,170; B 4 4 62,58; B 4 4 4,184; B 4 4 62,174; B 4 4 4,92; B 4 4 4,76; B 4 4 64,138; B 8 4 34,58; B 4 4 4,64; B 8 4 38,112; B 4 4 18,82; B 4 4 66,114; B 4 4 58,8; B 4 4 4,202; B 4 4 4,50; B 4 4 18,212; B 4 4 30,152; B 4 4 4,26; B 4 4 60,202; B 4 4 20,6; DF;

Synchronizer

DS 521 125/1; L NM; B 8 8 4,74; B 8 8 4,162; B 8 80 4,118; B 8 70 4,35; B 12 8 20,24; B 8 12 20,114; B 8 8 24,98; B 8 12 26,44; B 12 8 28,158; B 6 44 25,72; B 8 12 28,140; B 8 8 36,16; B 6 6 37,109; B 6 6 37,103; B 6 6 37,103; B 6 6 37,63; B 2 6 35,63; B 6 6 37,109; B 6 34 37,83; B 8 8 40,62; B 6 6 39,63; B 6 6 39,63; B 8 8 42,46; B 6 8 41,54; B 8 8 44,118; B 6 6 43,109; B 6 6 43,29; B 26 6 53,29; B 12 6 46,15; B 6 10 43,37; B 6 2 43,113; B 8 8 46,156; B 6 6 45,149; B 6 6 45,149; B 6 24 45,134; B 6 6 47,163; B 12 8 52,90; B 12 8 52,76; B 8 8 56,16; B 12 8 60,118; B 8 12 58,140; B 12 8 62,158; B 6 6 61,89; B 8 12 66,44; B 8 8 68,90; B 6 36 67,68; B 6 6 69,29; B 6 6 69,23; B 6 6 69,23; B 6 20 69,10; B 8 8 82,120; B 8 8 82,146; B 8 8 82,82; B 8 6 82,139; B 8 6 82,139; B 8 16 82,158; B 8 78 82,39; B 8 30 82,101; B 8 12 82,130; L ND; B 8 8 4,74; B 8 8 4,162; B 4 4 4,160; B 4 4 4,160; B 6 4 5,160; B 4 16 8,154; B 2 4 9,72; B 4 40 12,54; B 4 16 12,154; B 4 4 14,118; B 4 4 14,122; B 4 4 14,106; B 2 4 13,106; B 8 4 16,122; B 4 8 14,112; B 4 4 16,88; B 4 4 16,84; B 4 4 16,92; B 4 4 16,106; B 4 40 16,54; B 4 16 16,154; B 4 4 16,82; B 4 4 16,106; B 10 4 19,82; B 4 4 16,106; B 4 2 16,83; B 4 10 16,99; B 8 4 20,118; B 8 8 20,116; B 8 8 22,24; B 4 40 20,54; B 4 4 20,160; B 4 4 20,160; B 36 4 36,106; B 4 4 22,130; B 4 4 22,134; B 4 4 22,126; B 4 4 22,136; B 4 4 22,122; B 4 2 22,135; B 4 24 24,62; B 8 4 26,48; B 8 8 26,46; B 8 8 26,158; B 4 4 24,118; B 4 8 24,24; B 4 4 24,48; B 4 4 24,48; B 2 4 23,118; B 6 4 25,48; B 4 4 26,48; B 4 4 26,74; B 8 4 28,136; B 8 8 28,138; B 4 4 26,136; B 4 4 26,82; B 4 4 26,136; B 4 4 26,78; B 4 22 26,61; B 2 4 25,48; B 12 4 32,118; B 20 4 36,26; B 10 4 33,156; B 4 4 30,48; B 4 4 30,74; B 4 4 30,74; B 2 4 29,48; B 4 4 34,48; B 4 4 34,74; B 4 4 38,48; B 4 4 38,74; B 4 4 40,156; B 4 4 40,118; B 8 8 42,46; B 4 4 40,48; B 4 4 40,48; B 2 4 39,48; B 4 4 40,48; B 8 8 44,118; B 4 4 42,74; B 2 4 41,118; B 8 8 46,156; B 4 4 44,24; B 4 4 44,6; B 4 4 44,10; B 4 4 44,2; B 4 4 44,74; B 4 4 44,74; B 4 10 44,17; B 2 4 43,74; B 4 4 48,116; B 4 4 48,26; B 4 4 48,24; B 4 8 48,76; B 8 8 50,76; B 4 4 48,48; B 4 4 48,48; B 4 2 48,25; B 2 4 47,116; B 4 4 50,156; B 2 4 49,156; B 8 8 54,90; B 10 4 55,116; B 4 4 52,48; B 4 4 52,24; B 10 4 57,156; B 4 8 56,90; B 4 4 56,106; B 4 4 56,100; B 4 4 56,104; B 4 4 56,96; B 4 4 56,48; B 4 4 56,24; B 8 4 58,136; B 8 8 58,138; B 4 2 56,105; B 8 8 62,118; B 4 4 60,48; B 4 4 60,24; B 4 4 60,136; B 4 4 60,136; B 8 8 64,158; B 4 4 64,118; B 8 4 66,48; B 8 8 66,46; B 4 4 64,160; B 4 4 64,164; B 4 4 64,130; B 4 4 64,134; B 4 4 64,126; B 4 4 64,136; B 4 4 64,122; B 4 4 64,122; B 4 4 64,24; B 4 4 64,24; B 4 2 64,135; B 6 4 65,160; B 4 4 64,118; B 4 8 66,116; B 2 4 65,118; B 4 4 68,164; B 4 4 68,24; B 4 4 68,24; B 4 4 68,160; B 4 4 68,160; B 4 8 70,116; B 4 16 70,154; B 2 4 69,160; B 4 4 72,24; B 4 4 72,48; B 4 20 72,36; B 4 8 74,116; B 4 16 74,154; B 4 40 74,50; B 2 4 73,48; B 4 16 78,154; B 4 40 78,50; B 2 4 77,118; B 8 8 82,120; B 8 8 82,146; B 8 8 82,82; B 8 4 82,80; B 6 4 81,148; B 4 40 82,50; B 4 4 82,148; B 4 4 82,148; B 4 4 84,68; B 4 4 84,80; B 4 12 84,76; B 2 4 83,68; L NP; B 16 4 8,130; B 38 4 19,6; B 32 4 16,14; B 12 4 16,88; B 4 24 12,154; B 4 4 12,142; B 4 2 12,143; B 6 8 17,24; B 4 48 16,54; B 10 4 19,142; B 4 2 16,29; B 8 6 20,111; B 12 4 22,130; B 8 8 24,98; B 4 4 22,110; B 4 4 22,110; B 8 6 26,41; B 4 4 24,88; B 4 4 24,88; B 4 2 26,113; B 4 4 26,164; B 8 6 28,143; B 2 4 25,110; B 4 4 28,36; B 4 4 28,110; B 12 12 32,118; B 4 4 28,88; B 4 4 28,88; B 4 4 30,164; B 10 12 33,156; B 4 4 30,100; B 4 4 30,100; B 30 4 43,130; B 4 6 32,41; B 2 4 31,36; B 16 4 38,88; B 4 4 34,36; B 4 12 34,48; B 8 8 36,16; B 4 4 34,40; B 18 4 41,100; B 8 8 40,62; B 4 12 38,74; B 4 2 38,67; B 12 4 44,6; B 6 8 49,90; B 10 12 55,116; B 12 4 56,100; B 36 4 68,6; B 10 12 57,156; B 8 8 56,16; B 6 8 55,76; B 4 4 56,36; B 4 12 56,48; B 4 12 56,24; B 8 6 58,143; B 4 2 56,19; B 4 4 56,40; B 12 4 64,130; B 10 4 63,74; B 4 4 60,36; B 26 4 73,14; B 4 4 64,36; B 8 6 66,41; B 10 4 67,142; B 8 8 68,90; B 4 16 70,116; B 4 4 70,74; B 4 4 70,74; B 4 14 70,101; B 16 4 78,130; B 4 24 74,154; B 4 4 74,142; B 4 4 74,74; B 4 4 74,74; B 4 2 74,143; B 4 4 78,74; B 4 48 78,50; B 4 2 78,73; L NC; B 4 4 4,74; B 4 4 4,162; B 8 4 20,24; B 4 8 20,114; B 4 4 24,98; B 4 8 26,44; B 8 4 28,158; B 4 8 28,140; B 4 4 36,16; B 4 4 40,62; B 4 4 42,46; B 4 4 44,118; B 4 4 46,156; B 8 4 52,90; B 8 4 52,76; B 4 4 56,16; B 8 4 60,118; B 4 8 58,140; B 8 4 62,158; B 4 8 66,44; B 4 4 68,90; B 4 4 82,120; B 4 4 82,146; B 4 4 82,82; L NI; B 20 12 32,118; B 18 12 33,156; B 12 12 34,48; B 18 12 55,116; B 18 12 57,156; B 12 12 56,48; DF;

Cnt

DS 155 125/1; L NP; B 4 10 2,129; B 4 94 2,181; B 32 4 16,12; B 8 8 4,60; B 8 8 4,100; B 4 12 2,68; B

4 12 2,92; B 4 14 4,121; B 4 6 4,227; B 18 4 11,72; B 12 4 8,88; B 4 22 4,113; B 10 4 9,154; B 10 4 9,170; B 10 4 9,202; B 10 4 13,138; B 10 4 13,138; B 10 4 13,186; B 10 4 13,218; B 8 6 20,25; B 4 20 20,124; B 4 96 20,182; B 4 20 20,36; B 4 54 20,97; B 4 24 20,54; B 8 8 30,132; B 4 22 28,147; B 4 4 28,158; B 14 4 35,158; B 6 8 35,96; B 4 58 34,71; B 14 10 39,47; B 16 4 40,12; B 8 6 38,145; B 4 16 38,112; B 4 4 38,120; B 2 4 39,120; B 4 28 40,134; B 4 4 40,120; B 8 6 42,45; L NM; B 6 6 3,193; B 6 6 3,163; B 48 8 24,118; B 48 6 24,131; B 48 8 24,226; B 32 8 16,118; B 32 8 16,80; B 12 8 8,60; B 12 8 8,100; B 8 8 10,146; B 8 8 10,162; B 8 8 10,178; B 8 8 10,194; B 8 8 10,210; B 8 8 12,118; B 8 8 12,226; B 12 6 20,147; B 12 6 20,209; B 10 6 19,179; B 8 12 20,22; B 6 22 23,155; B 6 16 23,198; B 8 8 28,178; B 22 6 37,163; B 22 6 37,193; B 8 8 30,132; B 12 8 38,96; B 16 8 40,80; B 16 8 40,118; B 8 12 38,148; B 8 12 42,42; B 8 8 42,80; B 6 28 41,22; B 8 8 42,4; L NI; B 12 12 8,72; B 12 12 8,88; B 14 18 39,47; L NC; B 8 4 8,60; B 8 4 8,100; B 4 4 8,80; B 4 4 10,146; B 4 4 10,162; B 4 4 10,178; B 4 4 10,194; B 4 4 10,210; B 4 4 12,118; B 4 4 12,226; B 4 4 12,118; B 4 8 20,22; B 4 4 28,178; B 4 4 28,118; B 4 4 30,132; B 8 4 38,96; B 4 8 38,148; B 4 8 42,42; B 4 4 42,80; B 4 4 42,4; L ND; B 8 8 8,80; B 8 8 10,146; B 8 8 10,162; B 8 8 10,178; B 8 8 10,194; B 8 8 10,210; B 10 8 11,146; B 10 8 11,178; B 10 8 11,194; B 10 8 11,210; B 4 8 8,154; B 4 8 8,170; B 4 8 8,202; B 8 12 10,62; B 8 12 10,98; B 4 12 8,72; B 4 2 8,67; B 4 12 8,86; B 4 2 8,93; B 8 8 12,118; B 8 8 12,226; B 8 10 12,119; B 4 20 14,132; B 4 8 14,138; B 4 8 14,186; B 4 8 14,218; B 12 16 20,38; B 12 16 20,100; B 12 16 20,54; B 8 8 20,20; B 20 4 28,6; B 20 4 28,6; B 4 10 20,13; B 8 8 28,178; B 8 8 28,118; B 6 62 27,85; B 16 18 38,117; B 12 4 38,180; B 8 8 38,150; B 12 4 40,164; B 4 10 36,159; B 8 8 40,96; B 8 8 42,40; B 4 32 40,60; B 8 8 42,80; B 8 8 42,4; B 4 16 44,172; B 4 4 44,180; B 4 16 44,100; DF;

CntUD

DS 161 125/1; L NP; B 4 66 2,237; B 4 10 2,265; B 4 96 2,318; B 4 10 2,129; B 4 94 2,181; B 32 4 16,12; B 8 8 4,60; B 8 8 4,100; B 4 12 2,68; B 4 12 2,92; B 4 14 4,121; B 4 6 4,227; B 18 4 11,72; B 12 4 8,88; B 4 22 4,113; B 10 4 9,274; B 10 4 9,274; B 10 4 9,306; B 10 4 9,354; B 10 4 9,154; B 10 4 9,170; B 10 4 9,202; B 10 4 13,290; B 10 4 13,322; B 10 4 13,338; B 10 4 13,138; B 10 4 13,138; B 10 4 13,186; B 10 4 13,218; B 4 6 18,363; B 8 6 20,25; B 4 50 20,245; B 4 94 20,317; B 4 20 20,124; B 4 96 20,182; B 4 20 20,36; B 4 54 20,97; B 4 24 20,54; B 8 8 30,268; B 8 8 30,132; B 4 22 28,147; B 4 4 28,158; B 14 4 35,158; B 8 8 34,240; B 4 10 32,277; B 6 8 35,96; B 4 58 34,71; B 14 10 39,47; B 16 4 40,12; B 4 62 36,205; B 8 6 38,145; B 4 16 38,112; B 4 4 38,120; B 2 4 39,120; B 4 28 40,134; B 4 4 40,120; B 8 6 42,45; L NM; B 6 6 3,329; B 6 6 3,299; B 48 6 24,239; B 48 8 24,254; B 48 6 24,267; B 48 8 24,362; B 6 6 3,193; B 6 6 3,163; B 48 8 24,118; B 48 6 24,131; B 48 8 24,226; B 32 8 16,118; B 32 8 16,80; B 12 8 8,60; B 12 8 8,100; B 8 8 10,282; B 8 8 10,298; B 8 8 10,314; B 8 8 10,330; B 8 8 10,346; B 8 8 10,362; B 8 8 10,146; B 8 8 10,162; B 8 8 10,178; B 8 8 10,194; B 8 8 10,210; B 8 8 12,254; B 8 8 12,118; B 8 8 12,226; B 12 6 20,283; B 12 6 20,345; B 10 6 19,315; B 12 6 20,147; B 12 6 20,209; B 10 6 19,179; B 8 12 20,22; B 6 22 23,291; B 6 16 23,334; B 6 22 23,155; B 6 16 23,198; B 8 8 28,314; B 8 8 28,178; B 22 6 37,299; B 22 6 37,329; B 8 8 30,268; B 22 6 37,163; B 22 6 37,193; B 8 8 30,132; B 8 8 34,240; B 12 8 38,96; B 16 8 40,80; B 16 8 40,118; B 8 12 38,148; B 8 12 42,42; B 8 8 42,80; B 6 28 41,22; B 8 8 42,4; L NI; B 12 12 8,72; B 12 12 8,88; B 14 18 39,47; L NC; B 8 4 8,60; B 8 4 8,100; B 4 4 8,80; B 4 4 10,282; B 4 4 10,298; B 4 4 10,314; B 4 4 10,330; B 4 4 10,346; B 4 4 10,362; B 4 4 10,146; B 4 4 10,162; B 4 4 10,178; B 4 4 10,194; B 4 4 10,210; B 4 4 12,254; B 4 4 12,118; B 4 4 12,226; B 4 4 12,118; B 4 8 20,22; B 4 4 28,314; B 4 4 28,178; B 4 4 28,118; B 4 4 30,268; B 4 4 30,132; B 4 4 34,240; B 8 4 38,96; B 4 8 38,148; B 4 8 42,42; B 4 4 42,80; B 4 4 42,4; L ND; B 8 8 8,80; B 4 20 8,260; B 8 8 10,282; B 8 8 10,298; B 8 8 10,314; B 8 8 10,330; B 8 8 10,346; B 8 8 10,362; B 10 8 11,282; B 10 8 11,298; B 10 8 11,314; B 10 8 11,330; B 10 8 11,346; B 4 8 8,274; B 4 8 8,306; B 4 8 8,354; B 8 8 10,146; B 8 8 10,162; B 8 8 10,178; B 8 8 10,194; B 8 8 10,210; B 10 8 11,146; B 10 8 11,178; B 10 8 11,194; B 10 8 11,210; B 4 8 8,154; B 4 8 8,170; B 4 8 8,202; B 8 12 10,62; B 8 12 10,98; B 4 12 8,72; B 4 2 8,67; B 4 12 8,86; B 4 2 8,93; B 8 8 12,254; B 8 8 12,118; B 8 8 12,226; B 8 10 12,119; B 4 8 14,290; B 4 8 14,322; B 4 8 14,338; B 4 20 14,132; B 4 8 14,138; B 4 8 14,186; B 4 8 14,218; B 12 16 20,38; B 12 16 20,100; B 12 16 20,54; B 8 8 20,20; B 20 4 28,6; B 20 4 28,6; B 4 10 20,13; B 8 8 28,314; B 4 34 26,293; B 4 4 26,276; B 8 8 28,178; B 8 8 28,118; B 6 62 27,85; B 18 4 35,276; B 16 18 38,117; B 12 4 38,180; B 8 8 38,150; B 12 4 40,164; B 4 10 36,159; B 8 8 40,96; B 8 8 42,40; B 4 32 40,60; B 8 8 42,80; B 8 8 42,4; B 4 94 44,229; B 4 4 44,276; B 4 16 44,172; B 4 4 44,180; B 4 16 44,100; DF;

CntRestore

DS 162 125/1; L NP; B 4 10 2,129; B 4 94 2,181; B 32 4 16,12; B 8 8 4,60; B 8 8 4,100; B 4 12 2,68; B 4 12 2,92; B 4 14 4,121; B 4 6 4,227; B 18 4 11,72; B 12 4 8,88; B 4 22 4,113; B 10 4 9,154; B 10 4 9,170; B 10 4 9,202; B 10 4 13,138; B 10 4 13,138; B 10 4 13,186; B 10 4 13,218; B 8 6 20,25; B 4 20 20,124; B 4 96 20,182; B 4 20 20,36; B 4 54 20,97; B 4 24 20,54; B 8 8 30,132; B 4 22 28,147; B 4 4 28,158; B 14 4 35,158; B 6 8 35,96; B 4 58 34,71; B 14 10 39,47; B 16 4 40,12; B 8 6 38,145; B 4 16 38,112; B 4 4 38,120; B 2 4 39,120; B 4 28 40,134; B 4 4 40,120; B 8 6 42,45; B 40 4 68,12; B 8 8 62,164; B 4 50 60,193; B 4 24 60,142; B 4 4 60,154; B 12 4 66,154; B 8 8 70,192; B 4 34 72,171; B 4 4 72,154; B 6 8 73,208; B 10 12 75,206; B 6 8 73,144; B 10 12 75,142; L NM; B 6 6 3,193; B 6 6 3,163; B 48 8 24,118; B 48 6 24,131; B 48 8 24,226; B 32 8 16,118; B 32 8 16,80; B 12 8 8,60; B 12 8 8,100; B 8 8 10,146; B 8 8 10,162; B 8 8 10,178; B 8 8 10,194; B 8 8 10,210; B 8 8 12,118; B 8 8 12,226; B 12 6 20,147; B 12 6 20,209; B 10 6 19,179; B 8 12 20,22; B 6 22 23,155; B 6 16 23,198; B 8 8 28,178; B 22 6 37,163; B 22 6 37,193; B 8 8 30,132; B 12 8 38,96; B 16 8 40,80; B 16 8 40,118; B 8 12 38,148; B 8 12 42,42; B 8 8 42,80; B 6 28 41,22; B 8 8 42,4; B 40 8 68,80; B 40 8 68,226; B 40 8 68,118; B 10 6 53,163; B 18 6 57,193; B 40 6 68,131; B 8 8 56,118; B 8 8 62,164; B 12 8 70,208; B 12 8 70,144; B 8 8 70,192; B 6 18 79,151; B 12 6 82,163; B 10 6 81,207; B 8 8 82,226; B 6 14 83,203; B 8 6 84,193; L NI; B 12 12 8,72; B 12 12 8,88; B 14 18 39,47; B 18 12 75,206; B 18 12 75,142; L NC; B 8 4 8,60; B 8 4

8,100; B 4 4 8,80; B 4 4 10,146; B 4 4 10,162; B 4 4 10,178; B 4 4 10,194; B 4 4 10,210; B 4 4 12,118;
B 4 4 12,226; B 4 4 12,118; B 4 8 20,22; B 4 4 28,178; B 4 4 28,118; B 4 4 30,132; B 8 4 38,96; B 4 8
38,148; B 4 8 42,42; B 4 4 42,80; B 4 4 42,4; B 4 4 56,118; B 4 4 62,164; B 8 4 70,208; B 8 4 70,144; B
4 4 70,192; B 4 4 82,226; L ND; B 8 8 8,80; B 8 8 10,146; B 8 8 10,162; B 8 8 10,178; B 8 8 10,194; B 8
8 10,210; B 10 8 11,146; B 10 8 11,178; B 10 8 11,194; B 10 8 11,210; B 4 8 8,154; B 4 8 8,170; B 4 8
8,202; B 8 12 10,62; B 8 12 10,98; B 4 12 8,72; B 4 2 8,67; B 4 12 8,86; B 4 2 8,93; B 8 8 12,118; B 8 8
12,226; B 8 10 12,119; B 4 20 14,132; B 4 8 14,138; B 4 8 14,186; B 4 8 14,218; B 12 16 20,38; B 12
16 20,100; B 12 16 20,54; B 8 8 20,20; B 20 4 28,6; B 20 4 28,6; B 4 10 20,13; B 8 8 28,178; B 8 8
28,118; B 6 62 27,85; B 16 18 38,117; B 12 4 38,180; B 8 8 38,150; B 12 4 40,164; B 4 10 36,159; B 8
8 40,96; B 8 8 42,40; B 4 32 40,60; B 8 8 42,80; B 8 8 42,4; B 4 16 44,172; B 4 4 44,180; B 4 16
44,100; B 8 8 56,118; B 4 92 54,168; B 10 16 61,206; B 10 16 61,142; B 8 8 68,208; B 8 8 68,144; B
10 4 77,206; B 10 4 77,142; B 8 8 82,226; B 4 82 84,181; DF;

CntUDRestore

DS 163 125/1; L NP; B 4 66 2,237; B 4 10 2,265; B 4 96 2,318; B 4 10 2,129; B 4 94 2,181; B 32 4
16,12; B 8 8 4,60; B 8 8 4,100; B 4 12 2,68; B 4 12 2,92; B 4 14 4,121; B 4 6 4,227; B 18 4 11,72; B 12
4 8,88; B 4 22 4,113; B 10 4 9,274; B 10 4 9,274; B 10 4 9,306; B 10 4 9,354; B 10 4 9,154; B 10 4
9,170; B 10 4 9,202; B 10 4 13,290; B 10 4 13,322; B 10 4 13,338; B 10 4 13,138; B 10 4 13,138; B 10
4 13,186; B 10 4 13,218; B 4 6 18,363; B 8 6 20,25; B 4 50 20,245; B 4 94 20,317; B 4 20 20,124; B 4
96 20,182; B 4 20 20,36; B 4 54 20,97; B 4 24 20,54; B 8 8 30,268; B 8 8 30,132; B 4 22 28,147; B 4 4
28,158; B 14 4 35,158; B 8 8 34,240; B 4 10 32,277; B 6 8 35,96; B 4 58 34,71; B 14 10 39,47; B 16 4
40,12; B 4 62 36,205; B 8 6 38,145; B 4 16 38,112; B 4 4 38,120; B 2 4 39,120; B 4 28 40,134; B 4 4
40,120; B 8 6 42,45; B 40 4 68,12; B 8 8 62,300; B 4 50 60,329; B 4 24 60,278; B 4 4 60,290; B 8 8
62,164; B 4 50 60,193; B 4 24 60,142; B 4 4 60,154; B 12 4 66,290; B 12 4 66,154; B 8 8 70,328; B 8
8 70,192; B 4 34 72,307; B 4 4 72,290; B 6 8 73,344; B 10 12 75,342; B 6 8 73,280; B 10 12 75,278; B
4 34 72,171; B 4 4 72,154; B 6 8 73,208; B 10 12 75,206; B 6 8 73,144; B 10 12 75,142; L NM; B 6 6
3,329; B 6 6 3,299; B 48 6 24,239; B 48 8 24,254; B 48 6 24,267; B 48 8 24,362; B 6 6 3,193; B 6 6
3,163; B 48 8 24,118; B 48 6 24,131; B 48 8 24,226; B 32 8 16,118; B 32 8 16,80; B 12 8 8,60; B 12 8
8,100; B 8 8 10,282; B 8 8 10,298; B 8 8 10,314; B 8 8 10,330; B 8 8 10,346; B 8 8 10,362; B 8 8
10,146; B 8 8 10,162; B 8 8 10,178; B 8 8 10,194; B 8 8 10,210; B 8 8 12,254; B 8 8 12,118; B 8 8
12,226; B 12 6 20,283; B 12 6 20,345; B 10 6 19,315; B 12 6 20,147; B 12 6 20,209; B 10 6 19,179; B
8 12 20,22; B 6 22 23,291; B 6 16 23,334; B 6 22 23,155; B 6 16 23,198; B 8 8 28,314; B 8 8 28,178;
B 22 6 37,299; B 22 6 37,329; B 8 8 30,268; B 22 6 37,163; B 22 6 37,193; B 8 8 30,132; B 8 8
34,240; B 12 8 38,96; B 16 8 40,80; B 16 8 40,118; B 8 12 38,148; B 8 12 42,42; B 8 8 42,80; B 6 28
41,22; B 8 8 42,4; B 40 6 68,239; B 40 8 68,362; B 40 8 68,254; B 10 6 53,299; B 18 6 57,329; B 40 6
68,267; B 40 8 68,80; B 40 8 68,226; B 40 8 68,118; B 10 6 53,163; B 18 6 57,193; B 40 6 68,131; B 8
8 56,254; B 8 8 56,118; B 8 8 62,300; B 8 8 62,164; B 12 8 70,344; B 12 8 70,280; B 12 8 70,208; B
12 8 70,144; B 8 8 70,328; B 8 8 70,192; B 6 18 79,287; B 12 6 82,299; B 10 6 81,343; B 6 18 79,151;
B 12 6 82,163; B 10 6 81,207; B 8 8 82,362; B 8 8 82,226; B 6 14 83,339; B 8 6 84,329; B 6 14
83,203; B 8 6 84,193; L NI; B 12 12 8,72; B 12 12 8,88; B 14 18 39,47; B 18 12 75,342; B 18 12
75,278; B 18 12 75,206; B 18 12 75,142; L NC; B 8 4 8,60; B 8 4 8,100; B 4 4 8,80; B 4 4 10,282; B 4 4
10,298; B 4 4 10,314; B 4 4 10,330; B 4 4 10,346; B 4 4 10,362; B 4 4 10,146; B 4 4 10,162; B 4 4
10,178; B 4 4 10,194; B 4 4 10,210; B 4 4 12,254; B 4 4 12,118; B 4 4 12,226; B 4 4 12,118; B 4 8
20,22; B 4 4 28,314; B 4 4 28,178; B 4 4 28,118; B 4 4 30,268; B 4 4 30,132; B 4 4 34,240; B 8 4
38,96; B 4 8 38,148; B 4 8 42,42; B 4 4 42,80; B 4 4 42,4; B 4 4 56,254; B 4 4 56,118; B 4 4 62,300; B
4 4 62,164; B 8 4 70,344; B 8 4 70,280; B 8 4 70,208; B 8 4 70,144; B 4 4 70,328; B 4 4 70,192; B 4 4
82,362; B 4 4 82,226; L ND; B 8 8 8,80; B 4 20 8,260; B 8 8 10,282; B 8 8 10,298; B 8 8 10,314; B 8 8
10,330; B 8 8 10,346; B 8 8 10,362; B 10 8 11,282; B 10 8 11,298; B 10 8 11,314; B 10 8 11,330; B 10
8 11,346; B 4 8 8,274; B 4 8 8,306; B 4 8 8,354; B 8 8 10,146; B 8 8 10,162; B 8 8 10,178; B 8 8
10,194; B 8 8 10,210; B 10 8 11,146; B 10 8 11,178; B 10 8 11,194; B 10 8 11,210; B 4 8 8,154; B 4 8
8,170; B 4 8 8,202; B 8 12 10,62; B 8 12 10,98; B 4 12 8,72; B 4 2 8,67; B 4 12 8,86; B 4 2 8,93; B 8 8
12,254; B 8 8 12,118; B 8 8 12,226; B 8 10 12,119; B 4 8 14,290; B 4 8 14,322; B 4 8 14,338; B 4 20
14,132; B 4 8 14,138; B 4 8 14,186; B 4 8 14,218; B 12 16 20,38; B 12 16 20,100; B 12 16 20,54; B 8
8 20,20; B 20 4 28,6; B 20 4 28,6; B 4 10 20,13; B 8 8 28,314; B 4 34 26,293; B 4 4 26,276; B 8 8
28,178; B 8 8 28,118; B 6 62 27,85; B 18 4 35,276; B 16 18 38,117; B 12 4 38,180; B 8 8 38,150; B 12
4 40,164; B 4 10 36,159; B 8 8 40,96; B 8 8 42,40; B 4 32 40,60; B 8 8 42,80; B 8 8 42,4; B 4 94
44,229; B 4 4 44,276; B 4 16 44,172; B 4 4 44,180; B 4 16 44,100; B 8 8 56,254; B 4 92 54,304; B 8 8
56,118; B 4 92 54,168; B 10 16 61,342; B 10 16 61,278; B 10 16 61,206; B 10 16 61,142; B 8 8
68,344; B 8 8 68,280; B 8 8 68,208; B 8 8 68,144; B 10 4 77,342; B 10 4 77,278; B 10 4 77,206; B 10
4 77,142; B 8 8 82,362; B 8 8 82,226; B 4 82 84,317; B 4 82 84,181; DF;

BasicDesignFrame

DS 901 125/1; L NM; B 32 6 16,3; B 16 6 24,29; B 1900 32 950,22; B 32 6 1884,3; B 32 6 374,3; DF;
DS 902 125/1; L NC; B 4 4 4,4; L ND; B 8 8 4,4; L NM; B 8 8 4,4; DF; DS 903 125/1; L ND; B 4 12 6,6; L
NP; B 12 4 6,6; L NI; B 12 12 6,6; DF; DS 904 125/1; L NC; B 4 4 4,4; L NP; B 8 8 4,4; L NM; B 8 8 4,4; DF;
DS 905 125/1; L ND; B 8 12 8,6; L NP; B 16 4 8,6; L NI; B 16 12 8,6; DF; DS 906 125/1; L ND; B 16 12
12,6; L NP; B 24 4 12,6; DF; DS 907 125/1; L ND; B 32 12 20,6; L NP; B 40 4 20,6; DF; DS 908 125/1; L
ND; B 4 16 6,8; L NP; B 12 8 6,8; L NI; B 12 16 6,8; DF; DS 909 125/1; L ND; B 8 12 8,6; L NP; B 16 4 8,6;
DF; DS 910 125/1; C 902 R 0,-1 T 32,46; C 902 R 0,-1 T 78,14; C 903 T 30,8; C 902 R 0,-1 T 32,28; C
904 R 0,-1 T 42,20; C 902 R 0,-1 MX T 68,28; C 905 R -1,0 MX T 56,20; C 902 R 0,-1 T 26,8; C 906 T
24,28; C 907 T 46,28; C 902 R 0,-1 T 60,46; C 902 R 0,-1 T 60,8; C 904 T 16,20; C 902 T 6,0; C 902 T
4,38; C 902 T 4,20; C 908 T 4,6; C 909 T 0,28; L ND; B 4 8 36,4; B 48 8 52,42; B 8 6 82,3; L NP; B 6 4

```
53,14; B 4 6 18,17; B 4 8 22,32; L NM; B 8 6 82,17; B 10 8 45,24; B 18 8 77,24; B 4 8 14,24; DF; DS 911
125/1; L ND; B 200 12 104,6; L NP; B 208 4 104,6; DF; DS 912 125/1; C 904 R -1,0 T 238,268; C 911 T
28,246; C 902 T 48,256; C 902 T 80,256; C 902 T 112,256; C 902 T 144,256; C 902 T 176,256; C 902
T 208,256; C 902 T 36,24; C 902 T 68,24; C 902 T 100,24; C 902 T 132,24; C 902 T 164,24; C 902 T
196,24; C 902 T 228,24; C 902 T 198,40; C 902 T 260,24; C 902 T 4,24; C 902 T 230,240; C 902 T
258,240; L ND; B 8 256 4,160; B 272 16 136,24; B 8 202 268,133; B 4 204 34,142; B 162 4 117,42; B
236 8 126,284; B 200 4 132,246; B 196 6 130,261; B 20 8 254,230; B 8 54 248,261; B 4 40 262,268;
L NP; B 4 10 236,255; L NM; B 8 26 202,61; B 140 140 136,144; B 272 32 136,272; B 272 32 136,16; B
20 8 248,244; L NG; B 124 124 136,144; DF; DS 913 125/1; C 908 R 0,1 MX T 224,46; C 904 T 228,40;
C 902 MX T 246,48; C 902 T 214,48; C 902 T 214,18; C 902 MX T 246,18; C 904 T 228,10; C 908 R 0,1
MX T 224,16; C 909 R 0,1 T 200,26; C 908 T 176,42; C 904 R 0,-1 T 170,52; C 902 R 0,1 T 208,30; C
902 R 0,1 T 188,30; C 902 T 178,58; C 902 T 154,42; C 902 R 0,1 T 164,16; C 902 R 0,1 T 180,16; C
904 R 0,-1 T 146,36; C 908 T 152,28; C 909 R 0,1 T 174,12; C 904 T 90,56; C 902 MX T 28,38; C 902
MX T 44,38; C 904 MX T 70,26; C 902 MX T 60,34; C 908 R 0,1 MX T 58,32; C 909 R 0,1 MX T 26,34; C
909 R 0,1 MX T 42,34; C 909 R 0,1 T 118,32; C 909 R 0,1 T 134,32; C 908 R 0,1 T 102,32; C 902 T
100,32; C 902 T 76,34; C 904 T 90,26; C 902 T 116,36; C 902 T 132,36; C 904 T 22,6; C 902 T 0,6; C
902 T 252,6; L ND; B 2 4 223,52; B 2 4 223,22; B 4 4 182,40; B 4 4 158,26; B 2 4 75,38; B 2 4 85,38; B
4 6 256,3; L NP; B 36 4 210,42; B 76 4 140,54; B 198 4 129,12; B 60 4 76,50; B 34 4 17,50; B 70 4
35,28; B 4 10 112,27; B 20 4 136,30; B 90 4 45,58; B 114 4 57,20; L NM; B 8 44 218,48; B 10 8
233,52; B 10 8 233,22; B 8 30 204,15; B 2 8 179,48; B 2 8 155,32; B 8 14 184,45; B 8 4 182,68; B 8
20 158,60; B 8 16 176,8; B 8 12 160,30; B 2 8 61,38; B 2 8 99,36; B 42 8 49,56; B 8 38 40,19; B 8 18
66,43; B 8 14 24,53; B 8 24 136,52; B 8 24 94,46; B 8 28 80,56; B 8 38 120,19; B 34 8 115,60; B 14 8
15,10; B 6 8 249,52; B 8 42 256,35; DF; DS 914 125/1; C 902 T 344,6; C 910 R 0,1 T 358,416; C 904 T
330,400; C 904 T 330,502; C 910 R 0,1 T 358,314; C 912 R 0,-1 T 0,272; C 912 R 0,-1 T 0,544; C 913
R 0,-1 T 288,266; L ND; B 8 4 292,282; B 6 8 355,10; B 4 18 298,275; L NP; B 24 4 314,414; B 20 4
320,406; B 4 24 308,278; B 4 32 316,282; B 16 4 298,288; B 16 4 306,296; B 30 4 321,304; B 4 102
308,357; B 4 220 292,400; B 36 4 312,508; B 4 40 338,286; B 4 48 346,290; B 22 4 333,312; B 4 118
300,357; L NM; B 24 8 300,480; B 24 8 300,322; B 24 8 300,378; B 24 8 300,452; B 24 8 300,350; B
24 8 300,424; DF; DS 915 125/1; L NM; B 272 32 136,16; B 140 140 136,144; B 32 42 82,53; B 32 42
190,53; L NG; B 124 124 136,144; DF; DS 916 125/1; C 910 R 0,1 T 358,152; C 910 R 0,1 T 358,50; C
914 T 0,1360; C 914 T 0,816; C 914 T 0,272; C 912 R 0,-1 T 0,272; C 915 R 0,-1 T 0,2176; L ND; B 32 8
374,282; B 32 8 374,1858; B 32 8 374,1756; B 32 8 374,1212; B 32 8 374,1314; B 32 8 374,770; B 32
8 374,668; B 32 8 374,234; B 4 12 290,6; B 32 8 374,132; B 32 8 374,826; B 32 8 374,1370; L NP; B 4
152 306,76; B 18 4 317,150; B 4 50 324,25; L NM; B 24 8 300,188; B 24 8 300,58; B 32 2170
374,1091; B 24 8 300,114; B 24 8 300,216; B 24 8 300,86; B 24 8 300,160; DF; DS 917 125/1; L NM; B
272 96 136,378; B 140 140 136,136; B 272 32 136,16; B 48 124 90,268; B 48 124 182,268; L NG; B
124 124 136,136; DF; DS 918 125/1; L ND; B 160 12 84,6; L NP; B 168 4 84,6; DF; DS 919 125/1; L ND;
B 220 12 114,6; L NP; B 228 4 114,6; DF; DS 920 125/1; L ND; B 194 12 101,6; L NP; B 202 4 101,6; DF;
DS 921 125/1; L ND; B 174 12 91,6; L NP; B 182 4 91,6; DF; DS 922 125/1; L ND; B 224 12 116,6; L NP;
B 232 4 116,6; DF; DS 923 125/1; L ND; B 24 12 16,6; L NP; B 32 4 16,6; L NI; B 32 12 16,6; DF; DS 924
125/1; L ND; B 96 12 52,6; L NP; B 104 4 52,6; DF; DS 925 125/1; C 918 MX T 220,230; C 919 R 0,1 T
26,36; C 919 R 0,1 MX T 246,36; C 920 R 0,1 MX T 230,54; C 921 R 0,1 MX T 46,54; C 920 R 0,1 MX T
30,54; C 921 R 0,1 T 226,54; C 922 R -1,0 T 252,42; C 923 R 0,-1 T 212,304; C 923 R 0,-1 T 48,304; C
905 R 0,-1 T 212,328; C 905 R 0,-1 T 48,328; C 924 T 84,290; C 924 T 84,306; C 907 R -1,0 T
218,344; C 902 T 240,68; C 902 T 208,68; C 902 T 208,100; C 902 T 240,100; C 902 T 240,132; C
902 T 208,132; C 902 T 208,164; C 902 T 240,164; C 902 T 240,196; C 902 T 208,196; C 902 MX T
64,196; C 902 MX T 32,196; C 902 MX T 32,164; C 902 MX T 64,164; C 902 MX T 64,132; C 902 MX T
32,132; C 902 MX T 32,100; C 902 MX T 64,100; C 902 MX T 64,68; C 902 MX T 32,68; C 902 T
52,240; C 902 T 84,240; C 902 T 116,240; C 902 T 148,240; C 902 T 180,240; C 902 T 212,240; C
902 T 132,300; C 902 T 154,300; C 902 T 110,300; C 902 R -1,0 T 68,350; C 902 R -1,0 T 88,350; C
902 R -1,0 T 212,350; C 902 R -1,0 T 192,350; C 902 T 88,316; C 902 T 110,316; C 902 R 0,-1 T
176,292; C 904 R 0,-1 T 176,282; C 904 R 0,-1 T 190,282; C 902 R 0,-1 MX T 68,292; C 904 R 0,-1 MX
T 82,282; C 902 R 0,-1 T 154,292; C 902 R 0,-1 T 88,292; C 904 R 0,-1 T 88,282; C 902 R 0,-1 T
110,292; C 902 T 68,224; C 902 T 100,224; C 902 T 132,224; C 902 T 164,224; C 902 T 196,224; C
902 T 212,24; C 902 T 180,24; C 902 T 148,24; C 902 T 116,24; C 902 T 84,24; C 902 T 52,24; C 902
T 244,24; C 902 T 20,24; C 902 R -1,0 T 76,48; C 902 R -1,0 T 108,48; C 902 R -1,0 T 140,48; C 902 R
-1,0 T 172,48; C 902 R -1,0 T 204,48; C 904 T 50,304; C 904 T 214,304; C 902 T 58,316; C 902 T
206,316; C 902 T 176,316; C 902 T 154,316; C 902 T 88,300; C 902 T 176,300; C 907 R -1,0 T
94,344; C 902 T 40,210; C 902 T 224,210; L ND; B 48 56 24,296; B 156 8 136,226; B 16 236 8,150; B
16 236 264,150; B 222 8 137,46; B 180 12 136,246; B 4 210 28,155; B 4 210 244,155; B 6 194
45,155; B 6 194 227,155; B 8 8 210,328; B 8 8 62,328; B 48 56 248,296; B 124 8 150,288; B 96 4
136,304; B 272 16 136,24; B 6 164 61,140; B 6 164 211,140; B 96 6 136,321; B 32 6 74,347; B 32 6
198,347; L NP; B 4 4 52,236; B 4 4 220,236; B 22 4 65,304; B 4 36 78,312; B 30 4 203,312; B 58 4
47,264; B 62 4 223,264; B 56 4 62,256; B 56 4 210,256; B 4 8 194,270; B 4 20 180,264; B 4 12
78,268; B 4 20 92,264; B 20 4 228,54; B 20 4 44,54; B 4 4 252,36; B 4 4 20,36; B 4 86 136,383; B 100
4 130,328; B 4 4 82,296; B 4 10 178,335; B 44 4 116,338; L NM; B 8 8 72,288; B 36 8 92,320; B 14 8
169,288; B 24 90 136,285; B 140 140 136,136; B 272 32 136,16; B 24 8 228,72; B 2 8 207,72; B 2 8
207,104; B 24 8 228,104; B 24 8 228,136; B 2 8 207,136; B 2 8 207,168; B 24 8 228,168; B 24 8
228,200; B 2 8 207,200; B 2 8 65,200; B 24 8 44,200; B 24 8 44,168; B 2 8 65,168; B 2 8 65,136; B 24
8 44,136; B 24 8 44,104; B 2 8 65,104; B 2 8 65,72; B 24 8 44,72; B 96 8 136,304; B 8 34 78,299; B
14 8 103,288; B 8 18 72,215; B 8 18 104,215; B 8 18 136,215; B 8 18 168,215; B 8 18 200,215; B 8
18 72,57; B 8 18 104,57; B 8 18 136,57; B 8 18 168,57; B 8 18 200,57; B 8 12 218,318; B 8 12
```

```
54,318; B 192 28 136,254; B 8 34 194,299; B 36 8 180,320; B 8 2 180,283; B 8 2 92,283; B 272 96
136,378; B 8 26 228,231; B 8 26 44,231; L NG; B 124 124 136,136; DF; DS 926 125/1; C 917 R 0,1 T
426,0; C 925 R 0,1 T 426,272; C 925 R 0,1 T 426,544; C 925 R 0,1 T 426,816; C 925 R 0,1 T 426,1088;
C 925 R 0,1 T 426,1360; C 925 R 0,1 T 426,1632; C 925 R 0,1 T 426,1904; DF; DS 927 125/1; L ND; B 4
20 6,10; L NP; B 12 12 6,10; L NI; B 12 20 6,10; DF; DS 928 125/1; L ND; B 4 12 6,6; L NP; B 12 4 6,6;
DF; DS 929 125/1; L ND; B 22 12 15,6; L NP; B 30 4 15,6; DF; DS 930 125/1; C 902 T 20,210; C 904 T
6,202; C 902 T 34,204; C 904 T 40,194; C 902 T 34,180; C 904 T 28,224; C 903 T 6,154; C 904 T
4,176; C 902 T 10,166; C 902 T 6,148; C 927 T 30,186; C 928 R 0,1 T 40,214; C 907 R 0,1 T 28,166; C
906 R 0,1 T 28,104; C 902 T 8,116; C 903 T 4,136; C 902 T 26,132; C 904 T 38,116; C 904 T 38,14; C
928 T 28,60; C 904 T 28,80; C 902 T 36,106; C 902 T 28,70; C 928 T 4,82; C 928 T 4,66; C 928 T
10,50; C 928 T 4,34; C 928 T 10,18; C 928 R 0,-1 T 16,104; C 902 T 8,92; C 902 T 8,76; C 902 T 8,60;
C 902 T 8,44; C 902 T 8,28; C 904 T 6,106; C 902 T 28,54; C 929 R 0,1 T 38,86; C 902 T 12,0; L ND; B
8 4 24,220; B 4 10 40,217; B 8 54 30,147; B 4 12 10,130; B 4 8 16,48; B 4 6 16,33; B 4 6 16,65; B 4
10 16,13; L NP; B 2 4 43,192; B 10 4 19,206; B 4 4 46,230; B 4 76 46,190; B 8 4 20,160; B 4 18 6,167;
B 4 2 30,227; B 4 32 42,140; B 4 30 22,143; B 4 106 4,91; B 4 46 42,45; B 4 70 22,57; L NM; B 20 8
16,214; B 8 10 38,199; B 8 24 38,168; B 8 10 14,179; B 48 12 24,154; B 48 6 24,227; B 48 12 24,136;
B 10 2 41,115; B 6 16 45,86; B 8 2 32,79; B 48 8 24,4; B 48 6 24,19; B 18 6 33,31; B 6 22 45,39; B 32
6 32,97; B 8 6 4,81; B 28 6 22,61; B 8 6 4,47; B 8 6 20,31; B 10 2 11,115; DF; DS 931 125/1; C 904 T
20,50; C 904 T 6,58; C 902 T 32,0; C 903 R 0,1 T 36,34; C 904 T 26,28; C 902 T 12,30; C 902 R -1,0
MX T 16,86; C 904 R -1,0 MX T 26,96; C 903 R 0,-1 MX T 36,90; C 907 R 0,1 T 14,8; C 907 R 0,1 T
18,72; C 902 T 4,114; L ND; B 10 4 19,40; B 4 88 2,56; B 4 78 38,47; B 6 46 5,99; L NP; B 4 12 24,64; B
16 4 18,72; B 4 10 8,53; L NM; B 22 8 23,32; B 28 6 14,47; B 6 22 37,39; B 16 10 32,83; B 40 12
20,136; B 8 8 8,126; B 40 6 20,19; B 40 12 20,154; B 40 8 20,4; B 40 6 20,227; B 6 26 3,71; DF; DS
932 125/1; C 930 T 1024,0; C 930 T 48,0; C 930; C 930 T 696,0; C 930 T 744,0; C 930 T 792,0; C 930
T 840,0; C 931 T 888,0; C 930 T 464,0; C 930 T 512,0; C 930 T 560,0; C 930 T 608,0; C 931 T 656,0; C
930 T 232,0; C 930 T 280,0; C 930 T 328,0; C 930 T 376,0; C 931 T 424,0; C 930 T 96,0; C 930 T
144,0; C 931 T 192,0; C 930 T 928,0; C 930 T 976,0; DF; DS 933 125/1; C 908 R -1,0 MX T 14,38; C
902 T 204,42; C 904 T 4,58; C 902 R 0,-1 MX T 22,52; C 902 R 0,1 T 22,72; C 902 T 28,32; C 902 T
28,52; C 902 T 46,84; C 904 T 52,42; C 904 T 52,58; C 902 R 0,-1 T 16,16; C 909 R -1,0 T 26,70; C
909 R 0,-1 T 34,44; C 909 R 0,-1 T 34,64; C 904 T 68,42; C 904 T 66,26; C 908 R 0,-1 MX T 102,40; C
909 R 0,-1 T 68,64; C 903 R 0,1 T 210,60; C 902 R 0,1 T 200,60; C 902 R 0,1 T 200,24; C 903 R 0,1 T
210,20; C 906 R -1,0 T 184,62; C 902 R 0,1 T 172,42; C 906 R -1,0 T 184,42; C 902 R 0,1 T 158,42; C
902 R 0,1 T 128,42; C 902 R 0,1 T 112,32; C 902 R 0,1 T 112,52; C 908 R -1,0 T 156,66; C 908 R -1,0 T
156,42; C 909 R -1,0 T 132,64; C 909 R -1,0 T 132,40; C 908 R 0,1 T 102,52; C 904 R 0,-1 MX T 16,34;
C 902 T 80,66; C 904 T 90,66; C 902 T 80,18; C 904 T 90,18; C 902 T 134,24; C 904 T 136,34; C 904
T 136,50; C 902 T 134,60; C 909 R 0,-1 T 50,44; C 904 T 214,24; C 904 T 214,60; L ND; B 8 2 18,71; B
4 6 20,41; B 4 6 20,19; B 2 4 85,58; B 2 4 103,58; B 2 4 103,34; B 2 4 85,34; B 8 10 18,57; B 4 52
48,58; B 20 4 60,54; B 22 4 73,38; B 8 4 80,58; B 4 8 150,46; B 32 4 136,26; B 32 4 136,66; B 28 4
178,62; B 28 4 178,30; B 8 4 124,40; B 8 4 124,50; B 4 14 82,59; B 4 10 82,31; B 4 44 210,46; L NP; B
16 4 34,28; B 4 12 70,20; B 4 2 40,65; B 4 12 56,72; B 8 2 94,27; B 4 2 74,51; B 152 4 144,12; B 166
4 137,80; B 108 4 152,20; B 4 14 116,29; B 4 14 116,63; B 108 4 152,72; B 4 4 158,56; B 4 4 158,36;
B 8 2 94,65; B 4 10 218,19; B 4 10 218,73; B 16 4 34,64; L NM; B 4 8 2,62; B 18 8 195,46; B 8 42
182,29; B 8 18 18,35; B 8 4 18,82; B 6 12 33,26; B 68 6 64,17; B 6 12 33,66; B 2 8 89,70; B 6 8 71,38;
B 8 10 56,53; B 8 46 108,31; B 8 42 168,63; B 8 34 154,25; B 8 42 124,63; B 222 8 111,4; B 222 8
111,88; B 6 2 139,33; B 6 2 139,59; B 2 8 89,22; B 68 6 64,75; B 14 8 207,28; B 14 8 207,64; DF; DS
934 125/1; L ND; B 4 24 6,12; L NP; B 12 16 6,12; L NI; B 12 24 6,12; DF; DS 935 125/1; L ND; B 4 8
6,4; L NP; B 12 4 6,4; DF; DS 936 125/1; C 909 T 20,34; C 934 T 22,6; C 902 R 0,1 MX T 10,20; C 904 T
32,18; C 904 R -1,0 MX T 10,38; C 902 T 24,28; C 902 T 24,44; C 902 T 24,0; C 935 R 0,1 T 8,20; L ND;
B 2 4 9,26; B 14 4 39,30; B 4 4 44,26; L NP; B 10 4 15,40; B 4 20 4,42; B 4 18 36,9; L NM; B 8 10 28,23;
B 8 2 14,29; B 46 8 23,48; B 46 8 23,4; DF; DS 937 125/1; C 902 T 266,0; C 904 T 8,28; C 904 T
24,14; C 904 T 40,0; C 902 T 0,66; C 902 T 16,66; C 902 T 32,86; C 909 R 0,1 T 18,62; C 909 R 0,1 T
34,62; C 909 R 0,1 T 50,62; C 908 T 50,52; C 904 T 60,56; C 902 T 52,66; C 902 T 52,42; C 904 T
116,114; C 904 T 208,114; C 904 T 300,114; C 904 T 254,100; C 904 T 162,100; C 904 T 70,100; C
904 T 332,28; C 904 T 240,14; C 904 T 148,0; C 936 T 116,42; C 936 T 70,42; C 936 T 162,42; C 936
T 208,42; C 936 T 300,42; C 936 T 254,42; C 904 T 274,28; C 904 T 172,0; L ND; B 10 4 65,68; B 8 20
36,76; B 4 8 52,70; B 4 2 56,51; L NP; B 4 54 44,35; B 4 40 28,42; B 4 26 12,49; B 4 20 120,104; B 4
20 212,104; B 4 20 304,104; B 4 6 258,97; B 4 6 166,97; B 4 6 74,97; B 4 6 336,39; B 4 20 244,32; B
4 34 152,25; B 4 44 12,100; B 4 44 44,100; B 4 28 278,14; L NM; B 8 12 4,80; B 8 28 342,108; B 28 8
38,70; B 8 10 56,61; B 308 8 154,118; B 320 8 160,104; B 208 8 136,18; B 316 8 174,32; B 70 8
35,46; B 70 8 35,90; B 8 22 324,111; B 72 8 310,4; B 124 8 110,4; DF; DS 938 125/1; L ND; B 4 14 6,7;
L NP; B 12 6 6,7; DF; DS 939 125/1; C 904 T 22,6; C 902 T 16,20; C 902 T 18,62; C 904 T 26,44; C 902
T 20,34; C 909 R 0,1 T 34,16; C 934 R -1,0 T 28,64; C 902 T 12,6; C 938 R 0,1 T 14,4; L ND; B 4 4
30,36; B 4 30 34,23; L NP; B 4 2 28,15; B 6 4 7,2; B 6 60 7,46; L NM; B 4 8 2,10; B 18 22 27,45; B 26 8
23,10; B 36 8 18,66; B 36 8 18,24; DF; DS 940 125/1; C 904 T 14,72; C 902 T 0,46; C 904 R -1,0 T
66,78; C 902 R -1,0 T 72,64; C 902 R -1,0 T 70,22; C 904 R -1,0 T 62,40; C 902 R -1,0 T 68,50; C 909
R 0,-1 T 54,68; C 908 T 60,28; C 902 R -1,0 T 76,78; C 939 R -1,0 T 232,252; C 939 R -1,0 T 196,252;
C 939 R -1,0 T 160,252; C 939 R -1,0 T 124,252; C 939 R -1,0 T 88,252; C 939 T 52,92; C 939 T 88,92;
C 939 T 124,92; C 939 T 160,92; C 939 T 196,92; C 902 T 14,266; C 939 R -1,0 T 124,84; C 939 R -1,0
T 160,84; C 939 R -1,0 T 196,84; C 939 R -1,0 T 232,84; C 939 T 196,260; C 939 T 160,260; C 939 T
124,260; C 939 T 88,260; C 939 T 52,260; C 904 T 56,252; C 904 T 76,252; C 904 T 92,252; C 904 T
112,252; C 904 T 128,252; C 904 T 148,252; C 904 T 164,252; C 904 T 184,252; C 904 T 200,252; C
904 T 220,252; C 904 T 220,168; C 904 T 200,168; C 904 T 184,168; C 904 T 164,168; C 904 T
```

```
148,168; C 904 T 128,168; C 904 T 112,168; C 904 T 92,168; C 904 T 76,168; C 904 T 56,168; C 904
T 56,84; C 904 T 92,84; C 904 T 112,84; C 904 T 128,84; C 904 T 148,84; C 904 T 164,84; C 904 T
184,84; C 904 T 200,84; C 904 T 220,84; C 904 T 114,0; C 904 T 150,0; C 904 T 186,0; C 904 T
222,0; C 902 T 50,0; L ND; B 4 6 66,25; B 12 4 82,74; B 52 4 34,48; B 4 14 54,57; B 4 32 242,88; B 4
130 40,165; B 4 32 242,256; B 30 4 37,270; B 8 4 236,242; B 8 4 236,270; B 8 4 236,102; B 8 4
236,74; B 10 4 47,102; B 10 4 47,228; B 50 4 25,4; L NP; B 36 4 40,74; B 4 6 60,67; L NM; B 8 186
18,173; B 4 8 86,74; B 18 22 61,39; B 26 8 65,74; B 18 8 79,18; B 36 8 70,60; B 6 8 235,74; B 8 8
48,102; B 6 8 235,242; B 8 8 48,270; B 6 8 235,18; B 16 330 246,165; B 6 8 235,158; B 44 8 66,60; B
16 288 36,144; B 8 8 48,116; B 8 8 48,284; B 8 8 48,228; B 6 8 235,186; B 6 8 235,326; B 8 282
4,195; B 172 8 142,256; B 172 8 142,172; B 172 8 142,88; B 28 16 14,8; B 172 8 144,4; DF; DS 941
125/1; L NP; B 4 20 2,10; B 6 4 7,2; B 4 20 12,10; B 6 4 7,10; B 4 20 32,10; B 6 4 37,2; B 4 12 42,6; B 6
4 37,10; B 6 4 55,10; B 4 20 60,10; B 6 4 55,2; B 4 20 50,10; B 4 20 68,10; B 8 4 74,2; B 8 4 74,10; B 8
4 74,18; B 12 4 88,2; B 4 16 88,12; B 6 4 105,10; B 4 20 100,10; B 4 20 110,10; B 10 4 125,10; B 4 20
274,10; B 4 4 286,10; B 4 12 282,6; B 4 4 286,2; B 4 20 290,10; B 4 20 306,10; B 4 4 302,2; B 4 20
298,10; B 4 4 302,10; B 4 4 302,18; B 8 4 320,18; B 4 4 318,10; B 6 4 317,2; B 4 12 314,14; B 4 12
322,6; B 4 4 20,18; B 4 4 158,18; B 8 4 208,18; B 8 4 208,2; B 4 20 202,10; B 4 20 184,10; B 4 20
194,10; B 6 4 189,18; B 6 4 175,10; B 6 4 175,2; B 4 20 170,10; B 6 4 145,10; B 4 20 150,10; B 4 20
140,10; B 4 20 228,10; B 4 20 218,10; B 6 4 223,10; B 4 12 236,6; B 4 12 244,14; B 6 4 241,2; B 4 4
240,10; B 8 4 238,18; B 10 4 259,10; DF; DS 942 125/1; C 910 T 574,32; C 902 T 662,64; C 910 T
676,32; C 904 T 780,254; C 904 T 782,32; C 904 T 780,108; C 932 T 790,32; C 933 T 442,220; C 937
T 396,92; C 904 T 748,254; C 940 R 0,-1 T 38,306; C 941 T 38,32; L ND; B 4 30 290,329; B 4 8 42,310;
B 248 4 164,316; B 8 20 666,82; L NP; B 56 4 1366,278; B 4 28 674,78; B 4 16 1396,272; B 4 30
306,329; B 4 22 324,333; B 4 40 1580,284; B 24 8 768,258; B 4 68 784,74; B 4 24 1444,276; B 4 40
1068,284; B 4 32 1116,280; B 168 4 1154,302; B 136 4 1186,294; B 4 24 1164,276; B 104 4
1218,286; B 4 16 1212,272; B 72 4 1250,278; B 4 44 1240,322; B 4 52 1256,318; B 4 60 1272,314; B
4 8 1300,268; B 4 68 1288,310; B 4 76 1304,306; B 4 76 1320,306; B 4 68 1336,310; B 4 60
1352,314; B 4 44 1384,322; B 4 52 1368,318; B 4 36 1400,326; B 4 28 1416,330; B 4 20 1432,334; B
88 4 1398,286; B 192 4 1482,302; B 4 108 440,268; B 4 100 408,264; B 102 4 359,316; B 116 4
384,324; B 396 4 1660,342; B 4 80 1860,304; B 360 4 1630,334; B 4 72 1812,300; B 4 64 1764,296;
B 328 4 1598,326; B 4 56 1676,292; B 256 4 1546,318; B 4 48 1628,288; B 224 4 1514,310; B 160 4
1450,294; B 4 32 1532,280; B 4 48 980,288; B 240 4 1102,310; B 4 12 1448,338; B 4 36 1224,326; B
4 8 1348,268; B 24 4 1334,270; B 4 28 572,78; L NM; B 10 8 665,56; B 8 4 666,62; B 2 6 789,259; B 6
8 745,96; B 1824 32 950,328; B 60 8 694,284; B 8 66 720,247; B 28 6 776,51; B 16 6 782,79; B 2 6
789,113; B 16 12 782,168; B 12 236 768,194; B 16 6 46,309; B 6 16 35,60; B 12 196 384,130; B 32
344 16,172; B 32 344 1884,172; B 1836 32 950,16; B 6 12 1865,186; B 8 162 752,173; B 8 30
674,229; B 10 8 395,138; B 56 8 414,282; B 4 8 376,302; B 14 8 671,248; B 8 98 738,263; B 52 8
416,224; B 8 28 382,292; B 184 8 670,82; DF; DS 943 125/1; C 901 T 0,2520; C 916 T 0,344; C 926 T
1474,344; C 942; DF; DS 944 125/1; C 943 T 0,-2558 MY; DF;
```

BasicDesignFramePC

```
DS 626 1270/2; C 625 T 1,-7620 MY; DF; DS 601 1270/2; L NM; B 120 8 100,220; B 112 8 96,212; B
120 8 92,204; B 112 8 88,196; B 112 8 88,188; B 112 8 80,180; B 112 8 80,172; B 24 8 36,164; B 72
8 52,156; B 96 8 64,148; B 112 8 72,140; B 128 8 72,132; B 136 8 76,124; B 136 8 76,116; B 152 8
76,108; B 88 8 108,100; B 72 8 124,92; B 64 8 128,84; B 64 8 128,76; B 56 8 132,68; B 48 8 128,60;
B 48 8 128,52; B 40 8 124,44; B 40 8 116,36; B 40 8 108,28; B 24 8 20,44; B 48 8 24,36; B 64 8
32,28; B 120 8 60,20; B 96 8 56,12; B 56 8 52,4; DF; DS 602 1270/2; L NM; B 80 80 68,196; B 80 80
68,68; B 80 80 196,68; B 80 80 196,196; B 144 72 132,36; B 72 144 228,132; B 72 144 36,132; B 144
72 132,228; DF; DS 603 1270/2; C 602 T 268,1068; C 602 T 268,668; C 602 T 268,268; L NP; B 800
1600 400,800; DF; DS 604 1270/2; L NM; B 96 16 5646,454; B 160 16 5646,438; B 208 16 5654,422;
B 224 16 5646,406; B 48 16 5750,390; B 144 16 5606,390; B 128 16 5582,374; B 128 16 5582,358; B
112 16 5574,342; B 128 16 5566,326; B 128 16 5566,310; B 112 16 5574,262; B 288 16 5646,278; B
288 16 5646,294; B 112 16 5574,246; B 112 16 5574,230; B 96 16 5582,214; B 96 16 5582,198; B
96 16 5598,182; B 176 16 5654,166; B 112 16 5654,150; B 80 16 5718,182; B 80 16 5734,198; B 80
16 5734,214; B 96 16 5742,230; B 32 16 5758,374; B 96 16 5742,246; B 96 16 5742,262; B 160 16
5338,180; B 128 16 5370,212; B 128 16 5370,228; B 128 16 5370,244; B 128 16 5370,324; B 128 16
5370,308; B 128 16 5370,292; B 128 16 5370,276; B 128 16 5370,260; B 128 16 5370,340; B 128 16
5370,356; B 128 16 5370,372; B 128 16 5370,388; B 128 16 5370,404; B 160 16 5370,420; B 144 16
5362,196; B 144 16 5346,164; B 112 16 5346,148; B 48 16 5346,132; B 48 16 5170,132; B 112 16
5170,148; B 144 16 5170,164; B 160 16 5194,196; B 160 16 5194,420; B 128 16 5194,404; B 128 16
5194,388; B 128 16 5194,372; B 128 16 5194,356; B 128 16 5194,340; B 128 16 5194,260; B 128 16
5194,276; B 128 16 5194,292; B 128 16 5194,308; B 128 16 5194,324; B 128 16 5194,244; B 128 16
5194,228; B 128 16 5194,212; B 144 16 5010,164; B 160 16 5002,148; B 288 16 5098,180; B 144 16
5026,196; B 128 16 5018,212; B 128 16 5018,228; B 128 16 5018,244; B 128 16 5018,324; B 128 16
5018,308; B 128 16 5018,292; B 128 16 5018,276; B 128 16 5018,260; B 128 16 5018,340; B 128 16
5018,356; B 128 16 5018,372; B 128 16 5018,388; B 128 16 5018,404; B 160 16 5018,420; B 544 16
5194,436; B 96 16 4794,452; B 48 16 4642,452; B 144 16 4802,436; B 112 16 4642,436; B 320 16
4730,420; B 272 16 4706,404; B 112 16 4626,388; B 96 16 4618,372; B 96 16 4618,356; B 96 16
4634,340; B 80 16 4642,324; B 80 16 4658,308; B 128 16 4778,388; B 128 16 4778,372; B 128 16
4778,308; B 128 16 4778,324; B 128 16 4778,356; B 128 16 4778,340; B 192 16 4746,292; B 160 16
4762,276; B 128 16 4778,244; B 128 16 4778,260; B 128 16 4778,228; B 128 16 4778,212; B 112 16
4770,196; B 128 16 4762,180; B 64 16 4634,260; B 96 16 4634,244; B 112 16 4626,228; B 112 16
4626,212; B 96 16 4618,196; B 80 16 4626,180; B 208 16 4706,164; B 144 16 4706,148; B 64 16
```

```
4474,150; B 96 16 4474,166; B 48 16 4482,214; B 160 16 4330,150; B 144 16 4338,166; B 240 16
4402,182; B 240 16 4402,198; B 144 16 4354,214; B 128 16 4346,230; B 128 16 4346,246; B 128 16
4346,326; B 128 16 4346,310; B 128 16 4346,294; B 128 16 4346,278; B 128 16 4346,262; B 128 16
4346,342; B 128 16 4346,358; B 128 16 4346,374; B 128 16 4346,390; B 128 16 4346,406; B 160 16
4346,422; B 192 16 4346,438; B 64 16 4102,246; B 64 16 4102,198; B 48 16 4110,262; B 48 16
4110,182; B 32 16 4118,278; B 32 16 4118,294; B 32 16 4118,310; B 32 16 4118,166; B 32 16
4118,150; B 32 16 4118,134; B 32 16 4202,102; B 48 16 4194,86; B 48 16 4194,70; B 80 16 4174,54;
B 240 16 3966,438; B 176 16 3966,422; B 144 16 3966,246; B 144 16 3966,262; B 144 16 3966,278;
B 144 16 3966,294; B 144 16 3966,358; B 144 16 3966,342; B 144 16 3966,326; B 144 16 3966,310;
B 144 16 3966,374; B 144 16 3966,390; B 144 16 3966,406; B 240 16 4014,230; B 240 16 4014,214;
B 144 16 3966,198; B 144 16 3966,182; B 144 16 3966,118; B 144 16 3966,134; B 144 16 3966,150;
B 144 16 3966,166; B 144 16 3966,102; B 144 16 3966,86; B 144 16 3966,70; B 144 16 3966,54; B
336 16 4046,38; B 368 16 4030,22; B 48 16 3690,136; B 112 16 3690,152; B 144 16 3690,168; B
144 16 3706,200; B 160 16 3714,424; B 128 16 3714,408; B 128 16 3714,392; B 128 16 3714,376; B
128 16 3714,360; B 128 16 3714,344; B 128 16 3714,264; B 128 16 3714,280; B 128 16 3714,296; B
128 16 3714,312; B 128 16 3714,328; B 128 16 3714,248; B 128 16 3714,232; B 128 16 3714,216; B
144 16 3530,168; B 160 16 3522,152; B 304 16 3626,184; B 144 16 3546,200; B 128 16 3538,216; B
128 16 3538,232; B 128 16 3538,248; B 128 16 3538,328; B 128 16 3538,312; B 128 16 3538,296; B
128 16 3538,280; B 128 16 3538,264; B 128 16 3538,344; B 128 16 3538,360; B 128 16 3538,376; B
128 16 3538,392; B 128 16 3538,408; B 160 16 3538,424; B 368 16 3626,440; B 112 16 3126,244; B
112 16 3126,228; B 112 16 3126,212; B 112 16 3126,260; B 112 16 3126,276; B 96 16 3134,292; B
96 16 3134,196; B 112 16 3142,180; B 288 16 3246,164; B 112 16 3206,148; B 160 16 3310,180; B
112 16 3302,260; B 112 16 3302,244; B 112 16 3302,228; B 96 16 3294,212; B 96 16 3294,196; B
96 16 3294,276; B 96 16 3294,292; B 96 16 3278,308; B 112 16 3142,308; B 192 16 3214,324; B
128 16 3214,340; B 64 16 3150,356; B 96 16 3134,372; B 240 16 3190,388; B 288 16 3214,404; B
304 16 3222,420; B 144 16 3206,564; B 240 16 3206,548; B 112 16 3302,532; B 96 16 3326,516; B
96 16 3342,500; B 112 16 3334,484; B 112 16 3126,532; B 80 16 3110,516; B 80 16 3110,500; B 80
16 3126,484; B 272 16 3254,468; B 304 16 3238,452; B 320 16 3230,436; B 128 16 2934,24; B 128
16 2934,72; B 128 16 2934,56; B 128 16 2934,40; B 96 16 2934,8; B 96 16 2934,88; B 160 16
2918,152; B 144 16 2926,168; B 128 16 2934,184; B 128 16 2934,200; B 128 16 2934,216; B 128 16
2934,232; B 128 16 2934,248; B 128 16 2934,328; B 128 16 2934,312; B 128 16 2934,296; B 128 16
2934,280; B 128 16 2934,264; B 128 16 2934,344; B 128 16 2934,360; B 128 16 2934,376; B 128 16
2934,392; B 128 16 2934,408; B 160 16 2934,424; B 192 16 2934,440; B 16 16 2586,358; B 32 16
2594,374; B 32 16 2594,390; B 48 16 2602,406; B 16 16 2586,454; B 96 16 2690,454; B 192 16
2674,438; B 208 16 2682,422; B 112 16 2730,406; B 112 16 2746,390; B 128 16 2738,374; B 128 16
2738,358; B 144 16 2730,342; B 160 16 2706,326; B 176 16 2698,310; B 176 16 2682,294; B 176 16
2666,278; B 160 16 2658,262; B 160 16 2642,246; B 128 16 2626,230; B 112 16 2618,214; B 96 16
2626,198; B 112 16 2634,182; B 32 16 2770,230; B 32 16 2770,214; B 48 16 2762,198; B 64 16
2754,182; B 192 16 2690,166; B 32 16 2770,150; B 128 16 2674,150; B 64 16 2674,134; B 96 16
2374,452; B 160 16 2374,436; B 208 16 2382,420; B 224 16 2374,404; B 48 16 2478,388; B 144 16
2334,388; B 128 16 2310,372; B 128 16 2310,356; B 112 16 2302,340; B 128 16 2294,324; B 128 16
2294,308; B 112 16 2302,260; B 288 16 2374,276; B 288 16 2374,292; B 112 16 2302,244; B 112 16
2302,228; B 96 16 2310,212; B 96 16 2310,196; B 96 16 2326,180; B 176 16 2382,164; B 112 16
2382,148; B 80 16 2446,180; B 80 16 2462,196; B 80 16 2462,212; B 96 16 2470,228; B 32 16
2486,372; B 96 16 2470,244; B 96 16 2470,260; B 352 16 1866,440; B 352 16 1898,424; B 144 16
1810,248; B 144 16 1810,264; B 144 16 1810,280; B 144 16 1810,296; B 144 16 1810,360; B 144 16
1810,344; B 144 16 1810,328; B 144 16 1810,312; B 144 16 1810,376; B 144 16 1810,392; B 160 16
1818,408; B 144 16 1810,232; B 144 16 1810,216; B 144 16 1810,200; B 144 16 1810,184; B 144 16
1810,120; B 144 16 1810,136; B 144 16 1810,152; B 144 16 1810,168; B 144 16 1810,104; B 144 16
1810,88; B 144 16 1810,72; B 160 16 1818,56; B 352 16 1898,40; B 336 16 1858,24; B 144 16
2034,56; B 144 16 2050,72; B 144 16 2066,88; B 144 16 2066,104; B 144 16 2082,120; B 160 16
2090,136; B 160 16 2090,152; B 144 16 2098,168; B 144 16 2098,184; B 160 16 2106,200; B 160 16
2106,216; B 160 16 2106,248; B 160 16 2106,264; B 144 16 2098,280; B 144 16 2098,296; B 160 16
2090,312; B 160 16 2090,328; B 144 16 2082,344; B 144 16 2066,360; B 144 16 2066,376; B 144 16
2050,392; B 144 16 2034,408; B 160 16 2106,232; B 48 16 1600,260; B 80 16 1600,244; B 112 16
1600,228; B 112 16 1600,212; B 112 16 1600,196; B 96 16 1592,180; B 128 16 1528,148; B 192 16
1528,164; B 96 16 1464,180; B 112 16 1456,196; B 112 16 1440,212; B 112 16 1440,228; B 112 16
1440,244; B 128 16 1432,292; B 128 16 1432,276; B 128 16 1432,260; B 128 16 1432,308; B 128 16
1432,324; B 128 16 1432,340; B 144 16 1440,356; B 128 16 1448,372; B 160 16 1464,388; B 32 16
1624,388; B 240 16 1520,404; B 208 16 1520,420; B 176 16 1520,436; B 96 16 1512,452; B 128 16
1234,24; B 128 16 1234,72; B 128 16 1234,56; B 128 16 1234,40; B 96 16 1234,8; B 96 16 1234,88;
B 160 16 1218,152; B 144 16 1226,168; B 128 16 1234,184; B 128 16 1234,200; B 128 16 1234,216;
B 128 16 1234,232; B 128 16 1234,248; B 128 16 1234,328; B 128 16 1234,312; B 128 16 1234,296;
B 128 16 1234,280; B 128 16 1234,264; B 128 16 1234,344; B 128 16 1234,360; B 128 16 1234,376;
B 128 16 1234,392; B 128 16 1234,408; B 160 16 1234,424; B 192 16 1234,440; B 16 16 884,358; B
32 16 892,374; B 32 16 892,390; B 48 16 900,406; B 16 16 884,454; B 96 16 988,454; B 192 16
972,438; B 208 16 980,422; B 112 16 1028,406; B 112 16 1044,390; B 128 16 1036,374; B 128 16
1036,358; B 144 16 1028,342; B 160 16 1004,326; B 176 16 996,310; B 176 16 980,294; B 176 16
964,278; B 160 16 956,262; B 160 16 940,246; B 128 16 924,230; B 112 16 916,214; B 96 16
924,198; B 112 16 932,182; B 32 16 1068,230; B 32 16 1068,214; B 48 16 1060,198; B 64 16
1052,182; B 192 16 988,166; B 32 16 1068,150; B 128 16 972,150; B 64 16 972,134; B 96 16
734,456; B 48 16 582,456; B 144 16 742,440; B 112 16 582,440; B 320 16 670,424; B 272 16
```

```
646,408; B 112 16 566,392; B 96 16 558,376; B 96 16 558,360; B 96 16 574,344; B 80 16 582,328; B
80 16 598,312; B 128 16 718,392; B 128 16 718,376; B 128 16 718,312; B 128 16 718,328; B 128 16
718,360; B 128 16 718,344; B 192 16 686,296; B 160 16 702,280; B 128 16 718,248; B 128 16
718,264; B 128 16 718,232; B 128 16 718,216; B 112 16 710,200; B 128 16 702,184; B 64 16
574,264; B 96 16 574,248; B 112 16 566,232; B 112 16 566,216; B 96 16 558,200; B 80 16 566,184;
B 208 16 646,168; B 144 16 646,152; B 128 16 368,120; B 144 16 360,104; B 144 16 360,88; B 144
16 344,72; B 144 16 328,56; B 384 16 192,440; B 384 16 224,424; B 144 16 120,248; B 144 16
120,264; B 144 16 120,280; B 144 16 120,296; B 144 16 120,360; B 144 16 120,344; B 144 16
120,328; B 144 16 120,312; B 144 16 120,376; B 144 16 120,392; B 144 16 120,408; B 304 16
200,232; B 272 16 184,216; B 144 16 120,200; B 144 16 120,184; B 144 16 120,120; B 144 16
120,136; B 144 16 120,152; B 144 16 120,168; B 144 16 120,104; B 144 16 120,88; B 144 16 120,72;
B 144 16 120,56; B 352 16 208,40; B 352 16 176,24; B 176 16 344,408; B 160 16 368,392; B 160 16
368,376; B 160 16 384,360; B 160 16 384,344; B 160 16 384,328; B 144 16 376,312; B 160 16
368,296; B 144 16 360,280; B 144 16 344,264; B 144 16 328,248; B 112 16 312,200; B 128 16
336,184; B 128 16 352,168; B 128 16 352,152; B 128 16 368,136; DF; DS 605 1270/2; C 602 T
1600,2400; C 602 T 800,2400; C 602 T 400,2400; C 602 T 0,2400; C 602 T 1200,2400; C 602 T
2000,2400; C 602 T 2400,2400; C 602 T 2800,2400; C 602 T 3200,2400; C 602 T 3600,2400; C 602 T
4000,2400; C 602 T 4400,2400; C 602 T 4800,2400; C 602 T 5200,2400; C 602 T 5600,2400; C 602 T
6000,2400; C 602 T 6400,2400; C 602 T 6800,2400; C 602 T 7200,2400; C 602 T 7600,2400; C 602 T
7600,0; C 602 T 7200,0; C 602 T 6800,0; C 602 T 6400,0; C 602 T 6000,0; C 602 T 5600,0; C 602 T
5200,0; C 602 T 4800,0; C 602 T 4400,0; C 602 T 4000,0; C 602 T 3600,0; C 602 T 3200,0; C 602 T
2800,0; C 602 T 2400,0; C 602 T 2000,0; C 602 T 1200,0; C 602; C 602 T 400,0; C 602 T 800,0; C 602
T 1600,0; L NP; B 7600 2400 3932,1332; DF; DS 606 1270/2; C 602 T 2400,0; C 602 T 2000,0; C 602
T 1200,0; C 602; C 602 T 400,0; C 602 T 800,0; C 602 T 1600,0; C 602 T 1600,1200; C 602 T
800,1200; C 602 T 400,1200; C 602 T 0,1200; C 602 T 1200,1200; C 602 T 2000,1200; C 602 T
2400,1200; L NP; B 2400 1200 1332,732; DF; DS 607 1270/2; C 602 T 68,0; C 602 T 68,400; L NP; B
400 400 200,332; DF; DS 608 1270/2; C 602; C 602 T 0,1600; L NP; B 200 1600 132,932; DF; DS 609
1270/2; L NM; B 184 8 92,212; B 184 8 108,204; B 72 8 60,196; B 72 8 60,188; B 72 8 60,180; B 72 8
60,172; B 72 8 60,140; B 72 8 60,148; B 72 8 60,156; B 72 8 60,164; B 72 8 60,132; B 72 8 60,124; B
72 8 60,116; B 144 8 96,108; B 72 8 60,20; B 72 8 60,28; B 72 8 60,36; B 72 8 60,68; B 72 8 60,60; B
72 8 60,52; B 72 8 60,44; B 72 8 60,76; B 72 8 60,84; B 152 8 100,92; B 152 8 100,100; B 88 8 60,12;
B 120 8 60,4; B 80 8 168,196; B 72 8 180,188; B 72 8 180,180; B 72 8 188,172; B 72 8 188,164; B 72
8 188,156; B 64 8 184,148; B 64 8 184,140; B 64 8 176,132; B 56 8 172,124; B 56 8 156,116; B 80 8
144,84; B 80 8 152,76; B 80 8 152,68; B 80 8 160,60; B 80 8 168,52; B 80 8 168,44; B 80 8 192,12; B
80 8 184,20; B 80 8 184,28; B 80 8 176,36; B 88 8 204,4; DF; DS 610 1270/2; L NM; B 48 8 120,100; B
48 8 120,108; B 16 8 128,44; B 48 8 120,116; B 40 8 116,124; B 40 8 116,132; B 40 8 108,140; B 56
8 76,156; B 88 8 76,148; B 48 8 48,140; B 48 8 40,132; B 48 8 40,124; B 56 8 36,116; B 56 8 36,108;
B 144 8 72,84; B 144 8 72,92; B 56 8 36,100; B 64 8 32,76; B 64 8 32,68; B 56 8 36,60; B 64 8 40,52;
B 64 8 40,44; B 72 8 52,36; B 24 8 124,36; B 112 8 72,28; B 104 8 76,20; B 80 8 72,12; B 48 8 72,4;
DF; DS 611 1270/2; L NM; B 160 8 80,68; B 152 8 84,60; B 136 8 92,52; B 40 8 28,44; B 40 8 20,36; B
40 8 20,28; B 56 8 28,20; B 56 8 132,44; B 48 8 136,36; B 48 8 128,28; B 56 8 116,20; B 120 8 68,12;
B 72 8 68,4; B 152 8 76,76; B 144 8 72,84; B 120 8 60,92; B 48 8 32,100; B 32 8 40,108; B 64 8
72,116; B 96 8 72,124; B 56 8 36,132; B 48 8 104,132; B 48 8 112,140; B 48 8 112,148; B 48 8
112,188; B 48 8 112,180; B 56 8 116,172; B 56 8 116,164; B 56 8 116,156; B 80 8 120,196; B 56 8
68,212; B 144 8 88,204; B 56 8 36,196; B 48 8 32,188; B 48 8 32,140; B 56 8 28,148; B 56 8 28,156;
B 56 8 28,180; B 56 8 28,172; B 56 8 28,164; DF; DS 612 1270/2; L NM; B 80 8 192,212; B 48 8
192,204; B 32 8 192,196; B 24 8 188,180; B 24 8 188,188; B 24 8 172,132; B 24 8 172,140; B 24 8
172,148; B 24 8 180,172; B 24 8 180,164; B 24 8 180,156; B 24 8 164,108; B 24 8 164,116; B 24 8
164,124; B 24 8 156,100; B 24 8 156,92; B 24 8 156,84; B 48 8 112,4; B 64 8 112,12; B 64 8 112,20;
B 72 8 108,28; B 80 8 112,36; B 80 8 112,44; B 88 8 108,52; B 96 8 112,60; B 96 8 112,68; B 104 8
108,76; B 80 8 96,84; B 72 8 92,92; B 80 8 88,100; B 80 8 88,108; B 72 8 84,116; B 80 8 80,124; B 80
8 80,132; B 72 8 76,140; B 80 8 72,148; B 72 8 68,156; B 72 8 68,164; B 80 8 64,172; B 72 8 60,180;
B 80 8 56,188; B 80 8 56,196; B 88 8 52,204; B 104 8 52,212; DF; DS 613 1270/2; L NM; B 64 8
128,36; B 64 8 128,52; B 64 8 128,44; B 64 8 128,76; B 64 8 128,84; B 64 8 128,68; B 64 8 128,60; B
64 8 128,92; B 64 8 128,100; B 64 8 128,116; B 64 8 128,108; B 64 8 128,124; B 72 8 124,132; B 144
8 88,140; B 152 8 92,20; B 144 8 88,28; B 72 8 44,36; B 64 8 40,44; B 64 8 40,52; B 64 8 32,60; B 64
8 32,68; B 64 8 32,76; B 64 8 32,84; B 64 8 32,92; B 64 8 32,100; B 64 8 32,108; B 64 8 40,116; B 64
8 40,124; B 72 8 44,132; B 64 8 56,148; B 40 8 60,156; B 64 8 128,212; B 64 8 128,196; B 64 8
128,204; B 64 8 128,188; B 64 8 128,180; B 64 8 128,148; B 64 8 128,156; B 64 8 128,172; B 64 8
128,164; B 72 8 124,220; B 80 8 120,228; B 80 8 136,12; B 24 8 108,4; B 56 8 52,12; B 24 8 52,4; DF;
DS 614 1270/2; L NM; B 184 8 92,4; B 80 8 48,12; B 64 8 48,20; B 64 8 48,28; B 64 8 48,36; B 64 8
48,44; B 64 8 48,52; B 64 8 48,92; B 64 8 48,84; B 64 8 48,76; B 64 8 48,68; B 64 8 48,60; B 64 8
48,100; B 64 8 48,108; B 64 8 48,116; B 72 8 52,124; B 152 8 92,132; B 80 8 40,148; B 72 8 44,140;
B 64 8 136,116; B 64 8 136,108; B 64 8 136,100; B 64 8 136,60; B 64 8 136,68; B 64 8 136,76; B 64 8
136,84; B 64 8 136,92; B 64 8 136,52; B 64 8 136,44; B 64 8 136,36; B 64 8 136,28; B 64 8 136,20; B
80 8 136,12; B 72 8 132,124; B 72 8 124,140; B 56 8 124,148; B 24 8 124,156; DF; DS 615 1270/2; L
NM; B 72 8 132,228; B 120 8 132,220; B 72 8 92,212; B 72 8 188,212; B 16 8 216,220; B 8 8 220,228;
B 48 8 200,204; B 40 8 204,196; B 32 8 208,188; B 24 8 212,180; B 16 8 216,172; B 16 8 216,164; B
8 8 220,156; B 72 8 76,204; B 72 8 60,188; B 80 8 56,180; B 80 8 56,172; B 72 8 68,196; B 80 8
48,164; B 80 8 48,156; B 80 8 48,148; B 88 8 44,140; B 80 8 40,116; B 80 8 40,124; B 80 8 40,132; B
80 8 40,108; B 80 8 40,100; B 80 8 40,92; B 80 8 48,84; B 80 8 48,76; B 80 8 48,68; B 72 8 52,60; B
72 8 52,52; B 72 8 60,44; B 64 8 64,36; B 64 8 72,28; B 72 8 84,20; B 72 8 196,52; B 72 8 196,44; B
```

72 8 196,60; B 72 8 196,68; B 72 8 196,76; B 72 8 196,84; B 88 8 196,92; B 120 8 196,100; B 72 8 196,28; B 72 8 196,36; B 72 8 188,20; B 144 8 136,12; B 96 8 136,4; DF; DS 616 1270/2; L NM; B 88 8 44,124; B 88 8 44,116; B 88 8 44,108; B 88 8 44,100; B 88 8 44,92; B 88 8 44,132; B 88 8 44,140; B 80 8 48,148; B 80 8 48,156; B 88 8 52,164; B 80 8 56,172; B 80 8 56,180; B 72 8 60,188; B 72 8 68,196; B 64 8 72,204; B 64 8 88,212; B 64 8 88,20; B 64 8 72,28; B 72 8 68,36; B 72 8 60,44; B 80 8 56,52; B 80 8 56,60; B 88 8 52,68; B 80 8 48,76; B 80 8 48,84; B 80 8 224,84; B 80 8 224,76; B 88 8 220,68; B 80 8 216,60; B 80 8 216,52; B 72 8 212,44; B 72 8 204,36; B 64 8 200,28; B 64 8 184,20; B 64 8 184,212; B 64 8 200,204; B 72 8 204,196; B 72 8 212,188; B 80 8 216,180; B 80 8 216,172; B 88 8 220,164; B 80 8 224,156; B 80 8 224,148; B 88 8 228,140; B 88 8 228,132; B 88 8 228,92; B 88 8 228,100; B 88 8 228,108; B 88 8 228,116; B 88 8 228,124; B 128 8 136,220; B 80 8 136,228; B 80 8 136,4; B 128 8 136,12; DF; DS 617 1270/2; L NM; B 72 8 36,188; B 72 8 36,180; B 80 8 40,172; B 80 8 40,164; B 96 8 48,156; B 104 8 52,148; B 112 8 64,140; B 128 8 72,132; B 136 8 84,124; B 144 8 96,116; B 144 8 104,108; B 136 8 116,100; B 128 8 128,92; B 112 8 136,84; B 104 8 148,76; B 96 8 152,68; B 88 8 156,60; B 80 8 152,52; B 64 8 40,196; B 72 8 44,204; B 72 8 52,212; B 112 8 88,220; B 56 8 84,228; B 72 8 140,212; B 8 8 172,220; B 48 8 152,204; B 32 8 160,196; B 24 8 164,188; B 16 8 168,180; B 16 8 168,172; B 8 8 172,164; B 72 8 156,44; B 72 8 156,36; B 64 8 152,28; B 64 8 144,20; B 136 8 100,12; B 88 8 100,4; B 16 8 8,12; B 64 8 32,20; B 48 8 24,28; B 40 8 20,36; B 32 8 16,44; B 24 8 12,52; B 24 8 12,60; B 16 8 8,68; B 16 8 8,76; B 8 8 4,84; DF; DS 618 1270/2; L NM; B 16 8 8,164; B 24 8 12,172; B 32 8 24,180; B 32 8 80,220; B 72 8 76,212; B 96 8 80,204; B 112 8 80,196; B 120 8 76,188; B 88 8 100,180; B 72 8 108,172; B 72 8 108,164; B 64 8 112,156; B 64 8 112,148; B 64 8 112,140; B 56 8 108,132; B 64 8 104,124; B 56 8 100,116; B 48 8 96,108; B 48 8 88,100; B 40 8 84,92; B 40 8 76,84; B 32 8 72,76; B 32 8 64,68; B 32 8 56,60; B 8 8 156,60; B 120 8 100,52; B 128 8 96,44; B 128 8 88,36; B 136 8 84,28; B 144 8 80,20; B 144 8 72,12; B 144 8 72,4; DF; DS 619 1270/2; L NM; B 216 8 108,212; B 216 8 108,204; B 72 8 108,196; B 72 8 108,188; B 72 8 108,180; B 72 8 108,172; B 72 8 108,140; B 72 8 108,148; B 72 8 108,156; B 72 8 108,164; B 72 8 108,132; B 72 8 108,124; B 72 8 108,116; B 72 8 108,108; B 72 8 108,20; B 72 8 108,28; B 72 8 108,36; B 72 8 108,68; B 72 8 108,60; B 72 8 108,52; B 72 8 108,44; B 72 8 108,76; B 72 8 108,84; B 72 8 108,92; B 72 8 108,100; B 88 8 108,12; B 120 8 108,4; B 40 8 20,196; B 24 8 12,188; B 24 8 12,180; B 16 8 8,172; B 16 8 8,164; B 16 8 208,164; B 16 8 208,172; B 24 8 204,180; B 24 8 204,188; B 40 8 196,196; DF; DS 620 1270/2; L NM; B 96 8 48,4; B 80 8 48,12; B 64 8 48,20; B 64 8 48,28; B 64 8 48,36; B 64 8 48,44; B 64 8 48,52; B 64 8 48,92; B 64 8 48,84; B 64 8 48,76; B 64 8 48,68; B 64 8 48,60; B 64 8 48,100; B 64 8 48,108; B 72 8 52,116; B 120 8 76,124; B 120 8 76,132; B 72 8 44,140; B 80 8 40,148; B 24 8 116,116; B 48 8 112,140; B 32 8 112,148; DF; DS 621 1270/2; L NM; B 96 8 48,4; B 80 8 48,12; B 64 8 48,20; B 64 8 48,28; B 64 8 48,36; B 64 8 48,44; B 64 8 48,52; B 64 8 48,92; B 64 8 48,84; B 64 8 48,76; B 64 8 48,68; B 64 8 48,60; B 64 8 48,100; B 64 8 48,108; B 64 8 48,116; B 64 8 48,124; B 64 8 48,132; B 72 8 44,140; B 80 8 40,148; B 48 8 48,180; B 48 8 48,220; B 64 8 48,204; B 64 8 48,196; B 64 8 48,188; B 64 8 48,212; DF; DS 622 1270/2; L NM; B 272 8 136,4; B 80 8 48,12; B 64 8 48,20; B 64 8 48,28; B 64 8 48,36; B 64 8 48,44; B 64 8 48,52; B 64 8 48,92; B 64 8 48,84; B 64 8 48,76; B 64 8 48,68; B 64 8 48,60; B 64 8 48,100; B 64 8 48,108; B 64 8 48,116; B 72 8 52,124; B 144 8 88,132; B 80 8 40,148; B 72 8 44,140; B 64 8 136,116; B 64 8 136,108; B 64 8 136,100; B 64 8 136,60; B 64 8 136,68; B 64 8 136,76; B 64 8 136,84; B 64 8 136,92; B 64 8 136,52; B 64 8 136,44; B 64 8 136,36; B 64 8 136,28; B 64 8 136,20; B 80 8 136,12; B 80 8 136,124; B 72 8 124,140; B 56 8 124,148; B 24 8 124,156; B 24 8 212,156; B 56 8 212,148; B 72 8 212,140; B 72 8 220,124; B 80 8 224,12; B 64 8 224,20; B 64 8 224,28; B 64 8 224,36; B 64 8 224,44; B 64 8 224,52; B 64 8 224,92; B 64 8 224,84; B 64 8 224,76; B 64 8 224,68; B 64 8 224,60; B 64 8 224,100; B 64 8 224,108; B 64 8 224,116; B 80 8 208,132; DF; DS 623 1270/2; L NM; B 32 8 80,212; B 56 8 68,204; B 80 8 56,196; B 96 8 48,188; B 64 8 64,180; B 64 8 64,172; B 64 8 64,164; B 64 8 64,140; B 64 8 64,148; B 64 8 64,156; B 64 8 64,132; B 64 8 64,124; B 64 8 64,116; B 64 8 64,92; B 64 8 64,100; B 64 8 64,108; B 64 8 64,84; B 64 8 64,76; B 64 8 64,68; B 64 8 64,44; B 64 8 64,52; B 64 8 64,60; B 64 8 64,36; B 64 8 64,28; B 64 8 64,20; B 80 8 64,12; B 128 8 64,4; DF; DS 624 1270/2; L NM; B 64 8 56,4; B 104 8 60,12; B 128 8 64,20; B 64 8 32,28; B 48 8 24,36; B 24 8 20,44; B 40 8 116,28; B 48 8 120,36; B 48 8 128,44; B 48 8 128,52; B 56 8 132,60; B 56 8 132,68; B 64 8 128,76; B 64 8 128,84; B 72 8 124,92; B 72 8 116,100; B 88 8 108,108; B 104 8 92,116; B 96 8 88,124; B 56 8 92,132; B 32 8 88,140; B 40 8 100,148; B 48 8 104,156; B 56 8 108,164; B 56 8 108,172; B 64 8 112,180; B 72 8 108,188; B 80 8 104,196; B 120 8 84,204; B 104 8 84,212; B 88 8 84,220; B 56 8 84,228; B 32 8 32,196; B 24 8 20,188; B 16 8 16,180; DF; DS 625 1270/2; C 601 R -1,0 MX T 468,804; C 602 T 7466,266; C 602 T 7466,1866; C 603 T 400,300; C 604 T 1864,3406; C 603 T 7200,5300; C 603 T 400,5300; C 603 T 2000,5300; C 603 T 4000,5300; C 603 T 6000,5300; C 605 T 268,2368; C 602 T 264,968; C 602 T 264,1968; C 606 T 2668,668; C 607 T 1800,268; C 607 T 2600,268; C 607 T 3400,268; C 607 T 4200,268; C 607 T 5000,268; C 607 T 5800,272; C 607 T 6600,268; C 608 T 1468,268; C 609 R -1,0 MX T 504,340; C 610 R -1,0 MX T 772,348; C 611 R -1,0 MX T 932,404; C 612 R -1,0 MX T 1708,4456; C 613 R -1,0 MX T 1952,4460; C 613 R -1,0 MX T 2140,4460; C 613 R -1,0 MX T 7644,3290; C 614 R -1,0 MX T 7448,3286; C 615 R -1,0 MX T 7180,3290; C 616 R -1,0 MX T 348,6984; C 614 R -1,0 MX T 632,6984; C 617 R -1,0 MX T 3548,6220; C 618 R -1,0 MX T 3760,6212; C 617 R -1,0 MX T 5580,6220; C 619 R -1,0 MX T 6950,6824; C 620 R -1,0 MX T 7178,6824; C 621 R -1,0 MX T 7326,6824; C 622 R -1,0 MX T 7434,6824; C 623 R -1,0 MX T 5792,6212; C 624 R -1,0 MX T 1760,6220; C 617 R -1,0 MX T 1548,6220; L NM; B 232 56 166,1356; B 232 56 166,1808; B 56 220 166,1810; B 36 120 386,696; B 132 36 386,694; B 84 36 3082,90; B 32 36 7520,170; B 32 36 7520,90; B 48 180 7480,162; B 48 180 7560,162; B 32 48 7696,228; B 48 180 7656,162; B 48 180 7736,162; B 600 200 7200,400; B 50 450 8374,274; B 500 50 8150,24; B 50 450 8374,6924; B 500 50 8150,7174; B 50 500 24,6950; B 450 50 274,7174; B 450 50 274,24; B 50 500 24,250; B 84 36 7860,2212; B 36 96 7908,2170; B 36 96 7908,2254; B 36 96 7812,2170; B 60 36 8052,2286; B 60 36

8052,2140; B 36 182 8100,2212; B 36 182 8004,2212; B 84 36 2282,90; B 36 96 2330,120; B 36 96 2330,204; B 36 60 2162,162; B 36 132 2146,162; B 48 48 2120,96; B 72 48 1988,96; B 72 48 1844,228; B 48 48 2120,228; B 48 180 2072,162; B 48 180 1928,162; B 72 48 1988,228; B 72 36 1988,162; B 48 180 1784,162; B 48 180 2584,162; B 72 36 2788,162; B 72 48 2788,228; B 48 180 2728,162; B 48 180 2872,162; B 48 48 2920,228; B 72 48 2644,228; B 72 48 2788,96; B 48 48 2920,96; B 36 132 2946,162; B 36 60 2962,162; B 36 96 3034,120; B 36 96 3034,204; B 36 96 3130,204; B 84 36 3082,234; B 84 36 3082,162; B 84 36 3882,162; B 108 36 3870,234; B 84 36 3882,90; B 36 96 3930,204; B 36 96 3834,120; B 36 60 3762,162; B 36 132 3746,162; B 48 48 3720,96; B 72 48 3588,96; B 72 48 3444,228; B 48 48 3720,228; B 48 180 3672,162; B 48 180 3528,162; B 72 48 3588,228; B 72 36 3588,162; B 48 180 3384,162; B 48 180 4184,162; B 72 36 4388,162; B 72 48 4388,228; B 48 180 4328,162; B 48 180 4472,162; B 48 48 4520,228; B 72 48 4244,228; B 72 48 4388,96; B 48 48 4520,96; B 36 132 4546,162; B 36 60 4562,162; B 36 96 4634,120; B 36 96 4730,204; B 36 96 4730,120; B 84 36 4682,162; B 84 36 5482,162; B 84 36 5482,234; B 84 36 5482,90; B 36 96 5530,120; B 36 96 5530,204; B 36 60 5362,162; B 36 132 5346,162; B 48 48 5320,96; B 72 48 5188,96; B 72 48 5044,228; B 48 48 5320,228; B 48 180 5272,162; B 48 180 5128,162; B 72 48 5188,228; B 72 36 5188,162; B 48 180 4984,162; B 48 180 5784,162; B 72 36 5988,162; B 72 48 5988,228; B 48 180 5928,162; B 48 180 6072,162; B 48 48 6120,228; B 72 48 5844,228; B 72 48 5988,96; B 48 48 6120,96; B 36 132 6146,162; B 36 60 6162,162; B 36 96 6234,204; B 36 96 6330,120; B 84 36 6282,90; B 84 36 6282,234; B 84 36 6282,162; B 36 96 7130,120; B 36 96 7130,204; B 36 60 6962,162; B 36 132 6946,162; B 48 48 6920,96; B 72 48 6788,96; B 72 48 6644,228; B 48 48 6920,228; B 48 180 6872,162; B 48 180 6728,162; B 72 48 6788,228; B 72 36 6788,162; B 48 180 6584,162; B 2400 200 6700,2900; B 200 200 5600,2700; B 650 200 1224,700; B 200 1450 1600,1224; B 100 300 1600,2250; B 200 200 600,1100; B 400 200 1100,1100; B 200 1200 1200,1800; B 200 800 800,2000; B 200 200 400,2300; B 200 2200 400,3700; B 400 200 500,6100; B 200 1000 400,5500; B 200 600 800,5300; B 200 2200 800,3700; B 200 2200 1200,3700; B 100 1900 1600,3550; B 6100 100 4600,4550; B 100 200 2800,4700; B 100 200 7600,4700; B 200 2000 8000,3800; B 100 300 5200,2250; B 100 300 4800,2250; B 100 650 4400,1574; B 500 100 4600,1200; B 100 250 4800,1024; B 100 350 4800,1724; B 1200 100 5450,1600; B 100 650 6000,1224; B 100 1050 6800,1424; B 2200 100 6400,2000; B 100 1000 4000,1400; B 100 300 2800,2250; B 100 650 3200,1574; B 500 100 3000,1200; B 100 250 2800,1024; B 200 200 3200,600; B 100 1000 3600,1400; B 600 200 6400,400; B 600 200 5600,400; B 600 200 4800,400; B 600 200 4000,400; B 600 200 3200,400; B 600 200 2400,400; B 250 200 1774,400; B 200 1900 1200,5950; B 7000 200 4600,6948; B 250 100 7824,6500; B 200 1850 8000,5924; B 100 600 7600,5300; B 1150 100 5724,5700; B 100 650 5200,5324; B 100 750 4800,5374; B 150 100 4924,5700; B 100 850 4700,6474; B 150 100 4574,6100; B 100 800 4950,6150; B 1300 100 5650,6500; B 350 100 4124,6500; B 100 1450 4000,5724; B 100 600 4400,5300; B 100 650 3600,5324; B 1150 100 3074,5700; B 100 350 3200,5174; B 1200 100 2550,5300; B 350 100 2124,6500; B 100 1100 2000,5900; B 350 100 2274,800; B 100 1100 2400,1400; B 350 100 2524,2000; B 200 100 5000,800; B 200 100 4200,800; B 100 750 2800,6524; B 350 100 2674,6100; B 350 100 6674,6100; B 100 750 6800,6524; B 100 300 3200,2250; B 100 300 3600,2250; B 100 300 4000,2250; B 100 300 4400,2250; B 100 250 5600,5124; B 1600 100 6350,5300; B 100 800 7200,5650; B 350 100 7324,6100; DF;

RAM

DS 501 125/1; L NM; B 8 8 4,24; B 6 26 5,13; B 6 34 5,39; B 8 12 18,34; B 8 8 30,52; B 8 8 30,4; B 6 6 31,3; B 6 52 31,28; B 6 6 31,53; L ND; B 8 8 4,24; B 14 4 11,24; B 8 8 18,36; B 12 4 22,24; B 12 4 24,38; B 4 4 20,38; B 4 12 28,44; B 8 8 30,52; B 4 12 28,12; B 8 8 30,4; B 4 6 28,51; B 4 4 28,52; B 4 4 28,24; B 4 10 28,21; B 4 6 28,5; B 4 4 28,4; B 4 4 28,38; L NP; B 26 4 13,44; B 26 4 13,12; B 8 6 18,31; B 4 6 20,31; B 6 4 21,32; B 4 12 22,24; B 4 8 22,30; B 12 4 28,44; B 12 4 28,12; B 4 4 32,44; B 4 4 32,12; B 2 4 35,12; B 2 4 35,44; B 2 4 36,12; B 2 4 36,44; L NC; B 4 4 4,24; B 4 8 18,34; B 4 4 30,52; B 4 4 30,4; DF;

Interface

DS 502 125/1; L NM; B 6 6 5,3; B 6 6 31,3; B 6 16 5,10; B 12 8 14,238; B 8 2 14,53; B 8 2 20,103; B 8 2 20,81; B 8 2 20,39; B 12 8 26,64; B 12 8 28,8; B 8 12 26,190; B 8 2 28,255; B 8 2 30,215; B 6 2 31,3; B 36 8 18,22; B 36 6 18,35; B 36 6 18,49; B 36 6 18,77; B 36 8 18,98; B 36 12 18,172; B 36 12 18,208; B 36 6 18,251; B 8 2 20,159; B 8 12 30,118; B 36 8 18,154; L NP; B 6 8 11,238; B 4 14 16,63; B 8 4 18,72; B 8 8 20,78; B 8 8 20,36; B 4 30 18,17; B 10 2 25,195; B 8 10 24,201; B 8 4 26,192; B 8 8 28,252; B 14 4 25,230; B 6 8 31,8; B 4 16 30,240; B 18 4 25,178; B 8 8 30,64; B 4 38 32,87; B 4 48 32,36; B 4 52 32,150; B 4 70 10,199; B 4 66 8,127; B 6 4 9,162; B 20 6 24,121; B 12 26 20,137; B 20 4 24,108; B 20 4 16,92; L ND; B 12 2 14,85; B 4 16 10,76; B 8 8 14,50; B 14 10 17,17; B 4 24 12,34; B 14 2 15,87; B 8 8 18,92; B 12 4 20,224; B 4 20 16,212; B 4 12 16,192; B 10 2 19,97; B 8 6 20,101; B 8 6 22,169; B 10 4 17,240; B 4 14 20,249; B 14 4 19,236; B 20 8 18,64; B 14 2 21,183; B 10 2 23,173; B 8 8 24,178; B 20 6 20,9; B 8 2 26,5; B 4 8 24,230; B 16 2 22,185; B 8 6 26,189; B 8 8 30,212; B 20 4 24,200; B 4 6 32,205; B 8 8 20,156; B 4 36 20,134; B 4 8 20,108; B 16 4 26,114; B 8 4 30,118; L NI; B 16 12 24,200; B 12 40 20,134; L NC; B 8 4 14,238; B 4 4 14,50; B 4 4 20,100; B 4 4 20,78; B 4 4 20,36; B 4 4 22,170; B 8 4 26,64; B 8 4 28,8; B 4 8 26,190; B 4 4 28,252; B 4 8 30,118; B 4 4 30,212; B 4 4 20,156; DF;

Address

DS 503 125/1; L NP; B 8 6 4,83; B 4 38 2,173; B 4 80 2,40; B 14 16 7,200; B 8 2 8,209; B 4 24 8,138; B

20 4 10,152; B 18 4 9,256; B 8 4 14,252; B 16 4 18,244; B 4 20 12,232; B 4 42 18,57; B 4 28 24,260; B 12 12 30,198; B 26 4 23,220; B 8 2 28,77; B 16 16 20,172; B 4 56 26,134; B 8 2 30,209; B 6 2 31,11; B 4 24 30,52; B 8 4 34,240; B 4 4 34,2; B 8 6 32,7; B 20 4 34,206; B 18 4 39,244; B 16 4 36,66; B 4 136 42,136; B 14 2 21,163; B 18 4 25,80; B 12 12 28,88; L ND; B 8 8 4,102; B 8 4 4,86; B 4 24 2,118; B 8 8 6,270; B 18 4 9,90; B 8 4 8,210; B 4 50 6,241; B 4 18 8,199; B 16 4 12,214; B 18 8 15,60; B 12 8 18,52; B 8 8 16,144; B 4 8 14,152; B 10 4 17,158; B 8 4 18,162; B 8 8 20,228; B 4 8 18,220; B 4 18 18,173; B 16 4 24,234; B 12 16 26,122; B 8 8 26,8; B 14 4 19,42; B 4 28 24,26; B 8 4 28,76; B 8 4 30,210; B 4 18 28,87; B 8 8 30,100; B 14 4 31,72; B 4 18 30,199; B 18 4 35,214; B 4 16 30,224; B 8 8 34,146; B 36 4 18,132; B 4 8 34,138; B 4 8 36,66; B 8 8 38,58; B 30 4 27,46; B 8 8 40,34; B 8 8 42,270; B 4 6 40,41; B 4 50 42,241; B 28 8 20,186; B 18 4 23,112; B 4 18 16,101; L NM; B 8 12 4,86; B 48 20 24,28; B 48 20 24,54; B 48 20 24,108; B 48 20 24,140; B 48 12 24,188; B 48 8 24,228; B 48 10 24,269; B 8 12 8,210; B 8 2 14,251; B 48 6 24,255; B 8 12 18,162; B 12 8 28,8; B 8 12 28,76; B 8 12 30,210; B 8 2 30,97; B 8 2 34,245; B 48 6 24,241; L NI; B 12 26 8,201; B 12 12 8,132; B 12 26 18,171; B 12 26 28,85; B 12 26 30,201; B 12 12 30,46; L NC; B 4 4 4,102; B 4 8 4,86; B 4 4 6,270; B 4 8 8,210; B 4 4 10,186; B 4 4 10,60; B 4 4 14,254; B 4 4 16,144; B 4 8 18,162; B 4 4 20,186; B 4 4 20,228; B 8 4 28,8; B 4 8 28,76; B 4 4 30,186; B 4 8 30,210; B 4 4 30,100; B 4 4 34,242; B 4 4 34,146; B 4 4 38,58; B 4 4 40,34; B 4 4 42,270; DF;

ShReg

DS 504 125/1; L NM; B 24 6 36,37; B 34 6 31,149; B 16 6 40,5; B 22 8 37,92; B 20 8 38,176; B 8 12 36,122; B 8 8 34,4; B 8 12 32,52; B 8 8 32,176; B 36 6 18,5; B 36 8 18,176; B 8 8 30,92; B 34 8 17,92; B 8 8 26,36; B 12 8 24,22; B 28 6 14,37; B 12 8 14,162; B 8 8 16,148; B 18 6 9,149; B 8 12 6,132; B 8 8 4,92; B 8 8 4,92; L ND; B 14 4 41,146; B 8 8 36,124; B 8 8 34,4; B 4 24 36,136; B 4 12 34,24; B 8 8 32,50; B 8 8 32,176; B 4 16 34,12; B 4 22 34,39; B 4 40 32,70; B 8 8 30,92; B 4 12 32,164; B 4 4 32,90; B 4 8 32,172; B 10 4 29,158; B 4 4 32,158; B 24 4 16,136; B 8 4 24,136; B 4 26 26,147; B 8 8 22,22; B 8 4 20,140; B 4 8 22,138; B 6 4 19,24; B 4 12 18,42; B 4 96 18,94; B 4 16 18,30; B 8 8 12,162; B 6 4 9,164; B 8 8 6,134; B 4 12 8,152; B 4 10 8,161; B 4 4 8,146; B 10 4 5,146; B 8 8 4,92; B 4 40 6,114; B 4 4 6,94; B 4 4 6,136; L NP; B 8 6 36,119; B 12 4 34,24; B 6 4 37,84; B 4 38 38,101; B 12 32 32,70; B 12 4 32,164; B 8 6 32,55; B 6 4 29,24; B 8 8 26,36; B 6 8 27,22; B 14 4 23,164; B 4 12 26,38; B 8 4 24,42; B 24 4 14,12; B 4 14 24,7; B 12 4 18,42; B 6 8 17,162; B 8 8 16,148; B 8 4 14,152; B 4 6 16,151; B 12 4 8,152; B 6 4 11,128; B 12 4 8,84; B 4 48 12,106; B 12 32 6,114; B 8 6 6,129; B 4 76 4,48; L NC; B 4 8 36,122; B 4 4 34,4; B 4 8 32,52; B 4 4 32,176; B 4 4 30,92; B 4 4 26,36; B 8 4 24,22; B 8 4 14,162; B 4 4 16,148; B 4 8 6,132; B 4 4 4,92; L NI; B 12 40 32,70; B 12 10 32,53; B 12 40 6,114; B 12 10 6,131; DF;

RamPair

DS 505 125/1; L NM; B 8 8 4,59; B 6 6 3,60; B 26 6 13,34; B 26 6 13,32; B 8 8 4,7; B 6 6 3,6; B 52 6 28,60; B 52 6 28,6; B 8 8 24,33; B 8 8 24,33; B 34 6 39,34; B 34 6 39,32; B 12 8 34,47; B 12 8 34,19; B 8 8 52,59; B 8 8 52,7; B 6 6 53,60; B 6 6 53,6; L ND; B 8 8 4,59; B 8 8 4,7; B 6 4 5,57; B 4 4 4,57; B 6 4 5,9; B 4 4 4,9; B 12 4 12,57; B 12 4 12,9; B 10 4 21,57; B 10 4 21,9; B 8 8 24,33; B 8 8 24,33; B 4 4 24,57; B 4 12 24,51; B 4 14 24,40; B 4 14 24,26; B 4 12 24,15; B 4 4 24,9; B 8 8 36,47; B 8 8 36,19; B 4 4 38,57; B 4 12 38,53; B 4 4 38,49; B 4 12 38,13; B 4 4 38,17; B 4 4 38,9; B 12 4 44,57; B 12 4 44,9; B 8 8 52,59; B 6 4 51,57; B 8 8 52,7; B 6 4 51,9; B 4 4 52,57; B 4 4 52,9; L NC; B 4 4 4,59; B 4 4 4,7; B 4 4 24,33; B 4 4 24,33; B 8 4 34,47; B 8 4 34,19; B 4 4 52,59; B 4 4 52,7; L NP; B 4 2 12,65; B 4 2 12,64; B 4 12 12,57; B 4 4 12,61; B 4 26 12,42; B 4 26 12,24; B 4 12 12,9; B 4 4 12,5; B 4 2 12,2; B 4 2 12,1; B 12 4 24,51; B 12 4 24,15; B 8 4 30,51; B 8 4 30,15; B 6 8 31,47; B 6 4 31,49; B 6 8 31,19; B 6 4 31,17; B 4 6 32,50; B 4 6 32,16; B 4 2 44,65; B 4 2 44,64; B 4 12 44,57; B 4 4 44,61; B 4 26 44,42; B 4 26 44,24; B 4 12 44,9; B 4 4 44,5; B 4 2 44,2; B 4 2 44,1; DF;

InterfacePair

DS 506 125/1; L NM; B 6 6 6,3; B 8 6 33,3; B 6 6 60,3; B 12 8 9,8; B 8 2 7,215; B 6 2 6,3; B 12 8 11,64; B 8 2 9,255; B 8 12 11,190; B 8 2 17,103; B 8 2 17,81; B 8 2 17,39; B 12 8 23,238; B 8 2 23,53; B 8 16 33,10; B 12 8 43,238; B 8 2 43,53; B 8 2 49,103; B 8 2 49,81; B 8 2 49,39; B 12 8 55,64; B 12 8 57,8; B 8 12 55,190; B 8 2 57,255; B 8 2 59,215; B 6 2 60,3; B 66 8 33,22; B 66 6 33,35; B 66 6 33,49; B 66 6 33,77; B 66 8 33,98; B 66 12 33,172; B 66 12 33,208; B 66 6 33,251; B 8 2 17,159; B 8 12 7,118; B 8 2 49,159; B 8 12 59,118; B 66 8 33,154; L NP; B 6 8 6,8; B 4 38 5,87; B 4 48 5,36; B 4 52 5,150; B 8 8 7,64; B 8 8 9,252; B 18 4 12,178; B 4 16 7,240; B 14 4 12,230; B 8 4 11,192; B 18 4 12,108; B 10 2 12,195; B 8 10 13,201; B 8 8 17,78; B 8 8 17,36; B 8 4 19,72; B 4 30 19,17; B 4 14 21,63; B 6 8 26,238; B 6 8 40,238; B 4 14 45,63; B 8 4 47,72; B 8 8 49,78; B 8 8 49,36; B 4 30 47,17; B 10 2 54,195; B 8 10 53,201; B 8 4 55,192; B 8 8 57,252; B 18 4 54,108; B 14 4 54,230; B 6 8 60,8; B 4 16 59,240; B 18 4 54,178; B 8 8 59,64; B 4 38 61,87; B 4 48 61,36; B 4 52 61,150; B 4 70 27,199; B 4 66 29,127; B 6 4 28,162; B 20 4 21,92; B 12 26 17,137; B 20 6 13,121; B 4 70 39,199; B 4 66 37,127; B 6 4 38,162; B 20 4 45,92; B 20 6 53,121; B 12 26 49,137; L ND; B 8 8 7,212; B 4 6 5,205; B 8 2 11,5; B 8 6 11,189; B 8 8 13,178; B 10 2 14,173; B 8 6 15,169; B 12 4 17,224; B 4 8 13,230; B 16 2 15,185; B 14 2 16,183; B 8 6 17,101; B 4 8 15,108; B 10 2 18,97; B 8 8 19,92; B 4 14 17,249; B 10 4 20,240; B 14 4 18,236; B 14 2 22,87; B 12 2 23,85; B 8 8 23,50; B 20 4 13,200; B 4 20 21,212; B 4 12 21,192; B 20 6 17,9; B 14 10 20,17; B 4 24 25,34; B 20 8 19,64; B 4 16 27,76; B 12 2 43,85; B 4 16 39,76; B 8 8 43,50; B 14 10 46,17; B 4 24 41,34; B 14 2 44,87; B 8 8 47,92; B 12 4 49,224; B 4 20 45,212; B 4 12 45,192; B 10 2 48,97; B 8 6 49,101; B 8 6 51,169; B 10 4 46,240; B 4 14 49,249; B 14 4 48,236; B 20 8 47,64; B 4 8 51,108; B 14 2 50,183; B 10 2 52,173; B 8 8 53,178; B 20 6 49,9; B 8 2 55,5; B 4 8 53,230; B 16 2 51,185; B 8 6 55,189; B 8 8 59,212; B 20 4 53,200; B 4 6 61,205; B 8 8 17,156; B 4 36

17,134; B 8 4 7,118; B 16 4 11,114; B 8 8 49,156; B 4 36 49,134; B 16 4 55,114; B 8 4 59,118; L NI; B 16 12 13,200; B 16 12 53,200; B 12 40 17,134; B 12 40 49,134; L NC; B 8 4 9,8; B 4 8 7,118; B 4 4 7,212; B 8 4 11,64; B 4 4 9,252; B 4 8 11,190; B 4 4 15,170; B 4 4 17,100; B 4 4 17,78; B 4 4 17,36; B 8 4 23,238; B 4 4 23,50; B 8 4 43,238; B 4 4 43,50; B 4 4 49,100; B 4 4 49,78; B 4 4 49,36; B 4 4 51,170; B 8 4 55,64; B 8 4 57,8; B 4 8 55,190; B 4 4 57,252; B 4 8 59,118; B 4 4 59,212; B 4 4 17,156; B 4 4 49,156; DF;

Mult

DS 104 125/1; L NP; B 8 4 20,186; B 8 4 20,290; B 70 4 51,182; B 70 4 51,286; B 8 4 62,262; B 80 4 40,58; B 8 2 76,55; B 80 4 40,384; B 8 4 76,388; B 8 4 76,158; B 80 4 40,240; B 8 4 76,236; B 80 4 40,344; B 8 4 76,340; B 258 4 201,162; B 272 4 194,266; B 4 24 84,196; B 4 24 84,300; B 8 8 90,134; B 8 8 90,96; B 8 8 90,36; B 8 8 90,454; B 8 2 90,433; B 8 2 90,427; B 8 2 90,395; B 8 2 90,389; B 8 8 90,242; B 8 8 90,216; B 8 8 90,346; B 8 8 90,320; B 4 12 92,86; B 4 12 92,46; B 4 24 92,20; B 4 14 92,145; B 4 6 92,127; B 22 4 101,122; B 4 10 92,463; B 4 10 92,251; B 4 44 92,190; B 4 10 92,355; B 4 30 92,301; B 22 4 97,430; B 22 4 97,392; B 8 2 106,107; B 8 2 106,133; B 6 4 109,26; B 6 4 109,84; B 6 8 109,14; B 6 4 109,46; B 6 6 109,147; B 6 4 109,460; B 6 4 109,442; B 6 4 109,418; B 6 4 109,404; B 6 4 109,380; B 6 2 109,175; B 6 2 109,169; B 6 4 109,248; B 6 2 109,229; B 6 2 109,223; B 6 4 109,204; B 6 2 109,191; B 6 2 109,185; B 6 4 109,352; B 6 2 109,333; B 6 2 109,327; B 6 4 109,308; B 6 2 109,295; B 6 2 109,289; B 4 4 110,8; B 28 4 116,110; B 8 4 110,88; B 4 2 112,91; B 100 4 152,136; B 24 4 118,42; B 24 4 118,30; B 216 4 214,438; B 216 4 214,400; B 90 4 151,172; B 216 4 214,226; B 24 4 118,252; B 80 4 146,188; B 216 4 214,330; B 24 4 118,356; B 24 4 118,304; B 80 4 146,292; B 10 4 111,422; B 10 4 111,384; B 20 4 120,94; B 18 4 121,426; B 24 4 118,464; B 18 4 121,388; B 26 2 119,151; B 24 2 120,153; B 18 4 137,148; B 4 4 142,114; B 4 4 142,90; B 4 4 142,46; B 46 4 163,464; B 4 4 142,460; B 46 4 163,426; B 4 4 142,422; B 46 4 163,388; B 4 4 142,384; B 46 4 163,252; B 4 4 142,248; B 46 4 163,356; B 4 4 142,352; B 46 4 163,304; B 4 4 142,308; B 8 10 160,121; B 8 10 160,83; B 8 10 160,19; B 8 10 160,53; B 4 10 158,415; B 4 10 158,377; B 64 2 214,203; B 44 4 214,176; B 162 4 279,144; B 62 4 171,30; B 4 6 200,35; B 4 4 200,140; B 30 4 213,40; B 72 4 176,68; B 8 2 208,71; B 8 2 208,65; B 4 10 216,27; B 8 4 226,194; B 8 4 226,318; B 18 4 235,28; B 6 4 229,32; B 24 4 234,190; B 8 4 240,412; B 4 18 242,39; B 8 4 254,374; B 4 4 258,106; B 18 4 265,102; B 4 44 258,40; B 4 10 262,211; B 10 4 261,64; B 4 26 264,79; B 90 4 227,460; B 6 4 269,464; B 90 4 227,422; B 6 4 269,426; B 90 4 227,384; B 6 4 269,388; B 90 4 227,248; B 6 4 269,252; B 88 4 226,308; B 6 8 269,314; B 90 4 227,352; B 6 4 269,356; B 6 4 273,62; B 4 36 272,82; B 4 28 274,46; B 4 22 284,75; B 6 4 285,62; B 4 20 286,50; B 8 8 290,454; B 8 8 290,416; B 8 8 290,378; B 8 8 290,242; B 8 8 290,204; B 8 8 290,346; B 8 8 290,308; B 4 10 292,463; B 4 10 292,425; B 4 10 292,387; B 4 10 292,251; B 4 10 292,213; B 4 10 292,355; B 4 10 292,317; B 12 8 292,98; B 4 12 296,108; B 4 20 296,84; B 4 20 298,58; B 64 4 328,46; B 8 2 308,57; B 6 4 309,76; B 6 4 309,460; B 6 4 309,422; B 6 4 309,384; B 6 4 309,248; B 6 4 309,210; B 6 4 309,352; B 6 4 309,314; B 56 4 332,54; B 32 4 322,214; B 32 4 322,318; B 24 4 318,80; B 24 4 318,426; B 24 4 318,464; B 24 4 318,388; B 24 4 318,252; B 24 4 318,356; B 4 2 320,441; B 4 2 320,403; B 4 2 320,229; B 4 2 320,333; B 4 28 328,178; B 4 28 328,282; B 16 10 336,125; B 16 10 336,31; B 16 10 336,415; B 16 10 336,453; B 16 10 336,377; B 16 10 336,241; B 16 10 336,345; B 8 10 340,203; B 8 10 340,307; B 4 4 342,100; B 38 4 359,96; B 4 4 342,76; B 38 4 359,80; B 4 4 342,24; B 4 4 342,384; B 46 4 363,252; B 4 4 342,248; B 46 4 363,356; B 4 4 342,352; B 4 26 358,129; B 8 10 360,107; B 8 10 360,69; B 8 10 360,31; B 4 4 358,42; B 4 4 358,58; B 8 10 360,377; B 30 2 385,57; B 18 4 391,384; B 4 8 384,246; B 18 4 391,240; B 4 8 384,350; B 18 4 391,344; B 4 4 142,26; B 18 8 135,18; B 16 2 136,23; B 16 2 136,13; B 40 4 198,20; B 4 2 180,17; B 22 2 171,11; B 26 2 169,13; B 4 6 142,63; B 18 8 135,54; B 16 2 136,59; B 16 2 136,49; B 16 2 168,61; B 20 2 166,59; B 8 2 172,57; B 40 4 160,42; B 18 8 135,82; B 16 2 136,87; B 16 2 136,77; B 90 2 213,79; B 90 2 213,125; B 102 2 207,127; B 102 2 207,77; B 16 2 168,75; B 16 2 168,129; B 126 4 203,94; B 120 4 200,110; B 18 8 135,122; B 16 2 136,127; B 16 2 136,117; B 106 2 193,201; B 4 4 142,204; B 46 2 163,199; B 24 4 118,200; B 18 8 135,212; B 16 2 136,217; B 16 2 136,207; B 108 2 210,217; B 96 2 216,219; B 18 8 135,240; B 16 2 136,245; B 16 2 136,235; B 18 8 135,316; B 16 2 136,321; B 16 2 136,311; B 18 8 135,344; B 16 2 136,349; B 16 2 136,339; B 74 2 193,321; B 62 2 199,323; B 18 8 135,376; B 16 2 136,381; B 16 2 136,371; B 8 2 168,373; B 102 4 207,370; B 18 8 135,414; B 16 2 136,419; B 16 2 136,409; B 8 2 168,411; B 88 4 200,408; B 18 8 135,452; B 16 2 136,457; B 16 2 136,447; B 8 2 322,443; B 14 2 325,445; B 26 2 331,447; B 8 2 322,405; B 14 2 325,407; B 26 2 331,409; B 8 2 372,373; B 40 2 380,369; B 44 2 378,371; B 46 4 363,388; B 14 2 325,369; B 26 2 331,371; B 8 2 322,367; B 8 2 322,335; B 14 2 325,337; B 26 2 331,339; B 8 2 330,297; B 14 2 333,299; B 18 2 335,301; B 8 2 322,231; B 14 2 325,233; B 26 2 331,235; B 8 2 330,193; B 14 2 333,195; B 18 2 335,197; B 14 2 325,133; B 8 2 322,135; B 26 2 331,131; B 22 4 319,96; B 4 6 310,101; B 6 8 309,108; B 18 8 335,108; B 16 2 336,113; B 16 2 336,103; B 18 8 335,68; B 16 2 336,73; B 16 2 336,63; B 8 2 320,35; B 60 2 314,37; B 48 2 308,39; B 206 4 211,4; B 20 4 320,20; B 4 12 312,12; B 20 4 366,38; B 8 2 372,35; B 40 4 360,20; B 34 2 383,59; B 18 4 365,62; B 18 4 365,114; B 8 2 370,117; B 18 10 165,211; B 18 10 165,315; B 14 2 175,15; L ND; B 8 2 20,177; B 8 2 20,171; B 8 2 34,149; B 8 2 34,143; B 8 2 48,17; B 8 2 48,11; B 8 4 62,32; B 8 8 76,52; B 8 4 76,90; B 8 4 76,458; B 8 4 76,246; B 8 4 76,192; B 8 4 76,350; B 8 4 76,296; B 4 18 78,213; B 4 18 78,317; B 36 4 90,86; B 8 2 104,89; B 8 2 104,83; B 64 4 76,14; B 8 2 104,17; B 8 2 104,11; B 50 4 83,28; B 8 2 104,31; B 8 2 104,25; B 36 4 90,46; B 8 4 104,42; B 78 4 69,146; B 8 4 104,150; B 4 12 102,430; B 4 12 102,392; B 36 4 90,462; B 8 2 104,465; B 8 2 104,459; B 8 8 104,440; B 8 8 104,420; B 8 8 104,402; B 8 8 104,382; B 92 4 62,174; B 8 4 104,170; B 36 4 90,250; B 8 2 104,253; B 8 2 104,247; B 32 4 92,224; B 8 4 104,228; B 32 4 92,202; B 8 2 104,205; B 8 2 104,199; B 36 4 90,188; B 8 2 104,191; B 8 2 104,185; B 36 4 90,354; B 8 2 104,357; B 8 2 104,351; B 32 4 92,328; B 8 4

```
104,332; B 32 4 92,306; B 8 2 104,309; B 8 2 104,303; B 36 4 90,292; B 8 2 104,295; B 8 2 104,289;
B 8 8 106,114; B 8 8 106,130; B 4 8 106,122; B 8 4 122,146; B 78 4 157,36; B 78 4 157,432; B 78 4
157,470; B 78 4 157,394; B 78 4 157,258; B 78 4 157,362; B 78 4 157,298; B 24 4 130,142; B 62 4
165,154; B 8 8 138,148; B 8 2 150,125; B 8 2 150,119; B 8 2 150,85; B 8 2 150,79; B 8 2 150,15; B 8
2 150,57; B 8 2 150,51; B 8 2 150,417; B 8 2 150,411; B 8 2 150,379; B 8 2 150,373; B 8 2 150,215; B
8 2 150,209; B 8 2 150,347; B 8 2 150,341; B 8 2 150,319; B 8 2 150,313; B 16 10 170,425; B 16 10
170,387; B 8 16 178,184; B 8 16 178,288; B 8 4 192,150; B 82 8 229,12; B 8 4 192,40; B 8 4 192,428;
B 8 4 192,466; B 8 4 192,390; B 8 4 192,254; B 8 2 192,197; B 8 2 192,191; B 8 4 192,358; B 8 2
192,301; B 8 2 192,295; B 50 4 217,182; B 4 6 194,187; B 32 8 216,176; B 62 8 235,134; B 8 4
208,30; B 8 16 208,122; B 8 24 208,94; B 20 2 214,25; B 8 26 222,77; B 8 4 222,56; B 8 24 222,110;
B 4 16 222,42; B 24 2 216,27; B 8 6 224,31; B 8 4 236,68; B 8 40 236,102; B 4 6 236,47; B 14 6
241,33; B 8 8 240,210; B 20 8 244,40; B 22 4 243,72; B 32 4 254,196; B 4 8 240,202; B 4 10 240,189;
B 8 16 250,82; B 8 16 250,122; B 56 2 228,171; B 8 4 252,174; B 28 2 242,169; B 8 4 254,208; B 4 20
264,120; B 8 6 266,19; B 18 4 271,108; B 4 12 268,204; B 8 2 274,465; B 8 2 274,459; B 8 2 274,427;
B 8 2 274,421; B 8 2 274,389; B 8 2 274,383; B 8 2 274,253; B 8 2 274,247; B 8 4 274,312; B 8 2
274,357; B 8 2 274,351; B 8 2 276,123; B 8 2 276,117; B 4 24 278,94; B 50 4 297,120; B 48 8 258,26;
B 20 8 272,40; B 48 4 258,52; B 36 4 264,56; B 4 6 280,47; B 4 6 280,33; B 4 8 290,114; B 6 4
291,68; B 4 8 290,74; B 4 2 292,65; B 20 4 298,108; B 8 2 304,111; B 8 2 304,105; B 32 4 292,80; B 8
4 304,76; B 38 4 289,462; B 8 2 304,465; B 8 2 304,459; B 38 4 289,424; B 8 2 304,427; B 8 2
304,421; B 38 4 289,386; B 8 2 304,389; B 8 2 304,383; B 38 4 289,250; B 8 2 304,253; B 8 2
304,247; B 58 4 279,212; B 8 2 304,215; B 8 2 304,209; B 38 4 289,354; B 8 2 304,357; B 8 2
304,351; B 38 4 289,316; B 8 2 304,319; B 8 2 304,313; B 22 4 301,62; B 8 4 308,58; B 78 4 357,432;
B 78 4 357,470; B 78 4 357,394; B 78 4 357,220; B 78 4 357,258; B 78 4 357,324; B 78 4 357,362; B
16 10 326,213; B 16 10 326,317; B 8 2 350,129; B 8 2 350,123; B 8 2 350,111; B 8 2 350,105; B 8 2
350,71; B 8 2 350,65; B 8 2 350,35; B 8 2 350,29; B 8 2 350,417; B 8 2 350,411; B 8 2 350,455; B 8 2
350,449; B 8 2 350,379; B 8 2 350,373; B 8 2 350,205; B 8 2 350,199; B 8 2 350,243; B 8 2 350,237;
B 8 2 350,309; B 8 2 350,303; B 8 2 350,347; B 8 2 350,341; B 8 4 392,428; B 8 4 392,466; B 8 4
392,390; B 8 4 392,216; B 8 4 392,254; B 8 4 392,320; B 8 4 392,358; B 8 2 150,21; B 10 2 123,15; B
10 2 123,21; B 8 12 122,28; B 8 2 172,15; B 8 14 172,27; B 58 4 147,18; B 10 2 123,57; B 10 2
123,51; B 8 12 122,44; B 58 2 197,51; B 8 2 172,57; B 8 12 172,44; B 108 2 172,53; B 58 2 147,55; B
10 2 123,79; B 10 2 123,85; B 8 12 122,92; B 8 2 172,125; B 8 14 172,113; B 58 4 147,82; B 8 14
172,91; B 8 2 172,79; B 78 8 157,102; B 10 2 123,125; B 10 2 123,119; B 8 12 122,112; B 58 4
147,122; B 78 4 157,194; B 10 2 123,215; B 10 2 123,209; B 8 12 122,202; B 62 4 149,212; B 8 4
176,208; B 8 2 150,243; B 8 2 150,237; B 10 2 123,243; B 10 2 123,237; B 36 4 136,240; B 8 12
122,250; B 10 2 123,319; B 10 2 123,313; B 8 12 122,306; B 10 2 123,347; B 10 2 123,341; B 36 4
136,344; B 8 12 122,354; B 62 4 149,316; B 8 4 176,312; B 10 2 123,379; B 10 2 123,373; B 8 12
122,386; B 54 4 145,376; B 8 4 168,380; B 8 2 168,373; B 10 2 123,417; B 10 2 123,411; B 8 12
122,424; B 54 4 145,414; B 8 4 168,418; B 8 2 168,411; B 8 2 150,455; B 8 2 150,449; B 10 2
123,455; B 10 2 123,449; B 8 12 122,462; B 36 4 136,452; B 36 4 336,452; B 8 4 322,448; B 8 14
322,461; B 36 4 336,414; B 8 4 322,410; B 8 14 322,423; B 8 14 372,385; B 8 2 372,373; B 58 4
347,376; B 8 4 322,372; B 8 14 322,385; B 36 4 336,344; B 8 4 322,340; B 8 14 322,353; B 28 4
340,306; B 8 4 330,302; B 8 4 330,310; B 36 4 336,240; B 8 4 322,236; B 8 14 322,249; B 28 4
340,202; B 8 4 330,198; B 8 4 330,206; B 36 4 336,126; B 8 4 322,130; B 4 2 320,123; B 10 2
323,111; B 10 2 323,105; B 8 12 322,98; B 78 8 357,88; B 10 2 323,71; B 10 2 323,65; B 8 2 322,63;
B 8 12 322,78; B 8 2 320,35; B 80 8 356,12; B 8 14 320,23; B 60 4 346,32; B 8 2 372,35; B 8 14
372,23; B 56 4 346,68; B 8 4 370,64; B 8 14 370,77; B 56 4 346,108; B 8 4 370,112; B 8 14 370,99; L
NM; B 8 472 20,236; B 8 472 34,236; B 8 472 48,236; B 8 242 62,147; B 8 12 76,54; B 8 82 76,423; B
8 80 76,124; B 8 66 76,219; B 8 66 76,323; B 8 472 90,236; B 12 8 106,28; B 12 8 106,86; B 12 8
106,14; B 12 8 106,44; B 12 8 106,148; B 12 8 106,462; B 12 8 106,440; B 12 8 106,420; B 12 8
106,402; B 12 8 106,382; B 12 8 106,172; B 12 8 106,250; B 12 8 106,226; B 12 8 106,202; B 12 8
106,188; B 12 8 106,354; B 12 8 106,330; B 12 8 106,306; B 12 8 106,292; B 8 12 106,112; B 8 12
106,132; B 8 472 150,236; B 8 472 192,236; B 8 18 208,81; B 8 40 208,44; B 8 24 208,110; B 22 8
215,68; B 8 24 222,94; B 8 24 222,126; B 12 8 226,30; B 8 8 226,192; B 8 8 226,320; B 6 120
227,256; B 22 8 229,54; B 8 32 236,74; B 8 24 236,126; B 8 8 240,410; B 6 58 239,443; B 8 8
240,210; B 6 192 239,310; B 24 8 258,40; B 8 40 250,102; B 8 8 252,172; B 8 8 254,372; B 6 96
253,424; B 8 8 254,210; B 6 30 253,191; B 6 154 253,291; B 12 8 272,462; B 12 8 272,424; B 12 8
272,386; B 12 8 272,250; B 12 8 272,314; B 12 8 272,354; B 34 8 263,54; B 8 66 276,91; B 8 472
290,236; B 12 8 306,78; B 12 8 306,108; B 12 8 306,462; B 12 8 306,424; B 12 8 306,386; B 12 8
306,250; B 12 8 306,212; B 12 8 306,354; B 12 8 306,316; B 8 12 308,58; B 8 472 350,236; B 8 472
392,236; B 12 8 126,18; B 8 12 172,16; B 12 8 126,54; B 8 12 172,56; B 12 8 126,82; B 8 12 172,124;
B 8 12 172,80; B 14 8 125,122; B 8 22 122,137; B 12 8 126,212; B 8 30 178,191; B 14 8 175,210; B
12 8 126,240; B 12 8 126,316; B 12 8 126,344; B 8 30 178,295; B 14 8 175,314; B 12 8 126,376; B 8
12 168,374; B 12 8 126,414; B 8 12 168,412; B 12 8 126,452; B 8 12 322,448; B 8 12 322,410; B 8 12
372,374; B 8 12 322,372; B 8 12 322,340; B 8 12 330,302; B 8 12 322,236; B 8 12 330,198; B 8 12
322,130; B 12 8 326,108; B 12 8 326,68; B 8 12 320,34; B 8 12 372,34; B 8 12 370,64; B 8 12
370,112; L NI; B 16 12 160,18; B 16 12 160,54; B 12 12 158,376; B 24 12 336,126; B 24 12 336,108;
B 24 12 336,32; B 24 12 336,452; B 24 12 336,414; B 24 12 336,376; B 24 12 336,240; B 24 12
336,344; B 16 12 340,306; B 16 12 360,108; B 16 12 360,68; B 16 12 360,32; B 16 12 360,376; B 26
12 135,18; B 26 12 135,54; B 26 12 135,82; B 16 12 160,122; B 16 12 160,82; B 26 12 135,122; B 26
12 135,212; B 26 12 165,212; B 26 12 135,240; B 26 12 135,316; B 26 12 135,344; B 26 12 165,316;
B 26 12 135,376; B 26 12 135,414; B 12 12 158,414; B 26 12 135,452; B 16 12 340,202; B 24 12
336,68; L NC; B 4 4 20,174; B 4 4 20,184; B 4 4 20,288; B 4 4 34,146; B 4 4 48,14; B 4 4 62,30; B 4 4
```

62,264; B 4 8 76,54; B 4 4 76,88; B 4 4 76,386; B 4 4 76,460; B 4 4 76,160; B 4 4 76,248; B 4 4 76,238; B 4 4 76,190; B 4 4 76,352; B 4 4 76,342; B 4 4 76,294; B 4 4 90,134; B 4 4 90,96; B 4 4 90,36; B 4 4 90,454; B 4 4 90,430; B 4 4 90,392; B 4 4 90,242; B 4 4 90,216; B 4 4 90,346; B 4 4 90,320; B 8 4 106,28; B 8 4 106,86; B 8 4 106,14; B 8 4 106,44; B 8 4 106,148; B 4 8 106,112; B 4 8 106,132; B 8 4 106,462; B 8 4 106,440; B 8 4 106,420; B 8 4 106,402; B 8 4 106,382; B 8 4 106,172; B 8 4 106,250; B 8 4 106,226; B 8 4 106,202; B 8 4 106,188; B 8 4 106,354; B 8 4 106,330; B 8 4 106,306; B 8 4 106,292; B 4 4 122,144; B 4 4 150,122; B 4 4 150,82; B 4 4 150,18; B 4 4 150,54; B 4 4 150,414; B 4 4 150,376; B 4 4 150,452; B 4 4 150,240; B 4 4 150,212; B 4 4 150,344; B 4 4 150,316; B 4 4 178,180; B 4 4 178,284; B 4 4 192,152; B 4 4 192,102; B 4 4 192,38; B 4 4 192,12; B 4 4 192,430; B 4 4 192,468; B 4 4 192,392; B 4 4 192,256; B 4 4 192,194; B 4 4 192,360; B 4 4 192,298; B 4 4 208,68; B 4 4 208,28; B 4 4 208,102; B 4 4 208,118; B 4 4 208,86; B 4 4 222,68; B 4 4 222,54; B 4 4 222,86; B 4 4 222,102; B 4 4 222,118; B 4 4 222,134; B 8 4 226,30; B 4 4 226,192; B 4 4 226,320; B 4 4 236,70; B 4 4 236,118; B 4 4 236,86; B 4 4 236,134; B 4 4 240,410; B 4 4 240,210; B 4 4 250,54; B 4 4 250,40; B 4 4 250,118; B 4 4 250,86; B 4 4 252,172; B 4 4 254,372; B 4 4 254,210; B 4 4 266,40; B 8 4 272,462; B 8 4 272,424; B 8 4 272,386; B 8 4 272,250; B 8 4 272,314; B 8 4 272,354; B 4 4 276,120; B 4 4 290,98; B 4 4 290,454; B 4 4 290,416; B 4 4 290,378; B 4 4 290,242; B 4 4 290,204; B 4 4 290,346; B 4 4 290,308; B 8 4 306,78; B 8 4 306,108; B 4 8 308,58; B 8 4 306,462; B 8 4 306,424; B 8 4 306,386; B 8 4 306,250; B 8 4 306,212; B 8 4 306,354; B 8 4 306,316; B 4 4 350,126; B 4 4 350,108; B 4 4 350,68; B 4 4 350,32; B 4 4 350,414; B 4 4 350,452; B 4 4 350,376; B 4 4 350,202; B 4 4 350,240; B 4 4 350,306; B 4 4 350,344; B 4 4 392,88; B 4 4 392,12; B 4 4 392,430; B 4 4 392,468; B 4 4 392,392; B 4 4 392,218; B 4 4 392,256; B 4 4 392,322; B 4 4 392,360; B 8 4 126,18; B 4 8 172,16; B 8 4 126,54; B 4 8 172,56; B 8 4 126,82; B 4 8 172,124; B 4 8 172,80; B 8 4 126,122; B 8 4 126,212; B 8 4 174,210; B 8 4 126,240; B 8 4 126,316; B 8 4 126,344; B 8 4 174,314; B 8 4 126,376; B 4 8 168,374; B 8 4 126,414; B 4 8 168,412; B 8 4 126,452; B 4 8 322,448; B 4 8 322,410; B 4 8 372,374; B 4 8 322,372; B 4 8 322,340; B 4 8 330,302; B 4 8 322,236; B 4 8 330,198; B 4 8 322,130; B 8 4 326,108; B 8 4 326,68; B 4 8 320,34; B 4 8 372,34; B 4 8 370,64; B 4 8 370,112; DF;

DblMult

DS 105 125/1; L NP; B 8 4 20,650; B 8 4 20,546; B 8 4 20,186; B 8 4 20,290; B 70 4 51,654; B 70 4 51,550; B 70 4 51,182; B 70 4 51,286; B 8 4 62,574; B 8 4 62,262; B 80 4 40,778; B 8 2 76,781; B 8 4 76,678; B 80 4 40,596; B 8 4 76,600; B 80 4 40,492; B 8 4 76,496; B 80 4 40,58; B 8 2 76,55; B 80 4 40,384; B 8 4 76,388; B 8 4 76,158; B 80 4 40,240; B 8 4 76,236; B 80 4 40,344; B 8 4 76,340; B 258 4 201,674; B 272 4 194,570; B 258 4 201,162; B 272 4 194,266; B 4 24 84,640; B 4 24 84,536; B 4 24 84,196; B 4 24 84,300; B 8 8 90,702; B 8 8 90,740; B 8 8 90,800; B 8 8 90,594; B 8 8 90,620; B 8 8 90,490; B 8 8 90,516; B 8 8 90,134; B 8 8 90,96; B 8 8 90,36; B 8 8 90,454; B 8 2 90,433; B 8 2 90,427; B 8 2 90,395; B 8 2 90,389; B 8 8 90,242; B 8 8 90,216; B 8 8 90,346; B 8 8 90,320; B 4 12 92,750; B 4 24 92,816; B 4 12 92,790; B 4 6 92,709; B 4 14 92,691; B 22 4 101,714; B 4 10 92,585; B 4 44 92,646; B 4 10 92,481; B 4 30 92,535; B 4 12 92,86; B 4 12 92,46; B 4 24 92,20; B 4 14 92,145; B 4 6 92,127; B 22 4 101,122; B 4 10 92,463; B 4 10 92,251; B 4 44 92,190; B 4 10 92,355; B 4 30 92,301; B 22 4 97,430; B 22 4 97,392; B 8 2 106,729; B 8 2 106,703; B 8 2 106,107; B 8 2 106,133; B 6 4 109,810; B 6 4 109,752; B 6 8 109,822; B 6 4 109,790; B 6 6 109,689; B 6 2 109,667; B 6 2 109,661; B 6 4 109,588; B 6 2 109,613; B 6 2 109,607; B 6 4 109,632; B 6 2 109,651; B 6 2 109,645; B 6 4 109,484; B 6 2 109,509; B 6 2 109,503; B 6 4 109,528; B 6 2 109,547; B 6 2 109,541; B 6 4 109,26; B 6 4 109,84; B 6 8 109,14; B 6 4 109,46; B 6 6 109,147; B 6 4 109,460; B 6 4 109,442; B 6 4 109,418; B 6 4 109,404; B 6 4 109,380; B 6 2 109,175; B 6 2 109,169; B 6 4 109,248; B 6 2 109,229; B 6 2 109,223; B 6 4 109,204; B 6 2 109,191; B 6 2 109,185; B 6 4 109,352; B 6 2 109,333; B 6 2 109,327; B 6 4 109,308; B 6 2 109,295; B 6 2 109,289; B 206 4 211,832; B 4 4 110,828; B 28 4 116,726; B 206 4 211,4; B 4 4 110,8; B 28 4 116,110; B 8 4 110,748; B 4 2 112,745; B 100 4 152,700; B 24 4 118,794; B 24 4 118,806; B 90 4 151,664; B 216 4 214,610; B 24 4 118,584; B 24 4 118,636; B 80 4 146,648; B 216 4 214,506; B 24 4 118,480; B 24 4 118,532; B 80 4 146,544; B 8 4 110,88; B 4 2 112,91; B 100 4 152,136; B 24 4 118,42; B 24 4 118,30; B 216 4 214,438; B 216 4 214,400; B 90 4 151,172; B 216 4 214,226; B 24 4 118,252; B 24 4 118,200; B 80 4 146,188; B 216 4 214,330; B 24 4 118,356; B 24 4 118,304; B 80 4 146,292; B 10 4 111,422; B 10 4 111,384; B 20 4 120,742; B 20 4 120,94; B 18 4 121,426; B 24 4 118,464; B 18 4 121,388; B 26 2 119,685; B 24 2 120,683; B 16 10 136,715; B 16 10 136,783; B 26 2 119,151; B 24 2 120,153; B 16 10 136,121; B 16 10 136,83; B 16 10 136,53; B 16 10 136,377; B 18 4 137,688; B 18 4 137,148; B 4 4 142,722; B 4 4 142,746; B 4 4 142,810; B 4 4 142,790; B 4 6 142,773; B 46 4 163,584; B 4 4 142,588; B 46 2 163,637; B 4 4 142,632; B 46 4 163,480; B 4 4 142,484; B 46 4 163,532; B 4 4 142,528; B 4 4 142,114; B 4 4 142,90; B 4 4 142,26; B 4 4 142,46; B 4 6 142,63; B 46 4 163,464; B 4 4 142,460; B 46 4 163,426; B 4 4 142,422; B 46 4 163,388; B 4 4 142,384; B 46 4 163,252; B 4 4 142,248; B 46 2 163,199; B 4 4 142,204; B 46 4 163,356; B 4 4 142,352; B 46 4 163,304; B 4 4 142,308; B 8 10 160,715; B 8 10 160,753; B 8 10 160,817; B 8 10 160,783; B 8 10 160,121; B 8 10 160,83; B 8 10 160,19; B 8 10 160,53; B 4 10 158,415; B 4 10 158,377; B 106 2 193,635; B 64 2 214,633; B 106 2 193,201; B 64 2 214,203; B 44 4 214,660; B 44 4 214,176; B 162 4 279,692; B 4 6 200,801; B 4 4 200,696; B 162 4 279,144; B 4 6 200,35; B 4 4 200,140; B 30 4 213,796; B 30 4 213,40; B 72 4 176,768; B 8 2 208,771; B 8 2 208,765; B 72 4 176,68; B 8 2 208,71; B 8 2 208,65; B 4 10 216,809; B 4 10 216,27; B 8 4 226,642; B 8 4 226,518; B 8 4 226,194; B 8 4 226,318; B 18 4 235,808; B 6 4 229,804; B 18 4 235,28; B 6 4 229,32; B 24 4 234,646; B 24 4 234,190; B 8 4 240,412; B 4 18 242,797; B 4 18 242,39; B 8 4 254,374; B 4 4 258,730; B 18 4 265,734; B 4 44 258,796; B 4 4 258,106; B 18 4 265,102; B 4 44 258,40; B 4 10 262,625; B 4 10 262,211; B 10 4 261,772; B 4 26 264,757; B 10 4 261,64; B 4 26 264,79; B 90 4 227,588; B 6 4 269,584; B 88 4 226,528; B 6 8 269,522; B 90 4 227,484; B 6 4

269,480; B 90 4 227,460; B 6 4 269,464; B 90 4 227,422; B 6 4 269,426; B 90 4 227,384; B 6 4 269,388; B 90 4 227,248; B 6 4 269,252; B 88 4 226,308; B 6 8 269,314; B 90 4 227,352; B 6 4 269,356; B 4 36 272,754; B 4 36 272,82; B 6 4 273,774; B 4 28 274,790; B 6 4 273,62; B 4 28 274,46; B 4 22 284,761; B 4 22 284,75; B 6 4 285,774; B 4 20 286,786; B 6 4 285,62; B 4 20 286,50; B 8 8 290,594; B 8 8 290,632; B 8 8 290,490; B 8 8 290,528; B 8 8 290,454; B 8 8 290,416; B 8 8 290,378; B 8 8 290,242; B 8 8 290,204; B 8 8 290,346; B 8 8 290,308; B 4 10 292,585; B 4 10 292,623; B 4 10 292,481; B 4 10 292,519; B 4 10 292,463; B 4 10 292,425; B 4 10 292,387; B 4 10 292,251; B 4 10 292,213; B 4 10 292,355; B 4 10 292,317; B 12 8 292,738; B 4 20 296,752; B 4 12 296,728; B 12 8 292,98; B 4 12 296,108; B 4 20 296,84; B 4 20 298,778; B 64 4 328,790; B 4 20 298,58; B 64 4 328,46; B 8 2 308,779; B 8 2 308,57; B 6 4 309,760; B 6 4 309,726; B 6 4 309,588; B 6 4 309,626; B 6 4 309,484; B 6 4 309,522; B 6 4 309,76; B 6 4 309,110; B 6 4 309,460; B 6 4 309,422; B 6 4 309,384; B 6 4 309,248; B 6 4 309,210; B 6 4 309,352; B 6 4 309,314; B 56 4 332,782; B 56 4 332,54; B 4 12 312,824; B 8 4 310,730; B 4 6 312,735; B 32 4 322,622; B 32 4 322,518; B 4 12 312,12; B 8 4 310,106; B 4 6 312,101; B 32 4 322,214; B 32 4 322,318; B 20 4 320,740; B 24 4 318,756; B 24 4 318,584; B 24 4 318,480; B 20 4 320,96; B 24 4 318,80; B 20 4 320,20; B 24 4 318,426; B 24 4 318,464; B 24 4 318,388; B 24 4 318,252; B 24 4 318,356; B 4 2 320,607; B 4 2 320,503; B 4 2 320,441; B 4 2 320,403; B 4 2 320,229; B 4 2 320,333; B 4 28 328,658; B 4 28 328,554; B 4 30 328,179; B 4 30 328,283; B 16 10 336,711; B 16 10 336,805; B 16 10 336,595; B 16 10 336,491; B 16 10 336,125; B 16 10 336,31; B 16 10 336,415; B 16 10 336,453; B 16 10 336,377; B 16 10 336,241; B 16 10 336,345; B 8 10 340,633; B 8 10 340,529; B 8 10 340,203; B 8 10 340,307; B 4 4 342,736; B 38 4 359,740; B 4 4 342,760; B 38 4 359,756; B 4 4 342,812; B 38 4 359,816; B 46 4 363,584; B 4 4 342,588; B 46 4 363,480; B 4 4 342,484; B 4 4 342,100; B 4 4 342,76; B 4 4 342,24; B 46 4 363,388; B 4 4 342,384; B 46 4 363,252; B 4 4 342,248; B 46 4 363,356; B 4 4 342,352; B 4 26 358,707; B 8 10 360,729; B 8 10 360,767; B 8 10 360,805; B 4 4 358,794; B 4 4 358,778; B 4 26 358,129; B 8 10 360,107; B 8 10 360,69; B 8 10 360,31; B 4 4 358,42; B 4 4 358,58; B 8 10 360,377; B 30 2 385,779; B 30 4 385,58; B 4 8 384,590; B 18 4 391,596; B 4 8 384,486; B 18 4 391,492; B 18 4 391,384; B 4 8 384,246; B 18 4 391,240; B 4 8 384,350; B 18 4 391,344; B 18 8 135,18; B 16 2 136,23; B 16 2 136,13; B 14 2 125,61; B 8 2 122,63; B 26 2 131,59; B 8 2 172,57; B 62 4 171,30; B 16 2 168,61; B 20 2 166,59; B 40 4 160,42; B 22 2 171,11; B 26 2 169,13; B 40 4 198,20; B 4 2 180,17; B 90 2 213,79; B 102 2 207,77; B 16 2 168,75; B 126 4 203,94; B 14 2 125,75; B 26 2 131,77; B 8 2 122,73; B 28 2 130,127; B 8 2 120,125; B 16 2 124,129; B 90 2 213,125; B 102 2 207,127; B 16 2 168,129; B 120 4 200,110; B 18 8 135,212; B 16 2 136,217; B 16 2 136,207; B 18 8 135,240; B 16 2 136,245; B 16 2 136,235; B 108 2 210,217; B 96 2 216,219; B 18 8 135,316; B 16 2 136,321; B 16 2 136,311; B 74 2 193,321; B 62 2 199,323; B 18 8 135,344; B 16 2 136,349; B 16 2 136,339; B 14 2 125,369; B 26 2 131,371; B 8 2 122,367; B 8 2 168,373; B 102 4 207,370; B 18 8 135,414; B 16 2 136,419; B 16 2 136,409; B 8 2 168,411; B 88 4 200,408; B 18 8 135,452; B 16 2 136,457; B 16 2 136,447; B 18 8 135,492; B 16 2 136,497; B 16 2 136,487; B 18 8 135,520; B 16 2 136,525; B 16 2 136,515; B 74 2 193,515; B 62 2 199,513; B 18 8 135,596; B 16 2 136,601; B 16 2 136,591; B 18 8 135,624; B 16 2 136,629; B 16 2 136,619; B 108 2 210,619; B 96 2 216,617; B 28 2 130,709; B 8 2 120,711; B 16 2 124,707; B 90 2 213,711; B 102 2 207,709; B 16 2 168,707; B 120 4 200,726; B 18 8 135,754; B 16 2 136,759; B 16 2 136,749; B 90 2 213,757; B 102 2 207,759; B 16 2 168,761; B 126 4 203,742; B 14 2 125,775; B 26 2 131,777; B 8 2 122,773; B 8 2 172,779; B 62 4 171,806; B 16 2 168,775; B 20 2 166,777; B 40 4 160,794; B 22 2 171,825; B 26 2 169,823; B 40 4 198,816; B 4 2 180,819; B 18 8 135,818; B 16 2 136,823; B 16 2 136,813; B 8 2 320,35; B 60 2 314,37; B 48 2 308,39; B 20 4 366,38; B 8 2 372,35; B 40 4 360,20; B 18 8 335,68; B 16 2 336,73; B 16 2 336,63; B 20 4 366,62; B 8 2 372,65; B 40 4 360,80; B 18 8 335,108; B 16 2 336,113; B 16 2 336,103; B 20 4 366,114; B 8 2 372,111; B 40 4 360,96; B 28 2 330,131; B 8 2 320,129; B 16 2 324,133; B 20 2 334,197; B 8 2 328,199; B 16 2 332,195; B 8 2 322,231; B 14 2 325,233; B 26 2 331,235; B 20 2 334,301; B 8 2 328,303; B 16 2 332,299; B 8 2 322,335; B 14 2 325,337; B 26 2 331,339; B 14 2 325,369; B 26 2 331,371; B 8 2 322,367; B 44 2 378,371; B 40 2 380,369; B 8 2 370,367; B 8 2 322,405; B 14 2 325,407; B 26 2 331,409; B 8 2 322,443; B 14 2 325,445; B 26 2 331,447; B 8 2 322,501; B 14 2 325,499; B 26 2 331,497; B 8 2 330,539; B 14 2 333,537; B 18 2 335,535; B 8 2 322,605; B 14 2 325,603; B 26 2 331,601; B 8 2 330,643; B 14 2 333,641; B 18 2 335,639; B 14 2 325,703; B 26 2 331,705; B 8 2 322,701; B 18 8 335,728; B 16 2 336,733; B 16 2 336,723; B 18 4 365,722; B 8 2 370,719; B 18 8 335,768; B 16 2 336,773; B 16 2 336,763; B 34 2 383,777; B 18 4 365,774; B 8 2 320,801; B 60 2 314,799; B 48 2 308,797; B 18 4 319,816; B 18 4 365,798; B 8 2 370,795; B 18 10 165,211; B 18 10 165,315; B 18 10 165,521; B 18 10 165,625; B 14 2 175,15; B 14 2 175,821; L ND; B 8 2 20,665; B 8 2 20,659; B 8 2 20,177; B 8 2 20,171; B 8 2 34,693; B 8 2 34,687; B 8 2 34,149; B 8 2 34,143; B 8 2 48,825; B 8 2 48,819; B 8 2 48,17; B 8 2 48,11; B 8 4 62,804; B 8 4 62,32; B 8 8 76,784; B 8 4 76,746; B 8 4 76,590; B 8 4 76,644; B 8 4 76,486; B 8 4 76,540; B 8 8 76,52; B 8 4 76,90; B 8 4 76,458; B 8 4 76,246; B 8 4 76,192; B 8 4 76,350; B 8 4 76,296; B 4 18 78,623; B 4 18 78,519; B 4 18 78,213; B 4 18 78,317; B 36 4 90,750; B 8 2 104,753; B 8 2 104,747; B 64 4 76,822; B 8 2 104,825; B 8 2 104,819; B 50 4 83,808; B 8 2 104,811; B 8 2 104,805; B 36 4 90,790; B 8 4 104,794; B 78 4 69,690; B 8 4 104,686; B 92 4 62,662; B 8 4 104,666; B 36 4 90,586; B 8 2 104,589; B 8 2 104,583; B 32 4 92,612; B 8 4 104,608; B 32 4 92,634; B 8 2 104,637; B 8 2 104,631; B 36 4 90,648; B 8 2 104,651; B 8 2 104,645; B 36 4 90,482; B 8 2 104,485; B 8 2 104,479; B 32 4 92,508; B 8 4 104,504; B 32 4 92,530; B 8 2 104,533; B 8 2 104,527; B 36 4 90,544; B 8 2 104,547; B 8 2 104,541; B 36 4 90,86; B 8 2 104,89; B 8 2 104,83; B 64 4 76,14; B 8 2 104,17; B 8 2 104,11; B 50 4 83,28; B 8 2 104,31; B 8 2 104,25; B 36 4 90,46; B 8 4 104,42; B 78 4 69,146; B 8 4 104,150; B 4 12 102,430; B 4 12 102,392; B 36 4 90,462; B 8 2 104,465; B 8 2 104,459; B 8 8 104,440; B 8 8 104,420; B 8 8 104,402; B 8 8 104,382; B 92 4 62,174; B 8 4 104,170; B 36 4 90,250; B 8 2 104,253; B 8 2 104,247; B 32 4 92,224; B 8 4 104,228; B 32 4 92,202; B 8 2 104,205; B 8 2 104,199; B 36 4 90,188; B 8 2

104,191; B 8 2 104,185; B 36 4 90,354; B 8 2 104,357; B 8 2 104,351; B 32 4 92,328; B 8 4 104,332;
B 32 4 92,306; B 8 2 104,309; B 8 2 104,303; B 36 4 90,292; B 8 2 104,295; B 8 2 104,289; B 8 8
106,722; B 8 8 106,706; B 8 8 106,114; B 8 8 106,130; B 4 8 106,714; B 4 8 106,122; B 8 4 122,690;
B 78 4 157,578; B 78 4 157,642; B 78 4 157,538; B 8 4 122,146; B 78 4 157,432; B 78 8 157,472; B
78 4 157,394; B 78 4 157,258; B 78 4 157,194; B 78 4 157,362; B 78 4 157,298; B 24 4 130,694; B 24
4 130,142; B 8 8 138,688; B 8 8 138,148; B 62 4 165,682; B 62 4 165,154; B 8 2 150,717; B 8 2
150,711; B 8 2 150,757; B 8 2 150,751; B 8 2 150,821; B 8 2 150,815; B 8 2 150,785; B 8 2 150,779;
B 8 2 150,627; B 8 2 150,621; B 8 2 150,523; B 8 2 150,517; B 8 2 150,125; B 8 2 150,119; B 8 2
150,85; B 8 2 150,79; B 8 2 150,21; B 8 2 150,15; B 8 2 150,57; B 8 2 150,51; B 8 2 150,417; B 8 2
150,411; B 8 2 150,379; B 8 2 150,373; B 8 2 150,215; B 8 2 150,209; B 8 2 150,319; B 8 2 150,313;
B 16 10 170,425; B 16 10 170,387; B 8 16 178,652; B 8 16 178,548; B 8 16 178,184; B 8 16 178,288;
B 8 4 192,686; B 82 8 229,824; B 8 4 192,796; B 8 4 192,582; B 8 2 192,645; B 8 2 192,639; B 8 4
192,478; B 8 2 192,541; B 8 2 192,535; B 8 4 192,150; B 82 8 229,12; B 8 4 192,40; B 8 4 192,428; B
8 4 192,466; B 8 4 192,390; B 8 4 192,254; B 8 2 192,197; B 8 2 192,191; B 8 4 192,358; B 8 2
192,301; B 8 2 192,295; B 50 4 217,654; B 4 6 194,649; B 50 4 217,182; B 4 6 194,187; B 32 8
216,660; B 32 8 216,176; B 62 8 235,702; B 8 4 208,806; B 8 16 208,714; B 8 24 208,742; B 62 8
235,134; B 8 4 208,30; B 8 16 208,122; B 8 24 208,94; B 20 2 214,811; B 20 2 214,25; B 8 26
222,759; B 8 4 222,780; B 8 24 222,726; B 8 26 222,77; B 8 4 222,56; B 8 24 222,110; B 4 16
222,794; B 24 2 216,809; B 8 6 224,805; B 4 16 222,42; B 24 2 216,27; B 8 6 224,31; B 8 4 236,768;
B 8 40 236,734; B 8 4 236,68; B 8 40 236,102; B 4 6 236,789; B 14 6 241,803; B 4 6 236,47; B 14 6
241,33; B 8 8 240,626; B 8 8 240,210; B 20 8 244,796; B 22 4 243,764; B 4 10 240,647; B 4 8
240,634; B 20 8 244,40; B 22 4 243,72; B 4 8 240,202; B 4 10 240,189; B 32 4 254,640; B 32 4
254,196; B 8 16 250,754; B 8 16 250,714; B 8 16 250,82; B 8 16 250,122; B 56 2 228,665; B 28 2
242,667; B 8 4 252,662; B 56 2 228,171; B 8 4 252,174; B 28 2 242,169; B 8 4 254,628; B 8 4
254,208; B 4 20 264,716; B 8 6 266,817; B 4 20 264,120; B 8 6 266,19; B 18 4 271,728; B 18 4
271,108; B 4 12 268,632; B 4 12 268,204; B 8 2 274,589; B 8 2 274,583; B 8 4 274,524; B 8 2
274,485; B 8 2 274,479; B 8 2 274,465; B 8 2 274,459; B 8 2 274,427; B 8 2 274,421; B 8 2 274,389;
B 8 2 274,383; B 8 2 274,253; B 8 2 274,247; B 8 4 274,312; B 8 2 274,357; B 8 2 274,351; B 8 2
276,719; B 8 2 276,713; B 8 2 276,123; B 8 2 276,117; B 4 24 278,742; B 4 24 278,94; B 50 4
297,716; B 48 8 258,810; B 20 8 272,796; B 4 6 280,803; B 48 4 258,784; B 4 6 280,789; B 36 4
264,780; B 50 4 297,120; B 48 8 258,26; B 20 8 272,40; B 48 4 258,52; B 36 4 264,56; B 4 6 280,47;
B 4 6 280,33; B 4 8 290,722; B 4 8 290,762; B 4 8 290,114; B 4 8 290,74; B 6 4 291,768; B 4 2
292,771; B 6 4 291,68; B 4 2 292,65; B 20 4 298,728; B 8 2 304,731; B 8 2 304,725; B 32 4 292,756;
B 8 4 304,760; B 38 4 289,586; B 8 2 304,589; B 8 2 304,583; B 58 4 279,624; B 8 2 304,627; B 8 2
304,621; B 38 4 289,482; B 8 2 304,485; B 8 2 304,479; B 38 4 289,520; B 8 2 304,523; B 8 2
304,517; B 20 4 298,108; B 8 2 304,111; B 8 2 304,105; B 32 4 292,80; B 8 4 304,76; B 38 4 289,462;
B 8 2 304,465; B 8 2 304,459; B 38 4 289,424; B 8 2 304,427; B 8 2 304,421; B 38 4 289,386; B 8 2
304,389; B 8 2 304,383; B 38 4 289,250; B 8 2 304,253; B 8 2 304,247; B 58 4 279,212; B 8 2
304,215; B 8 2 304,209; B 38 4 289,354; B 8 2 304,357; B 8 2 304,351; B 38 4 289,316; B 8 2
304,319; B 8 2 304,313; B 22 4 301,774; B 8 4 308,778; B 22 4 301,62; B 8 4 308,58; B 78 8 357,748;
B 78 4 357,616; B 78 4 357,578; B 78 4 357,512; B 78 8 357,88; B 78 4 357,432; B 78 8 357,472; B
78 4 357,394; B 78 4 357,220; B 78 4 357,258; B 78 4 357,324; B 78 4 357,362; B 16 10 326,623; B
16 10 326,519; B 16 10 326,213; B 16 10 326,317; B 8 2 350,713; B 8 2 350,707; B 8 2 350,731; B 8
2 350,725; B 8 2 350,771; B 8 2 350,765; B 8 2 350,807; B 8 2 350,801; B 8 2 350,637; B 8 2
350,631; B 8 2 350,599; B 8 2 350,593; B 8 2 350,533; B 8 2 350,527; B 8 2 350,495; B 8 2 350,489;
B 8 2 350,129; B 8 2 350,123; B 8 2 350,111; B 8 2 350,105; B 8 2 350,71; B 8 2 350,65; B 8 2
350,35; B 8 2 350,29; B 8 2 350,417; B 8 2 350,411; B 8 2 350,455; B 8 2 350,449; B 8 2 350,379; B 8
2 350,373; B 8 2 350,205; B 8 2 350,199; B 8 2 350,243; B 8 2 350,237; B 8 2 350,309; B 8 2
350,303; B 8 2 350,347; B 8 2 350,341; B 8 4 392,620; B 8 4 392,582; B 8 4 392,516; B 8 4 392,478;
B 8 4 392,428; B 8 4 392,466; B 8 4 392,390; B 8 4 392,216; B 8 4 392,254; B 8 4 392,320; B 8 4
392,358; B 10 2 123,21; B 8 12 122,28; B 10 2 123,15; B 8 2 122,13; B 8 14 122,45; B 58 2 197,51; B
8 2 172,57; B 8 12 172,44; B 8 14 172,27; B 8 2 172,15; B 108 2 172,53; B 58 2 147,55; B 58 4
147,18; B 78 4 157,36; B 8 14 172,91; B 8 2 172,79; B 58 4 147,82; B 8 4 122,78; B 8 14 122,91; B 8
2 120,125; B 60 4 146,122; B 8 2 172,125; B 8 14 172,113; B 80 8 156,102; B 8 14 120,113; B 10 2
123,215; B 8 2 122,217; B 10 2 123,209; B 8 12 122,202; B 36 4 136,240; B 8 2 150,243; B 8 2
150,237; B 10 2 123,243; B 8 12 122,250; B 10 2 123,237; B 8 2 122,235; B 62 4 149,212; B 8 4
176,208; B 10 2 123,319; B 8 2 122,321; B 10 2 123,313; B 8 12 122,306; B 62 4 149,316; B 8 4
176,312; B 8 2 150,347; B 8 2 150,341; B 10 2 123,347; B 8 12 122,354; B 10 2 123,341; B 8 2
122,339; B 36 4 136,344; B 8 4 122,372; B 8 14 122,385; B 54 4 145,376; B 8 4 168,380; B 8 2
168,373; B 10 2 123,417; B 8 12 122,424; B 10 2 123,411; B 8 2 122,409; B 54 4 145,414; B 8 2
168,411; B 8 4 168,418; B 8 2 150,455; B 8 2 150,449; B 10 2 123,455; B 8 12 122,462; B 10 2
123,449; B 8 2 122,447; B 36 4 136,452; B 8 2 150,495; B 8 2 150,489; B 10 2 123,495; B 10 2
123,489; B 8 12 122,482; B 36 4 136,492; B 8 2 122,497; B 10 2 123,523; B 10 2 123,517; B 8 2
122,515; B 8 12 122,530; B 62 4 149,520; B 8 4 176,524; B 8 2 150,599; B 8 2 150,593; B 10 2
123,599; B 10 2 123,593; B 8 12 122,586; B 8 2 122,601; B 36 4 136,596; B 10 2 123,627; B 10 2
123,621; B 8 2 122,619; B 8 12 122,634; B 62 4 149,624; B 8 4 176,628; B 8 2 120,711; B 8 14
120,723; B 80 8 156,734; B 60 4 146,714; B 8 14 172,723; B 8 2 172,711; B 10 2 123,757; B 10 2
123,751; B 8 12 122,744; B 58 4 147,754; B 8 2 172,757; B 8 14 172,745; B 8 4 122,778; B 8 14
122,791; B 8 2 172,821; B 8 14 172,809; B 58 2 197,785; B 8 12 172,792; B 8 2 172,779; B 108 2
172,783; B 58 2 147,781; B 78 4 157,800; B 8 2 122,759; B 10 2 123,821; B 10 2 123,815; B 8 12
122,808; B 58 4 147,818; B 8 2 122,823; B 8 2 320,35; B 80 8 356,12; B 8 14 320,23; B 60 4 346,32;
B 8 2 372,35; B 8 14 372,23; B 10 2 323,71; B 8 12 322,78; B 10 2 323,65; B 8 2 322,63; B 58 4

347,68; B 8 14 372,77; B 8 2 372,65; B 10 2 323,111; B 10 2 323,105; B 8 12 322,98; B 58 4 347,108;
B 8 2 372,111; B 8 14 372,99; B 38 4 335,126; B 8 2 320,129; B 8 2 320,123; B 30 4 339,202; B 8 4
328,206; B 8 2 328,199; B 36 4 336,240; B 8 4 322,236; B 8 14 322,249; B 30 4 339,306; B 8 4
328,310; B 8 2 328,303; B 36 4 336,344; B 8 4 322,340; B 8 14 322,353; B 8 4 322,372; B 8 14
322,385; B 56 4 346,376; B 8 4 370,372; B 8 14 370,385; B 36 4 336,414; B 8 4 322,410; B 8 14
322,423; B 36 4 336,452; B 8 4 322,448; B 8 14 322,461; B 36 4 336,492; B 8 4 322,496; B 8 14
322,483; B 28 4 340,530; B 8 4 330,534; B 8 4 330,526; B 36 4 336,596; B 8 4 322,600; B 8 14
322,587; B 28 4 340,634; B 8 4 330,638; B 8 4 330,630; B 36 4 336,710; B 8 4 322,706; B 4 2
320,713; B 8 12 322,738; B 10 2 323,731; B 10 2 323,725; B 56 4 346,728; B 8 4 370,724; B 8 14
370,737; B 8 2 322,773; B 8 12 322,758; B 10 2 323,771; B 10 2 323,765; B 56 4 346,768; B 8 4
370,772; B 8 14 370,759; B 8 2 320,801; B 80 8 356,824; B 8 14 320,813; B 58 4 345,804; B 8 4
370,800; B 8 14 370,813; B 8 4 122,58; L NM; B 8 836 20,418; B 8 836 34,418; B 8 836 48,418; B 8
242 62,689; B 8 242 62,147; B 8 12 76,782; B 8 80 76,712; B 8 66 76,617; B 8 66 76,513; B 8 12
76,54; B 8 82 76,423; B 8 80 76,124; B 8 66 76,219; B 8 66 76,323; B 8 836 90,418; B 12 8 106,808;
B 12 8 106,750; B 12 8 106,822; B 12 8 106,792; B 12 8 106,688; B 12 8 106,664; B 12 8 106,586; B
12 8 106,610; B 12 8 106,634; B 12 8 106,648; B 12 8 106,482; B 12 8 106,506; B 12 8 106,530; B 12
8 106,544; B 12 8 106,28; B 12 8 106,86; B 12 8 106,14; B 12 8 106,44; B 12 8 106,148; B 12 8
106,462; B 12 8 106,440; B 12 8 106,420; B 12 8 106,402; B 12 8 106,382; B 12 8 106,172; B 12 8
106,250; B 12 8 106,226; B 12 8 106,202; B 12 8 106,188; B 12 8 106,354; B 12 8 106,330; B 12 8
106,306; B 12 8 106,292; B 8 12 106,724; B 8 12 106,704; B 8 12 106,112; B 8 12 106,132; B 8 836
150,418; B 8 836 192,418; B 8 40 208,792; B 8 18 208,755; B 8 24 208,726; B 8 18 208,81; B 8 40
208,44; B 8 24 208,110; B 22 8 215,768; B 22 8 215,68; B 8 24 222,742; B 8 24 222,710; B 8 24
222,94; B 8 24 222,126; B 12 8 226,806; B 12 8 226,30; B 8 8 226,644; B 8 8 226,516; B 8 8 226,192;
B 8 8 226,320; B 6 120 227,580; B 6 120 227,256; B 22 8 229,782; B 8 32 236,762; B 8 24 236,710; B
22 8 229,54; B 8 32 236,74; B 8 24 236,126; B 8 8 240,626; B 8 8 240,410; B 6 208 239,518; B 8 8
240,210; B 6 192 239,310; B 24 8 258,796; B 8 40 250,734; B 24 8 258,40; B 8 40 250,102; B 8 8
252,664; B 8 8 252,172; B 8 8 254,626; B 6 30 253,645; B 8 8 254,372; B 6 246 253,499; B 8 8
254,210; B 6 30 253,191; B 6 154 253,291; B 12 8 272,586; B 12 8 272,522; B 12 8 272,482; B 12 8
272,462; B 12 8 272,424; B 12 8 272,386; B 12 8 272,250; B 12 8 272,314; B 12 8 272,354; B 34 8
263,782; B 8 66 276,745; B 34 8 263,54; B 8 66 276,91; B 8 836 290,418; B 12 8 306,758; B 12 8
306,728; B 12 8 306,586; B 12 8 306,624; B 12 8 306,482; B 12 8 306,520; B 12 8 306,78; B 12 8
306,108; B 12 8 306,462; B 12 8 306,424; B 12 8 306,386; B 12 8 306,250; B 12 8 306,212; B 12 8
306,354; B 12 8 306,316; B 8 12 308,778; B 8 12 308,58; B 8 836 350,418; B 8 836 392,418; B 12 8
126,18; B 8 12 172,16; B 8 12 172,56; B 8 12 172,80; B 8 12 122,78; B 8 12 120,124; B 10 8 121,134;
B 8 10 122,143; B 8 12 172,124; B 12 8 126,212; B 12 8 126,240; B 8 30 178,191; B 14 8 175,210; B
12 8 126,316; B 8 30 178,295; B 14 8 175,314; B 12 8 126,344; B 8 12 122,372; B 8 12 168,374; B 12
8 126,414; B 8 12 168,412; B 12 8 126,452; B 12 8 126,492; B 12 8 126,520; B 8 30 178,541; B 14 8
175,522; B 12 8 126,596; B 12 8 126,624; B 8 30 178,645; B 14 8 175,626; B 8 12 172,712; B 12 8
126,754; B 8 12 172,756; B 8 12 122,778; B 8 12 172,820; B 8 12 172,780; B 12 8 126,818; B 8 12
320,34; B 8 12 372,34; B 12 8 326,68; B 8 12 372,66; B 12 8 326,108; B 8 12 372,110; B 8 12
320,128; B 8 12 328,200; B 8 12 322,236; B 8 12 328,304; B 8 12 322,340; B 8 12 322,372; B 8 12
370,372; B 8 12 322,410; B 8 12 322,448; B 8 12 322,496; B 8 12 330,534; B 8 12 322,600; B 8 12
330,638; B 8 12 322,706; B 12 8 326,728; B 8 12 370,724; B 12 8 326,768; B 8 12 370,772; B 8 12
320,802; B 8 12 370,800; B 8 12 122,58; B 8 10 122,693; B 8 12 120,712; B 10 8 121,702; L NI; B 24
12 136,714; B 24 12 136,782; B 24 12 136,122; B 24 12 136,82; B 24 12 136,54; B 24 12 136,376; B
16 12 160,714; B 16 12 160,754; B 16 12 160,818; B 16 12 160,782; B 16 12 160,122; B 16 12
160,82; B 16 12 160,18; B 16 12 160,54; B 12 12 158,414; B 12 12 158,376; B 24 12 336,710; B 24
12 336,804; B 24 12 336,596; B 24 12 336,492; B 24 12 336,126; B 24 12 336,32; B 24 12 336,452; B
24 12 336,414; B 24 12 336,376; B 24 12 336,240; B 24 12 336,344; B 16 12 340,634; B 16 12
340,530; B 16 12 340,202; B 16 12 340,306; B 16 12 360,728; B 16 12 360,768; B 16 12 360,804; B
16 12 360,108; B 16 12 360,68; B 16 12 360,32; B 16 12 360,376; B 26 12 135,18; B 26 12 135,212;
B 26 12 135,240; B 24 12 164,212; B 26 12 135,316; B 26 12 165,316; B 26 12 135,344; B 26 12
135,414; B 26 12 135,452; B 26 12 135,492; B 26 12 135,520; B 26 12 165,520; B 26 12 135,596; B
26 12 135,624; B 26 12 165,624; B 26 12 135,754; B 26 12 135,818; B 26 12 335,68; B 26 12
335,108; B 26 12 335,728; B 26 12 335,768; L NC; B 4 4 20,662; B 4 4 20,652; B 4 4 20,548; B 4 4
20,174; B 4 4 20,184; B 4 4 20,288; B 4 4 34,690; B 4 4 34,146; B 4 4 48,822; B 4 4 48,14; B 4 4
62,806; B 4 4 62,572; B 4 4 62,30; B 4 4 62,264; B 4 8 76,782; B 4 4 76,748; B 4 4 76,676; B 4 4
76,588; B 4 4 76,598; B 4 4 76,646; B 4 4 76,484; B 4 4 76,494; B 4 4 76,542; B 4 8 76,54; B 4 4
76,88; B 4 4 76,386; B 4 4 76,460; B 4 4 76,160; B 4 4 76,248; B 4 4 76,238; B 4 4 76,190; B 4 4
76,352; B 4 4 76,342; B 4 4 76,294; B 4 4 90,702; B 4 4 90,740; B 4 4 90,800; B 4 4 90,594; B 4 4
90,620; B 4 4 90,490; B 4 4 90,516; B 4 4 90,134; B 4 4 90,96; B 4 4 90,36; B 4 4 90,454; B 4 4
90,430; B 4 4 90,392; B 4 4 90,242; B 4 4 90,216; B 4 4 90,346; B 4 4 90,320; B 8 4 106,808; B 8 4
106,750; B 8 4 106,822; B 8 4 106,792; B 8 4 106,688; B 8 4 106,28; B 8 4 106,86; B 8 4 106,14; B 8
4 106,44; B 8 4 106,148; B 4 8 106,724; B 4 8 106,704; B 4 8 106,112; B 4 8 106,132; B 8 4 106,664;
B 8 4 106,586; B 8 4 106,610; B 8 4 106,634; B 8 4 106,648; B 8 4 106,482; B 8 4 106,506; B 8 4
106,530; B 8 4 106,544; B 8 4 106,462; B 8 4 106,440; B 8 4 106,420; B 8 4 106,402; B 8 4 106,382;
B 8 4 106,172; B 8 4 106,250; B 8 4 106,226; B 8 4 106,202; B 8 4 106,188; B 8 4 106,354; B 8 4
106,330; B 8 4 106,306; B 8 4 106,292; B 4 4 122,692; B 4 4 122,144; B 4 4 150,714; B 4 4 150,754;
B 4 4 150,818; B 4 4 150,782; B 4 4 150,596; B 4 4 150,624; B 4 4 150,492; B 4 4 150,520; B 4 4
150,122; B 4 4 150,82; B 4 4 150,18; B 4 4 150,54; B 4 4 150,414; B 4 4 150,376; B 4 4 150,452; B 4
4 150,240; B 4 4 150,212; B 4 4 150,344; B 4 4 150,316; B 4 4 178,656; B 4 4 178,552; B 4 4
178,180; B 4 4 178,284; B 4 4 192,684; B 4 4 192,734; B 4 4 192,798; B 4 4 192,824; B 4 4 192,580;

B 4 4 192,642; B 4 4 192,476; B 4 4 192,538; B 4 4 192,152; B 4 4 192,102; B 4 4 192,38; B 4 4 192,12; B 4 4 192,430; B 4 4 192,468; B 4 4 192,392; B 4 4 192,256; B 4 4 192,194; B 4 4 192,360; B 4 4 192,298; B 4 4 208,768; B 4 4 208,808; B 4 4 208,734; B 4 4 208,718; B 4 4 208,750; B 4 4 208,68; B 4 4 208,28; B 4 4 208,102; B 4 4 208,118; B 4 4 208,86; B 4 4 222,768; B 4 4 222,782; B 4 4 222,750; B 4 4 222,734; B 4 4 222,718; B 4 4 222,702; B 4 4 222,68; B 4 4 222,54; B 4 4 222,86; B 4 4 222,102; B 4 4 222,118; B 4 4 222,134; B 8 4 226,806; B 8 4 226,30; B 4 4 226,644; B 4 4 226,516; B 4 4 226,192; B 4 4 226,320; B 4 4 236,766; B 4 4 236,718; B 4 4 236,750; B 4 4 236,702; B 4 4 236,70; B 4 4 236,118; B 4 4 236,86; B 4 4 236,134; B 4 4 240,626; B 4 4 240,410; B 4 4 240,210; B 4 4 250,782; B 4 4 250,796; B 4 4 250,718; B 4 4 250,750; B 4 4 250,54; B 4 4 250,40; B 4 4 250,118; B 4 4 250,86; B 4 4 252,664; B 4 4 252,172; B 4 4 254,626; B 4 4 254,372; B 4 4 254,210; B 4 4 266,796; B 4 4 266,40; B 8 4 272,586; B 8 4 272,522; B 8 4 272,482; B 8 4 272,462; B 8 4 272,424; B 8 4 272,386; B 8 4 272,250; B 8 4 272,314; B 8 4 272,354; B 4 4 276,716; B 4 4 276,120; B 4 4 290,738; B 4 4 290,594; B 4 4 290,632; B 4 4 290,490; B 4 4 290,528; B 4 4 290,98; B 4 4 290,454; B 4 4 290,416; B 4 4 290,378; B 4 4 290,242; B 4 4 290,204; B 4 4 290,346; B 4 4 290,308; B 8 4 306,758; B 8 4 306,728; B 8 4 306,78; B 8 4 306,108; B 4 8 308,778; B 8 4 306,586; B 8 4 306,624; B 8 4 306,482; B 8 4 306,520; B 4 8 308,58; B 8 4 306,462; B 8 4 306,424; B 8 4 306,386; B 8 4 306,250; B 8 4 306,212; B 8 4 306,354; B 8 4 306,316; B 4 4 350,710; B 4 4 350,728; B 4 4 350,768; B 4 4 350,804; B 4 4 350,634; B 4 4 350,596; B 4 4 350,530; B 4 4 350,492; B 4 4 350,126; B 4 4 350,108; B 4 4 350,68; B 4 4 350,32; B 4 4 350,414; B 4 4 350,452; B 4 4 350,376; B 4 4 350,202; B 4 4 350,240; B 4 4 350,306; B 4 4 350,344; B 4 4 392,748; B 4 4 392,824; B 4 4 392,618; B 4 4 392,580; B 4 4 392,514; B 4 4 392,476; B 4 4 392,88; B 4 4 392,12; B 4 4 392,430; B 4 4 392,468; B 4 4 392,392; B 4 4 392,218; B 4 4 392,256; B 4 4 392,322; B 4 4 392,360; B 8 4 126,18; B 4 8 122,58; B 4 8 172,16; B 4 8 172,56; B 4 8 172,80; B 4 8 122,78; B 4 8 120,124; B 4 8 172,124; B 8 4 126,212; B 8 4 126,240; B 8 4 174,210; B 8 4 126,316; B 8 4 174,314; B 8 4 126,344; B 4 8 122,372; B 4 8 168,374; B 8 4 126,414; B 4 8 168,412; B 8 4 126,452; B 8 4 126,492; B 8 4 126,520; B 8 4 174,522; B 8 4 126,596; B 8 4 126,624; B 8 4 174,626; B 4 8 120,712; B 4 8 172,712; B 8 4 126,754; B 4 8 172,756; B 4 8 122,778; B 4 8 172,820; B 4 8 172,780; B 8 4 126,818; B 4 8 320,34; B 4 8 372,34; B 8 4 326,68; B 4 8 372,66; B 8 4 326,108; B 4 8 372,110; B 4 8 320,128; B 4 8 328,200; B 4 8 322,236; B 4 8 328,304; B 4 8 322,340; B 4 8 322,372; B 4 8 370,372; B 4 8 322,410; B 4 8 322,448; B 4 8 322,496; B 4 8 330,534; B 4 8 322,600; B 4 8 330,638; B 4 8 322,706; B 8 4 326,728; B 4 8 370,724; B 8 4 326,768; B 4 8 370,772; B 4 8 320,802; B 4 8 370,800; DF;

TimingDelay

DS 106 125/1; L NP; B 4 72 26,36; B 4 16 12,8; B 8 8 12,20; B 8 8 6,48; B 4 104 4,104; B 8 8 28,76; B 8 4 96,88; B 8 8 72,90; B 8 8 34,90; B 18 4 101,92; B 4 14 72,101; B 4 14 34,101; B 8 6 104,109; B 8 6 82,109; B 8 6 62,109; B 8 6 44,109; B 8 6 24,109; B 4 244 80,234; B 4 206 42,215; B 8 4 66,114; B 8 4 28,114; B 4 14 68,123; B 4 18 106,121; B 4 14 30,123; B 4 38 106,163; B 20 4 60,142; B 4 34 68,161; B 20 4 22,142; B 4 38 30,163; B 22 4 13,158; B 12 4 94,158; B 14 4 93,162; B 4 30 90,191; B 8 4 104,184; B 4 80 102,226; B 8 4 28,184; B 4 80 26,226; B 8 4 66,200; B 4 64 68,234; B 4 12 64,192; B 4 38 14,217; B 8 8 90,210; B 8 8 14,240; B 4 40 56,266; B 8 6 104,269; B 8 6 66,269; B 8 6 28,269; B 8 4 96,288; B 8 4 58,288; B 8 4 20,288; B 18 4 101,292; B 18 4 63,292; B 18 4 25,292; B 8 6 104,309; B 8 6 66,309; B 8 6 28,309; B 4 18 68,321; B 4 18 106,321; B 4 18 30,321; B 4 2 12,327; B 14 4 55,330; B 4 2 50,327; B 12 8 56,336; B 14 4 17,330; B 12 8 18,336; B 20 4 60,342; B 4 38 68,363; B 20 4 22,342; B 4 38 30,363; B 4 38 106,363; B 22 4 89,358; B 14 4 17,358; B 4 30 12,385; B 4 2 12,361; B 14 4 55,358; B 4 30 50,385; B 4 2 50,361; B 8 4 28,384; B 4 14 26,393; B 8 2 18,127; B 12 12 18,134; B 8 2 56,127; B 12 12 56,134; B 6 8 13,168; B 4 4 12,162; B 18 4 57,184; B 14 10 55,161; B 4 16 50,174; B 6 8 89,172; B 4 4 88,166; B 8 2 94,127; B 20 4 98,142; B 12 12 94,134; B 8 2 94,327; B 20 4 98,342; B 12 12 94,334; B 12 4 46,320; B 6 4 49,324; B 6 10 87,365; B 6 8 11,322; B 6 8 11,366; B 6 8 49,366; L ND; B 8 8 82,34; B 4 62 82,69; B 8 8 102,76; B 4 20 104,90; B 8 8 104,104; B 8 4 82,106; B 28 4 72,102; B 8 4 62,106; B 8 4 44,106; B 28 4 34,102; B 8 4 24,106; B 4 14 74,181; B 4 40 74,146; B 4 12 112,182; B 4 62 36,157; B 8 8 56,150; B 8 8 94,150; B 8 8 18,150; B 8 8 72,192; B 8 8 110,192; B 8 8 34,192; B 8 8 8,192; B 4 10 74,201; B 16 4 14,204; B 4 6 8,199; B 12 28 14,220; B 20 8 66,210; B 4 36 62,232; B 8 8 50,240; B 16 4 56,252; B 4 6 50,247; B 8 8 104,274; B 8 8 66,274; B 8 8 28,274; B 4 22 104,289; B 4 22 66,289; B 4 22 28,289; B 8 8 104,304; B 8 8 66,304; B 8 8 28,304; B 4 10 74,383; B 4 36 74,344; B 4 10 112,383; B 4 36 112,344; B 4 10 36,383; B 4 36 36,344; B 4 20 56,336; B 4 20 18,336; B 8 8 56,350; B 8 8 18,350; B 8 8 94,350; B 4 8 94,358; B 4 8 18,358; B 4 8 56,358; B 14 8 107,374; B 14 8 31,374; B 14 8 69,374; B 8 8 72,392; B 8 8 110,392; B 8 8 34,392; B 4 18 18,137; B 8 2 18,127; B 24 8 26,122; B 4 18 56,137; B 8 2 56,127; B 24 8 64,122; B 4 30 20,187; B 8 8 18,168; B 4 10 18,159; B 8 2 58,165; B 4 10 56,159; B 22 8 65,170; B 4 14 94,161; B 4 42 112,147; B 24 8 102,172; B 4 18 94,137; B 8 2 94,127; B 24 8 102,122; B 4 18 94,337; B 8 2 94,327; B 24 8 102,322; B 26 8 63,322; B 26 8 101,366; B 26 8 63,366; B 26 8 25,366; B 26 8 25,322; L NM; B 114 8 57,20; B 114 8 57,34; B 114 8 57,48; B 82 8 65,76; B 114 8 57,90; B 8 12 104,106; B 8 12 82,106; B 8 12 62,106; B 8 12 44,106; B 8 12 24,106; B 114 8 57,150; B 114 8 57,192; B 38 8 75,210; B 44 8 32,240; B 8 12 104,272; B 8 12 66,272; B 8 12 28,272; B 114 8 57,290; B 8 12 104,306; B 114 8 57,350; B 114 8 57,392; B 12 8 14,366; B 12 8 14,322; B 8 12 66,306; B 12 8 52,366; B 12 8 90,366; B 12 8 52,322; B 8 12 94,326; B 12 8 92,172; B 8 12 94,126; B 12 8 16,168; B 8 12 58,166; B 8 12 56,126; B 8 12 18,126; B 8 12 28,306; L NI; B 12 24 18,136; B 12 16 56,160; B 12 24 56,336; B 12 24 18,336; B 12 12 94,358; B 12 12 18,358; B 12 12 56,358; B 12 24 56,136; B 12 12 18,158; B 12 16 94,160; B 12 24 94,136; B 12 24 94,336; L NC; B 4 4 12,20; B 4 4 82,34; B 4 4 6,48; B 4 4 28,76; B 4 4 102,76; B 4 4 96,90; B 4 4 72,90; B 4 4 34,90; B 4 8 104,106; B 4 8 82,106; B 4 8 62,106; B 4 8 44,106; B 4 8 24,106; B 4 4 56,150; B 4 4 94,150; B 4 4 18,150; B 4 4 72,192; B 4 4 110,192; B 4 4 34,192; B 4 4 8,192; B 4 4 90,210; B 4 4 60,210; B 4 4 14,240; B 4 4 50,240; B 4 8

104,272; B 4 8 66,272; B 4 8 28,272; B 4 4 96,290; B 4 4 58,290; B 4 4 20,290; B 4 8 104,306; B 4 8 66,306; B 4 8 28,306; B 4 4 56,350; B 4 4 18,350; B 4 4 94,350; B 4 4 72,392; B 4 4 110,392; B 4 4 34,392; B 4 8 18,126; B 4 8 56,126; B 8 4 16,168; B 4 8 58,166; B 8 4 92,172; B 4 8 94,126; B 4 8 94,326; B 8 4 52,322; B 8 4 90,366; B 8 4 14,322; B 8 4 14,366; B 8 4 52,366; DF;

Recoder

DS 107 125/1; L NP; B 10 4 5,334; B 8 4 4,338; B 4 14 8,325; B 8 8 18,394; B 8 8 18,52; B 8 2 18,251; B 8 2 18,245; B 8 8 18,122; B 8 8 18,316; B 8 8 18,354; B 4 10 20,403; B 4 10 20,61; B 4 10 20,131; B 4 10 20,113; B 18 4 27,138; B 4 10 20,325; B 4 10 20,363; B 22 4 25,248; B 6 4 37,400; B 6 4 37,58; B 6 4 37,260; B 6 4 37,236; B 6 4 37,112; B 6 2 37,131; B 6 2 37,125; B 6 4 37,150; B 6 4 37,320; B 6 4 37,334; B 6 4 37,358; B 6 4 37,372; B 24 4 46,116; B 276 4 172,256; B 24 4 46,128; B 10 4 39,240; B 10 2 39,323; B 10 2 39,361; B 24 4 46,404; B 24 4 46,62; B 18 4 49,244; B 24 2 46,325; B 18 2 49,327; B 24 2 46,363; B 18 2 49,365; B 16 10 64,393; B 16 10 64,51; B 16 10 64,233; B 16 10 64,315; B 38 4 87,404; B 4 4 70,400; B 46 4 91,62; B 4 4 70,58; B 4 4 70,132; B 46 4 91,128; B 40 4 88,326; B 4 4 70,322; B 4 4 70,360; B 8 10 88,393; B 8 10 88,315; B 8 10 88,353; B 60 2 154,305; B 66 4 137,322; B 174 4 121,338; B 8 4 134,356; B 8 4 148,350; B 8 4 162,356; B 112 4 124,224; B 8 2 176,227; B 8 2 176,221; B 8 4 176,350; B 8 4 190,142; B 160 4 114,376; B 8 4 190,372; B 8 4 190,214; B 8 4 190,318; B 46 4 209,218; B 46 4 209,322; B 92 4 156,58; B 6 2 199,61; B 6 2 199,55; B 92 4 156,132; B 6 4 199,128; B 8 6 204,395; B 4 2 202,391; B 8 4 204,158; B 8 4 204,342; B 8 4 204,372; B 8 4 204,202; B 8 4 204,306; B 254 4 161,154; B 32 4 216,206; B 32 4 216,310; B 8 8 218,394; B 8 8 218,236; B 8 8 218,52; B 8 8 218,136; B 8 8 218,366; B 8 8 218,176; B 8 8 218,280; B 4 10 220,403; B 4 10 220,245; B 4 10 220,61; B 4 10 220,127; B 4 10 220,357; B 4 10 220,185; B 4 10 220,167; B 4 10 220,289; B 4 10 220,271; B 6 4 237,400; B 6 4 237,58; B 6 2 237,131; B 6 2 237,125; B 6 4 237,362; B 6 2 237,245; B 6 2 237,239; B 6 4 237,166; B 6 2 237,185; B 6 2 237,179; B 6 2 237,349; B 6 2 237,343; B 6 4 237,270; B 6 2 237,289; B 6 2 237,283; B 24 4 246,358; B 24 4 246,404; B 24 4 246,128; B 24 4 246,242; B 24 4 246,170; B 24 4 246,182; B 24 4 246,346; B 24 4 246,274; B 6 4 243,210; B 6 4 243,314; B 24 4 246,62; B 74 4 277,214; B 74 4 277,318; B 16 10 264,51; B 16 10 264,369; B 60 4 298,384; B 46 4 291,62; B 4 4 270,58; B 38 4 287,128; B 4 4 270,132; B 8 10 288,139; B 4 4 286,150; B 16 10 292,229; B 16 10 292,193; B 16 10 292,333; B 16 10 292,297; B 16 4 306,182; B 80 4 274,286; B 8 4 306,260; B 22 2 317,241; B 22 2 317,345; B 18 4 319,58; B 14 2 53,43; B 26 2 59,45; B 8 2 50,41; B 18 8 63,140; B 16 2 64,145; B 16 2 64,135; B 14 2 53,225; B 26 2 59,227; B 8 2 50,223; B 14 2 53,307; B 26 2 59,309; B 8 2 50,305; B 44 2 106,309; B 96 2 136,307; B 8 2 98,305; B 18 8 63,352; B 16 2 64,357; B 16 2 64,347; B 8 2 100,349; B 92 2 134,345; B 96 2 132,347; B 60 4 136,360; B 42 4 89,364; B 14 2 53,385; B 26 2 59,387; B 8 2 50,383; B 120 2 144,387; B 8 2 98,383; B 18 8 263,392; B 16 2 264,397; B 16 2 264,387; B 60 2 230,377; B 8 2 250,379; B 72 2 236,375; B 26 2 315,343; B 26 2 297,339; B 14 2 303,341; B 18 8 263,298; B 16 2 264,303; B 16 2 264,293; B 26 2 297,303; B 14 2 303,305; B 8 2 306,307; B 18 8 263,230; B 16 2 264,235; B 16 2 264,225; B 26 2 315,239; B 26 2 297,235; B 14 2 303,237; B 26 2 297,199; B 14 2 303,201; B 8 2 306,203; B 18 8 263,194; B 16 2 264,199; B 16 2 264,189; B 18 4 293,146; B 8 2 298,149; B 86 2 229,145; B 74 2 223,147; B 14 2 253,43; B 26 2 259,45; B 8 2 250,41; B 100 2 154,389; B 20 2 98,385; B 18 10 263,139; B 18 8 263,334; B 16 2 264,339; B 16 2 264,329; L ND; B 124 8 62,298; B 4 20 2,312; B 8 2 4,117; B 8 2 4,111; B 16 4 8,328; B 4 2 14,331; B 4 6 14,367; B 32 4 20,402; B 8 2 32,405; B 8 2 32,399; B 32 4 20,60; B 8 2 32,63; B 8 2 32,57; B 8 8 32,258; B 4 12 30,248; B 8 8 32,238; B 36 4 18,114; B 8 2 32,117; B 8 2 32,111; B 8 8 32,128; B 8 8 32,152; B 4 16 30,140; B 36 4 18,324; B 8 4 32,320; B 24 4 24,334; B 8 4 32,338; B 32 4 20,362; B 8 4 32,358; B 24 4 24,372; B 8 4 32,376; B 78 4 85,410; B 78 4 85,68; B 78 4 85,250; B 78 4 85,122; B 8 16 50,112; B 78 4 85,332; B 78 4 85,370; B 8 2 78,395; B 8 2 78,389; B 8 2 78,53; B 8 2 78,47; B 8 2 78,235; B 8 2 78,229; B 8 2 78,143; B 8 2 78,137; B 8 2 78,317; B 8 2 78,311; B 8 2 78,355; B 8 2 78,349; B 8 4 120,406; B 8 4 120,64; B 8 4 120,246; B 8 4 120,126; B 8 4 120,328; B 8 4 120,366; B 8 24 134,306; B 8 24 148,322; B 8 4 148,172; B 8 2 148,247; B 8 2 148,241; B 8 24 162,322; B 8 4 176,188; B 8 24 176,306; B 8 4 176,268; B 8 2 176,361; B 8 2 176,355; B 4 4 178,292; B 8 4 190,344; B 8 4 204,398; B 8 4 204,56; B 8 4 204,132; B 4 8 226,218; B 4 8 226,206; B 4 8 226,322; B 4 8 226,310; B 20 4 234,224; B 20 4 234,200; B 20 4 234,304; B 36 4 218,402; B 8 2 232,405; B 8 2 232,399; B 36 4 218,60; B 8 2 232,63; B 8 2 232,57; B 36 4 218,128; B 8 2 232,131; B 8 2 232,125; B 64 4 204,358; B 8 4 232,362; B 92 4 190,244; B 8 4 232,240; B 92 4 190,168; B 8 2 232,171; B 8 2 232,165; B 64 4 204,184; B 8 4 232,180; B 50 4 211,348; B 8 4 232,344; B 64 4 204,272; B 8 2 232,275; B 8 2 232,269; B 60 4 206,288; B 8 4 232,284; B 18 4 233,212; B 8 2 238,215; B 8 2 238,209; B 18 4 233,316; B 8 2 238,319; B 8 2 238,313; B 78 4 285,68; B 78 4 285,122; B 8 16 250,166; B 8 16 250,270; B 8 2 278,395; B 8 2 278,389; B 8 2 278,53; B 8 2 278,47; B 8 2 278,143; B 8 2 278,137; B 8 2 278,373; B 8 2 278,367; B 8 2 278,197; B 8 2 278,191; B 8 2 278,233; B 8 2 278,227; B 8 2 278,301; B 8 2 278,295; B 8 2 278,337; B 8 2 278,331; B 78 4 285,410; B 8 4 320,406; B 8 4 320,64; B 8 4 320,126; B 78 4 285,248; B 8 2 320,251; B 8 2 320,245; B 22 4 313,208; B 8 4 320,212; B 78 4 285,176; B 8 2 320,179; B 8 2 320,173; B 78 4 285,352; B 8 2 320,355; B 8 2 320,349; B 22 4 313,312; B 8 4 320,316; B 78 4 285,280; B 8 2 320,283; B 8 2 320,277; B 36 4 64,50; B 8 4 50,46; B 8 14 50,59; B 36 4 64,140; B 8 2 50,145; B 8 12 50,130; B 10 2 51,143; B 10 2 51,137; B 36 4 64,232; B 8 4 50,228; B 8 14 50,241; B 8 4 50,310; B 8 14 50,323; B 56 4 74,314; B 8 4 98,310; B 8 14 98,323; B 8 12 50,362; B 8 2 50,347; B 10 2 51,355; B 10 2 51,349; B 58 4 75,352; B 8 2 100,349; B 8 14 100,361; B 8 4 50,388; B 8 14 50,401; B 56 4 74,392; B 8 4 98,388; B 8 14 98,401; B 36 4 264,392; B 8 12 250,402; B 8 2 250,387; B 10 2 251,395; B 10 2 251,389; B 36 4 264,370; B 8 4 250,374; B 8 14 250,361; B 8 4 306,338; B 8 18 306,323; B 8 12 250,288; B 16 2 248,301; B 10 2 251,295; B 70 4 275,298; B 8 4 306,302; B 8 14 306,289; B 8 12 250,240; B 10 2 251,233; B 16 2 248,227; B 70 4 275,230; B 8 4 306,234; B 8 18 306,219; B 8 4 306,198; B 8 14 306,185; B 8 12 250,184; B 16 2

248,197; B 10 2 251,191; B 70 4 275,194; B 8 4 298,144; B 8 14 298,131; B 8 10 250,129; B 58 4 275,140; B 10 4 251,136; B 36 4 264,50; B 8 4 250,46; B 8 14 250,59; B 20 4 234,328; B 70 4 275,334; B 10 2 251,337; B 8 12 250,344; B 16 2 248,331; L NM; B 8 456 4,228; B 8 472 18,236; B 12 8 34,402; B 12 8 34,60; B 12 8 34,258; B 12 8 34,238; B 12 8 34,114; B 12 8 34,128; B 12 8 34,152; B 12 8 34,322; B 12 8 34,336; B 12 8 34,360; B 12 8 34,374; B 8 472 78,236; B 8 472 120,236; B 8 52 134,336; B 22 8 141,298; B 8 16 148,310; B 8 128 148,230; B 8 26 148,339; B 8 16 162,310; B 8 36 162,344; B 22 8 169,298; B 8 52 176,336; B 8 92 176,228; B 8 8 190,144; B 8 36 190,360; B 8 8 190,216; B 6 64 189,180; B 6 96 189,268; B 8 8 190,320; B 12 8 202,58; B 12 8 202,130; B 8 12 204,398; B 8 8 204,156; B 8 42 204,357; B 8 8 204,204; B 6 40 203,180; B 6 96 203,256; B 8 8 204,308; B 8 472 218,236; B 12 8 234,402; B 12 8 234,60; B 12 8 234,128; B 12 8 234,360; B 12 8 234,242; B 12 8 234,168; B 12 8 234,182; B 12 8 234,346; B 12 8 234,272; B 12 8 234,286; B 12 8 240,212; B 12 8 240,316; B 8 472 278,236; B 8 472 320,236; B 8 12 50,46; B 14 8 53,140; B 8 2 50,145; B 8 32 50,120; B 8 12 50,228; B 8 12 50,310; B 8 12 98,310; B 12 8 54,352; B 8 12 100,350; B 8 12 50,388; B 8 12 98,388; B 12 8 254,392; B 8 12 250,374; B 8 12 306,338; B 14 8 253,298; B 8 32 250,278; B 8 54 306,281; B 12 8 254,230; B 8 12 306,234; B 8 12 306,198; B 14 8 253,194; B 8 32 250,174; B 8 12 298,144; B 12 8 254,138; B 8 12 250,46; B 12 8 254,334; L NI; B 24 12 64,392; B 24 12 64,50; B 24 12 64,232; B 24 12 64,314; B 16 12 88,392; B 16 12 88,314; B 16 12 88,352; B 24 12 264,50; B 24 12 264,370; B 16 12 288,140; B 24 12 292,230; B 24 12 292,194; B 24 12 292,334; B 24 12 292,298; B 26 12 63,140; B 26 12 63,352; B 26 12 263,392; B 26 12 263,298; B 26 12 263,230; B 26 12 263,194; B 26 12 263,140; B 26 12 263,334; L NC; B 4 4 4,114; B 4 4 4,336; B 4 4 18,394; B 4 4 18,52; B 4 4 18,248; B 4 4 18,122; B 4 4 18,316; B 4 4 18,354; B 8 4 34,114; B 8 4 34,128; B 8 4 34,152; B 8 4 34,402; B 8 4 34,60; B 8 4 34,258; B 8 4 34,238; B 8 4 34,322; B 8 4 34,336; B 8 4 34,360; B 8 4 34,374; B 4 4 50,108; B 4 4 78,392; B 4 4 78,50; B 4 4 78,232; B 4 4 78,140; B 4 4 78,314; B 4 4 78,352; B 4 4 120,408; B 4 4 120,298; B 4 4 120,66; B 4 4 120,248; B 4 4 120,124; B 4 4 120,330; B 4 4 120,368; B 4 4 134,314; B 4 4 134,298; B 4 4 134,358; B 4 4 148,314; B 4 4 148,330; B 4 4 148,170; B 4 4 148,348; B 4 4 148,244; B 4 4 162,314; B 4 4 162,330; B 4 4 162,358; B 4 4 176,186; B 4 4 176,224; B 4 4 176,314; B 4 4 176,298; B 4 4 176,348; B 4 4 176,270; B 4 4 176,358; B 4 4 190,144; B 4 4 190,374; B 4 4 190,346; B 4 4 190,216; B 4 4 190,320; B 8 4 202,58; B 8 4 202,130; B 4 8 204,398; B 4 4 204,156; B 4 4 204,340; B 4 4 204,374; B 4 4 204,204; B 4 4 204,308; B 4 4 218,394; B 4 4 218,236; B 4 4 218,52; B 4 4 218,136; B 4 4 218,366; B 4 4 218,176; B 4 4 218,280; B 8 4 234,128; B 8 4 234,360; B 8 4 234,242; B 8 4 234,168; B 8 4 234,182; B 8 4 234,346; B 8 4 234,272; B 8 4 234,286; B 8 4 234,402; B 8 4 234,60; B 8 4 240,212; B 8 4 240,316; B 4 4 250,162; B 4 4 250,266; B 4 4 278,392; B 4 4 278,50; B 4 4 278,140; B 4 4 278,370; B 4 4 278,194; B 4 4 278,230; B 4 4 278,298; B 4 4 278,334; B 4 4 306,258; B 4 4 320,408; B 4 4 320,66; B 4 4 320,124; B 4 4 320,248; B 4 4 320,210; B 4 4 320,176; B 4 4 320,352; B 4 4 320,314; B 4 4 320,280; B 4 8 50,46; B 8 4 54,140; B 4 8 50,228; B 4 8 50,310; B 4 8 98,310; B 8 4 54,352; B 4 8 100,350; B 4 8 50,388; B 4 8 98,388; B 8 4 254,392; B 4 8 250,374; B 4 8 306,338; B 8 4 254,298; B 4 8 306,302; B 8 4 254,230; B 4 8 306,234; B 4 8 306,198; B 8 4 254,194; B 4 8 298,144; B 8 4 254,138; B 4 8 250,46; B 8 4 254,334; DF;

DblRecoder

DS 108 125/1; L NP; B 8 4 4,498; B 8 4 4,338; B 10 4 5,502; B 4 14 8,511; B 10 4 5,334; B 4 14 8,325; B 8 8 18,784; B 8 2 18,591; B 8 2 18,585; B 8 8 18,714; B 8 8 18,520; B 8 8 18,482; B 8 8 18,394; B 8 8 18,52; B 8 2 18,251; B 8 2 18,245; B 8 8 18,122; B 8 8 18,316; B 8 8 18,354; B 4 10 20,775; B 4 10 20,723; B 4 10 20,705; B 18 4 27,698; B 4 10 20,511; B 4 10 20,473; B 4 10 20,403; B 4 10 20,61; B 4 10 20,131; B 4 10 20,113; B 18 4 27,138; B 4 10 20,325; B 4 10 20,363; B 22 4 25,588; B 22 4 25,248; B 6 4 37,778; B 6 4 37,576; B 6 4 37,600; B 6 4 37,724; B 6 2 37,711; B 6 2 37,705; B 6 4 37,686; B 6 4 37,516; B 6 4 37,502; B 6 4 37,478; B 6 4 37,464; B 6 4 37,400; B 6 4 37,58; B 6 4 37,260; B 6 4 37,236; B 6 4 37,112; B 6 2 37,131; B 6 2 37,125; B 6 4 37,150; B 6 4 37,320; B 6 4 37,334; B 6 4 37,358; B 6 4 37,372; B 24 4 46,720; B 24 4 46,116; B 276 4 172,580; B 24 4 46,708; B 276 4 172,256; B 24 4 46,128; B 10 4 39,596; B 10 2 39,513; B 10 2 39,475; B 10 4 39,240; B 10 2 39,323; B 10 2 39,361; B 24 4 46,774; B 18 4 49,592; B 24 2 46,511; B 18 2 49,509; B 24 2 46,473; B 18 2 49,471; B 24 4 46,404; B 24 4 46,62; B 18 4 49,244; B 24 2 46,325; B 18 2 49,327; B 24 2 46,363; B 18 2 49,365; B 16 10 64,785; B 16 10 64,603; B 16 10 64,521; B 16 10 64,393; B 16 10 64,51; B 16 10 64,233; B 16 10 64,315; B 46 4 91,774; B 4 4 70,778; B 4 4 70,704; B 46 4 91,708; B 40 4 88,510; B 4 4 70,514; B 4 4 70,476; B 38 4 87,404; B 4 4 70,400; B 46 4 91,62; B 4 4 70,58; B 4 4 70,132; B 46 4 91,128; B 40 4 88,326; B 4 4 70,322; B 4 4 70,360; B 8 10 88,521; B 8 10 88,483; B 8 10 88,393; B 8 10 88,315; B 8 10 88,353; B 60 2 154,531; B 60 2 154,305; B 66 4 137,514; B 174 4 121,498; B 66 4 137,322; B 174 4 121,338; B 8 4 134,480; B 8 4 134,356; B 8 4 148,486; B 8 4 148,350; B 8 4 162,480; B 8 4 162,356; B 112 4 124,612; B 8 2 176,615; B 8 2 176,609; B 8 4 176,486; B 112 4 124,224; B 8 2 176,227; B 8 2 176,221; B 8 4 176,350; B 8 4 190,694; B 160 4 114,460; B 8 4 190,464; B 8 4 190,622; B 8 4 190,518; B 8 4 190,142; B 160 4 114,376; B 8 4 190,372; B 8 4 190,214; B 8 4 190,318; B 46 4 209,618; B 46 4 209,514; B 46 4 209,218; B 46 4 209,322; B 92 4 156,778; B 6 2 199,781; B 6 2 199,775; B 92 4 156,704; B 6 4 199,708; B 92 4 156,58; B 6 2 199,61; B 6 2 199,55; B 92 4 156,132; B 6 4 199,128; B 8 4 204,678; B 8 4 204,494; B 8 4 204,464; B 8 4 204,634; B 8 4 204,530; B 8 6 204,395; B 4 2 202,391; B 8 4 204,158; B 8 4 204,342; B 8 4 204,372; B 8 4 204,202; B 8 4 204,306; B 254 4 161,682; B 32 4 216,630; B 32 4 216,526; B 254 4 161,154; B 32 4 216,206; B 32 4 216,310; B 8 8 218,600; B 8 8 218,784; B 8 8 218,700; B 8 8 218,470; B 8 8 218,660; B 8 8 218,556; B 8 8 218,394; B 8 8 218,236; B 8 8 218,52; B 8 8 218,136; B 8 8 218,366; B 8 8 218,176; B 8 8 218,280; B 4 10 220,591; B 4 10 220,775; B 4 10 220,709; B 4 10 220,479; B 4 10 220,669; B 4 10 220,651; B 4 10 220,565; B 4 10 220,547; B 4 10 220,403; B 4 10 220,245; B 4 10 220,61; B 4 10 220,127; B 4 10 220,357; B 4 10 220,185; B 4 10

```
220,167; B 4 10 220,289; B 4 10 220,271; B 6 4 237,778; B 6 2 237,711; B 6 2 237,705; B 6 4
237,474; B 6 2 237,597; B 6 2 237,591; B 6 4 237,670; B 6 2 237,657; B 6 2 237,651; B 6 2 237,493;
B 6 2 237,487; B 6 4 237,566; B 6 2 237,553; B 6 2 237,547; B 6 4 237,400; B 6 4 237,58; B 6 2
237,131; B 6 2 237,125; B 6 4 237,362; B 6 2 237,245; B 6 2 237,239; B 6 4 237,166; B 6 2 237,185;
B 6 2 237,179; B 6 2 237,349; B 6 2 237,343; B 6 4 237,270; B 6 2 237,289; B 6 2 237,283; B 24 4
246,478; B 24 4 246,358; B 24 4 246,708; B 24 4 246,594; B 24 4 246,666; B 24 4 246,654; B 24 4
246,490; B 24 4 246,562; B 24 4 246,404; B 24 4 246,128; B 24 4 246,242; B 24 4 246,170; B 24 4
246,182; B 24 4 246,346; B 24 4 246,274; B 6 4 243,626; B 6 4 243,522; B 6 4 243,210; B 6 4
243,314; B 24 4 246,774; B 24 4 246,62; B 74 4 277,622; B 74 4 277,518; B 74 4 277,214; B 74 4
277,318; B 16 10 264,785; B 16 10 264,467; B 16 10 264,393; B 16 10 264,51; B 46 4 291,774; B 4 4
270,778; B 38 4 287,708; B 4 4 270,704; B 60 4 298,384; B 46 4 291,62; B 4 4 270,58; B 38 4
287,128; B 4 4 270,132; B 8 10 288,697; B 4 4 286,686; B 16 10 292,607; B 16 10 292,643; B 16 10
292,503; B 16 10 292,539; B 8 10 288,139; B 4 4 286,150; B 16 10 292,229; B 16 10 292,193; B 16
10 292,333; B 16 10 292,297; B 16 4 306,654; B 80 4 274,550; B 16 4 306,182; B 80 4 274,286; B 8 4
306,576; B 8 4 306,260; B 22 2 317,595; B 22 2 317,491; B 22 2 317,241; B 22 2 317,345; B 18 4
319,778; B 18 4 319,58; B 14 2 53,43; B 26 2 59,45; B 8 2 50,41; B 18 8 63,140; B 16 2 64,145; B 16
2 64,135; B 14 2 53,225; B 26 2 59,227; B 8 2 50,223; B 14 2 53,307; B 26 2 59,309; B 8 2 50,305; B
44 2 106,309; B 96 2 136,307; B 8 2 98,305; B 16 2 64,357; B 16 2 64,347; B 18 8 63,352; B 8 2
100,349; B 92 2 134,345; B 96 2 132,347; B 60 4 136,360; B 42 4 89,364; B 14 2 53,385; B 26 2
59,387; B 8 2 50,383; B 120 2 144,387; B 8 2 98,383; B 18 8 63,484; B 16 2 64,489; B 16 2 64,479; B
8 2 100,487; B 92 2 134,491; B 96 2 132,489; B 60 4 136,476; B 14 2 53,529; B 8 2 50,531; B 26 2
59,527; B 44 2 106,527; B 96 2 136,529; B 8 2 98,531; B 14 2 53,611; B 8 2 50,613; B 26 2 59,609; B
18 8 63,696; B 16 2 64,701; B 16 2 64,691; B 14 2 53,793; B 8 2 50,795; B 26 2 59,791; B 14 2
253,793; B 8 2 250,795; B 26 2 259,791; B 74 2 223,689; B 86 2 229,691; B 18 4 293,690; B 8 2
298,687; B 18 8 263,642; B 16 2 264,647; B 16 2 264,637; B 26 2 297,637; B 14 2 303,635; B 8 2
306,633; B 26 2 315,597; B 26 2 297,601; B 14 2 303,599; B 18 8 263,538; B 16 2 264,543; B 16 2
264,533; B 26 2 297,533; B 14 2 303,531; B 8 2 306,529; B 26 2 315,493; B 26 2 297,497; B 14 2
303,495; B 60 2 230,459; B 72 2 236,461; B 8 2 250,457; B 14 2 253,385; B 26 2 259,387; B 8 2
250,383; B 72 2 236,375; B 60 2 230,377; B 16 2 264,339; B 16 2 264,329; B 26 2 315,343; B 26 2
297,339; B 14 2 303,341; B 18 8 263,298; B 16 2 264,303; B 16 2 264,293; B 26 2 297,303; B 14 2
303,305; B 8 2 306,307; B 18 8 263,230; B 16 2 264,235; B 16 2 264,225; B 26 2 315,239; B 26 2
297,235; B 14 2 303,237; B 18 8 263,194; B 16 2 264,199; B 16 2 264,189; B 26 2 297,199; B 14 2
303,201; B 8 2 306,203; B 74 2 223,147; B 86 2 229,145; B 18 4 293,146; B 8 2 298,149; B 14 2
253,43; B 26 2 259,45; B 8 2 250,41; B 100 2 154,389; B 20 2 98,385; B 42 4 89,472; B 18 8 263,334;
B 18 10 263,139; B 18 10 263,369; B 18 10 263,697; B 16 2 264,507; B 16 2 264,497; B 18 8
263,502; B 18 8 263,606; B 16 2 264,611; B 16 2 264,601; L ND; B 124 8 62,538; B 4 20 2,524; B 8 2
4,725; B 8 2 4,719; B 124 8 62,298; B 4 20 2,312; B 8 2 4,117; B 8 2 4,111; B 16 4 8,508; B 4 2
14,505; B 4 6 14,469; B 16 4 8,328; B 4 2 14,331; B 4 6 14,367; B 32 4 20,776; B 8 2 32,779; B 8 2
32,773; B 8 8 32,578; B 4 12 30,588; B 8 8 32,598; B 36 4 18,722; B 8 2 32,725; B 8 2 32,719; B 8 8
32,708; B 8 8 32,684; B 4 16 30,696; B 36 4 18,512; B 8 4 32,516; B 24 4 24,502; B 8 4 32,498; B 32
4 20,474; B 8 4 32,478; B 24 4 24,464; B 8 4 32,460; B 32 4 20,402; B 8 2 32,405; B 8 2 32,399; B 32
4 20,60; B 8 2 32,63; B 8 2 32,57; B 8 8 32,258; B 4 12 30,248; B 8 8 32,238; B 36 4 18,114; B 8 2
32,117; B 8 2 32,111; B 8 8 32,128; B 8 8 32,152; B 4 16 30,140; B 36 4 18,324; B 8 4 32,320; B 24 4
24,334; B 8 4 32,338; B 32 4 20,362; B 8 4 32,358; B 24 4 24,372; B 8 4 32,376; B 78 4 85,768; B 78
4 85,586; B 78 4 85,714; B 8 16 50,724; B 78 4 85,504; B 78 4 85,466; B 78 4 85,410; B 78 4 85,68; B
78 4 85,250; B 78 4 85,122; B 8 16 50,112; B 78 4 85,332; B 78 4 85,370; B 8 2 78,789; B 8 2 78,783;
B 8 2 78,607; B 8 2 78,601; B 8 2 78,699; B 8 2 78,693; B 8 2 78,525; B 8 2 78,519; B 8 2 78,487; B 8
2 78,481; B 8 2 78,395; B 8 2 78,389; B 8 2 78,53; B 8 2 78,47; B 8 2 78,235; B 8 2 78,229; B 8 2
78,143; B 8 2 78,137; B 8 2 78,317; B 8 2 78,311; B 8 2 78,355; B 8 2 78,349; B 8 4 120,772; B 8 4
120,590; B 8 4 120,710; B 8 4 120,508; B 8 4 120,470; B 8 4 120,406; B 8 4 120,64; B 8 4 120,246; B
8 4 120,126; B 8 4 120,328; B 8 4 120,366; B 8 24 134,530; B 8 24 134,306; B 8 24 148,514; B 8 4
148,664; B 8 2 148,595; B 8 2 148,589; B 8 24 148,322; B 8 4 148,172; B 8 2 148,247; B 8 2 148,241;
B 8 24 162,514; B 8 24 162,322; B 8 4 176,648; B 8 24 176,530; B 8 4 176,568; B 8 2 176,481; B 8 2
176,475; B 8 4 176,188; B 8 24 176,306; B 8 4 176,268; B 8 2 176,361; B 8 2 176,355; B 4 4 178,544;
B 4 4 178,292; B 8 4 190,492; B 8 4 190,344; B 8 4 204,780; B 8 4 204,704; B 8 4 204,398; B 8 4
204,56; B 8 4 204,132; B 4 8 226,630; B 4 8 226,618; B 4 8 226,526; B 4 8 226,514; B 4 8 226,218; B
4 8 226,206; B 4 8 226,322; B 4 8 226,310; B 20 2 234,613; B 20 4 234,636; B 20 4 234,532; B 20 4
234,224; B 20 4 234,200; B 20 4 234,328; B 20 4 234,304; B 36 4 218,776; B 8 2 232,779; B 8 2
232,773; B 36 4 218,708; B 8 2 232,711; B 8 2 232,705; B 64 4 204,478; B 8 4 232,474; B 92 4
190,592; B 8 4 232,596; B 92 4 190,668; B 8 2 232,671; B 8 2 232,665; B 64 4 204,652; B 8 4
232,656; B 50 4 211,488; B 8 4 232,492; B 64 4 204,564; B 8 2 232,567; B 8 2 232,561; B 60 4
206,548; B 8 4 232,552; B 36 4 218,402; B 8 2 232,405; B 8 2 232,399; B 36 4 218,60; B 8 2 232,63;
B 8 2 232,57; B 36 4 218,128; B 8 2 232,131; B 8 2 232,125; B 64 4 204,358; B 8 4 232,362; B 92 4
190,244; B 8 4 232,240; B 92 4 190,168; B 8 2 232,171; B 8 2 232,165; B 64 4 204,184; B 8 4
232,180; B 50 4 211,348; B 8 4 232,344; B 64 4 204,272; B 8 2 232,275; B 8 2 232,269; B 60 4
206,288; B 8 4 232,284; B 18 4 233,624; B 8 2 238,627; B 8 2 238,621; B 18 4 233,520; B 8 2
238,523; B 8 2 238,517; B 18 4 233,212; B 8 2 238,215; B 8 2 238,209; B 18 4 233,316; B 8 2
238,319; B 8 2 238,313; B 78 4 285,768; B 78 4 285,714; B 8 16 250,670; B 8 16 250,566; B 78 4
285,68; B 78 4 285,122; B 8 16 250,166; B 8 16 250,270; B 8 2 278,789; B 8 2 278,783; B 8 2
278,699; B 8 2 278,693; B 8 2 278,469; B 8 2 278,463; B 8 2 278,645; B 8 2 278,639; B 8 2 278,609;
B 8 2 278,603; B 8 2 278,541; B 8 2 278,535; B 8 2 278,505; B 8 2 278,499; B 8 2 278,395; B 8 2
278,389; B 8 2 278,53; B 8 2 278,47; B 8 2 278,143; B 8 2 278,137; B 8 2 278,197; B 8 2 278,191; B 8
```

2 278,233; B 8 2 278,227; B 8 2 278,301; B 8 2 278,295; B 8 2 278,337; B 8 2 278,331; B 8 4 320,772; B 8 4 320,710; B 78 4 285,588; B 8 2 320,591; B 8 2 320,585; B 22 4 313,628; B 8 4 320,624; B 78 4 285,660; B 8 2 320,663; B 8 2 320,657; B 8 2 320,487; B 8 2 320,481; B 22 4 313,524; B 8 4 320,520; B 78 4 285,556; B 8 2 320,559; B 8 2 320,553; B 78 4 285,410; B 8 4 320,406; B 8 4 320,64; B 8 4 320,126; B 78 4 285,248; B 8 2 320,251; B 8 2 320,245; B 22 4 313,208; B 8 4 320,212; B 78 4 285,176; B 8 2 320,179; B 8 2 320,173; B 78 4 285,352; B 8 2 320,355; B 8 2 320,349; B 22 4 313,312; B 8 4 320,316; B 78 4 285,280; B 8 2 320,283; B 8 2 320,277; B 36 4 64,50; B 8 4 50,46; B 8 14 50,59; B 36 4 64,140; B 8 2 50,145; B 8 12 50,130; B 10 2 51,143; B 10 2 51,137; B 36 4 64,232; B 8 4 50,228; B 8 14 50,241; B 8 4 50,310; B 8 14 50,323; B 56 4 74,314; B 8 4 98,310; B 8 14 98,323; B 8 12 50,362; B 8 2 50,347; B 10 2 51,355; B 10 2 51,349; B 58 4 75,352; B 8 14 100,361; B 8 2 100,349; B 8 4 50,388; B 8 14 50,401; B 56 4 74,392; B 8 4 98,388; B 8 14 98,401; B 8 12 50,474; B 10 2 51,487; B 10 2 51,481; B 8 2 50,489; B 58 4 75,484; B 8 2 100,487; B 8 14 100,475; B 8 4 50,526; B 8 14 50,513; B 56 4 74,522; B 8 4 98,526; B 8 14 98,513; B 36 4 64,604; B 8 4 50,608; B 8 14 50,595; B 36 4 64,696; B 8 12 50,706; B 8 2 50,691; B 10 2 51,699; B 10 2 51,693; B 36 4 64,786; B 8 4 50,790; B 8 14 50,777; B 36 4 264,786; B 8 4 250,790; B 8 14 250,777; B 8 10 250,707; B 10 4 251,700; B 58 4 275,696; B 8 4 298,692; B 8 14 298,705; B 10 2 251,645; B 16 2 248,639; B 8 12 250,652; B 70 4 275,642; B 8 4 306,638; B 8 14 306,651; B 8 4 306,602; B 8 18 306,617; B 10 2 251,541; B 16 2 248,535; B 8 12 250,548; B 70 4 275,538; B 8 4 306,534; B 8 14 306,547; B 8 4 306,498; B 8 18 306,513; B 36 4 264,466; B 8 4 250,462; B 8 14 250,475; B 36 4 264,392; B 8 4 250,388; B 8 14 250,401; B 8 2 278,373; B 8 2 278,367; B 8 10 250,359; B 36 4 264,370; B 10 4 251,366; B 8 12 250,344; B 8 4 306,338; B 8 18 306,323; B 8 12 250,288; B 16 2 248,301; B 10 2 251,295; B 70 4 275,298; B 8 4 306,302; B 8 14 306,289; B 8 12 250,240; B 10 2 251,233; B 16 2 248,227; B 70 4 275,230; B 8 4 306,234; B 8 18 306,219; B 8 12 250,184; B 16 2 248,197; B 10 2 251,191; B 70 4 275,194; B 8 4 306,198; B 8 14 306,185; B 10 4 251,136; B 8 10 250,129; B 58 4 275,140; B 8 4 298,144; B 8 14 298,131; B 36 4 264,50; B 8 4 250,46; B 8 14 250,59; B 70 4 275,334; B 16 2 248,331; B 10 2 251,337; B 18 4 233,508; B 80 4 284,484; B 72 4 274,502; B 18 2 247,505; B 12 2 250,499; B 8 12 248,492; B 70 4 275,606; B 16 2 248,609; B 30 2 239,611; B 10 2 251,603; B 8 12 250,596; L NM; B 8 836 4,418; B 8 836 18,418; B 12 8 34,776; B 12 8 34,578; B 12 8 34,598; B 12 8 34,722; B 12 8 34,708; B 12 8 34,684; B 12 8 34,514; B 12 8 34,500; B 12 8 34,476; B 12 8 34,462; B 12 8 34,402; B 12 8 34,60; B 12 8 34,258; B 12 8 34,238; B 12 8 34,114; B 12 8 34,128; B 12 8 34,152; B 12 8 34,322; B 12 8 34,336; B 12 8 34,360; B 12 8 34,374; B 8 836 78,418; B 8 836 120,418; B 8 52 134,500; B 8 52 134,336; B 22 8 141,538; B 8 128 148,606; B 8 16 148,526; B 8 26 148,497; B 22 8 141,298; B 8 16 148,310; B 8 128 148,230; B 8 26 148,339; B 8 16 162,526; B 8 36 162,492; B 8 16 162,310; B 8 36 162,344; B 22 8 169,538; B 22 8 169,298; B 8 52 176,500; B 8 92 176,608; B 8 52 176,336; B 8 92 176,228; B 8 8 190,692; B 8 36 190,476; B 8 8 190,620; B 6 64 189,656; B 6 96 189,568; B 8 8 190,516; B 8 8 190,144; B 8 36 190,360; B 8 8 190,216; B 6 64 189,180; B 6 96 189,268; B 8 8 190,320; B 12 8 202,778; B 12 8 202,706; B 12 8 202,58; B 12 8 202,130; B 8 8 204,680; B 8 42 204,479; B 8 8 204,632; B 6 40 203,656; B 6 96 203,580; B 8 8 204,528; B 8 12 204,398; B 8 8 204,156; B 8 42 204,357; B 8 8 204,204; B 6 40 203,180; B 6 96 203,256; B 8 8 204,308; B 8 836 218,418; B 12 8 234,776; B 12 8 234,708; B 12 8 234,476; B 12 8 234,594; B 12 8 234,668; B 12 8 234,654; B 12 8 234,490; B 12 8 234,564; B 12 8 234,550; B 12 8 234,402; B 12 8 234,60; B 12 8 234,128; B 12 8 234,360; B 12 8 234,242; B 12 8 234,168; B 12 8 234,182; B 12 8 234,346; B 12 8 234,272; B 12 8 234,286; B 12 8 240,624; B 12 8 240,520; B 12 8 240,212; B 12 8 240,316; B 8 836 278,418; B 8 836 320,418; B 8 12 50,46; B 14 8 53,140; B 8 2 50,145; B 8 32 50,120; B 8 12 50,228; B 8 12 50,310; B 8 12 98,310; B 12 8 54,352; B 8 12 100,350; B 8 12 50,388; B 8 12 98,388; B 12 8 54,484; B 8 12 100,486; B 8 12 50,526; B 8 12 98,526; B 8 12 50,608; B 14 8 53,696; B 8 32 50,716; B 8 2 50,691; B 8 12 50,790; B 8 12 250,790; B 12 8 254,698; B 8 12 298,692; B 14 8 253,642; B 8 32 250,662; B 8 12 306,638; B 8 12 306,602; B 14 8 253,538; B 8 32 250,558; B 8 54 306,555; B 8 12 306,498; B 8 12 250,462; B 8 12 250,388; B 12 8 254,368; B 8 12 306,338; B 14 8 253,298; B 8 32 250,278; B 8 54 306,281; B 12 8 254,230; B 8 12 306,234; B 14 8 253,194; B 8 32 250,174; B 8 12 306,198; B 8 12 298,144; B 8 12 250,46; B 12 8 254,138; B 12 8 254,334; B 12 8 254,502; B 12 8 254,606; L NI; B 24 12 64,786; B 24 12 64,604; B 24 12 64,522; B 24 12 64,392; B 24 12 64,50; B 24 12 64,232; B 24 12 64,314; B 16 12 88,522; B 16 12 88,484; B 16 12 88,392; B 16 12 88,314; B 16 12 88,352; B 24 12 264,786; B 24 12 264,466; B 24 12 264,392; B 24 12 264,50; B 24 12 264,194; B 16 12 288,696; B 24 12 292,606; B 24 12 292,642; B 24 12 292,502; B 24 12 292,538; B 16 12 288,140; B 24 12 292,230; B 24 12 292,194; B 24 12 292,334; B 24 12 292,298; B 26 12 63,140; B 26 12 63,352; B 26 12 63,484; B 26 12 63,696; B 26 12 263,696; B 26 12 263,642; B 26 12 263,538; B 26 12 263,370; B 26 12 263,334; B 26 12 263,298; B 26 12 263,230; B 26 12 263,140; B 26 12 263,502; B 26 12 263,606; L NC; B 4 4 4,722; B 4 4 4,500; B 4 4 4,114; B 4 4 4,336; B 4 4 18,784; B 4 4 18,588; B 4 4 18,714; B 4 4 18,520; B 4 4 18,482; B 4 4 18,394; B 4 4 18,52; B 4 4 18,248; B 4 4 18,122; B 4 4 18,316; B 4 4 18,354; B 8 4 34,722; B 8 4 34,708; B 8 4 34,684; B 8 4 34,114; B 8 4 34,128; B 8 4 34,152; B 8 4 34,776; B 8 4 34,578; B 8 4 34,598; B 8 4 34,514; B 8 4 34,500; B 8 4 34,476; B 8 4 34,462; B 8 4 34,402; B 8 4 34,60; B 8 4 34,258; B 8 4 34,238; B 8 4 34,322; B 8 4 34,336; B 8 4 34,360; B 8 4 34,374; B 4 4 50,728; B 4 4 50,108; B 4 4 78,786; B 4 4 78,604; B 4 4 78,696; B 4 4 78,522; B 4 4 78,484; B 4 4 78,392; B 4 4 78,50; B 4 4 78,232; B 4 4 78,140; B 4 4 78,314; B 4 4 78,352; B 4 4 120,538; B 4 4 120,770; B 4 4 120,588; B 4 4 120,712; B 4 4 120,506; B 4 4 120,468; B 4 4 120,408; B 4 4 120,298; B 4 4 120,66; B 4 4 120,248; B 4 4 120,124; B 4 4 120,330; B 4 4 120,368; B 4 4 134,522; B 4 4 134,538; B 4 4 134,478; B 4 4 134,314; B 4 4 134,298; B 4 4 134,358; B 4 4 148,522; B 4 4 148,506; B 4 4 148,666; B 4 4 148,488; B 4 4 148,592; B 4 4 148,314; B 4 4 148,330; B 4 4 148,170; B 4 4 148,348; B 4 4 148,244; B 4 4 162,522; B 4 4 162,506; B 4 4 162,478; B 4 4 162,314; B 4 4 162,330; B 4 4 162,358; B 4 4 176,650; B 4 4 176,612; B 4 4 176,522; B 4 4 176,538; B 4 4 176,488; B 4 4 176,566; B 4 4 176,478; B 4 4

176,186; B 4 4 176,224; B 4 4 176,314; B 4 4 176,298; B 4 4 176,348; B 4 4 176,270; B 4 4 176,358; B 4 4 190,692; B 4 4 190,462; B 4 4 190,490; B 4 4 190,620; B 4 4 190,516; B 4 4 190,144; B 4 4 190,374; B 4 4 190,346; B 4 4 190,216; B 4 4 190,320; B 8 4 202,778; B 8 4 202,706; B 8 4 202,58; B 8 4 202,130; B 4 4 204,680; B 4 4 204,496; B 4 4 204,462; B 4 4 204,632; B 4 4 204,528; B 4 8 204,398; B 4 4 204,156; B 4 4 204,340; B 4 4 204,374; B 4 4 204,204; B 4 4 204,308; B 4 4 218,600; B 4 4 218,784; B 4 4 218,700; B 4 4 218,470; B 4 4 218,660; B 4 4 218,556; B 4 4 218,394; B 4 4 218,236; B 4 4 218,52; B 4 4 218,136; B 4 4 218,366; B 4 4 218,176; B 4 4 218,280; B 8 4 234,708; B 8 4 234,476; B 8 4 234,594; B 8 4 234,668; B 8 4 234,654; B 8 4 234,490; B 8 4 234,564; B 8 4 234,550; B 8 4 234,128; B 8 4 234,360; B 8 4 234,242; B 8 4 234,168; B 8 4 234,182; B 8 4 234,346; B 8 4 234,272; B 8 4 234,286; B 8 4 234,776; B 8 4 234,402; B 8 4 234,60; B 8 4 240,624; B 8 4 240,520; B 8 4 240,212; B 8 4 240,316; B 4 4 250,674; B 4 4 250,570; B 4 4 250,162; B 4 4 250,266; B 4 4 278,786; B 4 4 278,696; B 4 4 278,466; B 4 4 278,642; B 4 4 278,606; B 4 4 278,538; B 4 4 278,502; B 4 4 278,392; B 4 4 278,50; B 4 4 278,140; B 4 4 278,370; B 4 4 278,194; B 4 4 278,230; B 4 4 278,298; B 4 4 278,334; B 4 4 306,578; B 4 4 306,258; B 4 4 320,770; B 4 4 320,712; B 4 4 320,588; B 4 4 320,626; B 4 4 320,660; B 4 4 320,484; B 4 4 320,522; B 4 4 320,556; B 4 4 320,408; B 4 4 320,66; B 4 4 320,124; B 4 4 320,248; B 4 4 320,210; B 4 4 320,176; B 4 4 320,352; B 4 4 320,314; B 4 4 320,280; B 4 8 50,46; B 8 4 54,140; B 4 8 50,228; B 4 8 50,310; B 4 8 98,310; B 8 4 54,352; B 4 8 100,350; B 4 8 50,388; B 4 8 98,388; B 8 4 54,484; B 4 8 100,486; B 4 8 50,526; B 4 8 98,526; B 4 8 50,608; B 8 4 54,696; B 4 8 50,790; B 4 8 250,790; B 8 4 254,698; B 4 8 298,692; B 8 4 254,642; B 4 8 306,638; B 4 8 306,602; B 8 4 254,538; B 4 8 306,534; B 4 8 306,498; B 4 8 250,462; B 4 8 250,388; B 8 4 254,368; B 8 4 254,334; B 4 8 306,338; B 8 4 254,298; B 4 8 306,302; B 8 4 254,230; B 4 8 306,234; B 8 4 254,194; B 4 8 306,198; B 8 4 254,138; B 4 8 298,144; B 4 8 250,46; B 8 4 254,502; B 8 4 254,606; DF;

RecoderTime

DS 109 125/1; L NP; B 4 2 12,55; B 4 2 12,93; B 4 98 12,151; B 8 8 14,204; B 4 2 12,255; B 14 4 17,286; B 4 2 12,289; B 4 30 12,313; B 14 4 17,58; B 12 8 18,64; B 14 4 17,90; B 12 4 18,86; B 14 4 17,258; B 12 8 18,264; B 8 4 20,216; B 8 4 22,16; B 18 4 25,220; B 18 4 27,20; B 8 6 28,37; B 8 6 28,237; B 4 14 26,321; B 4 18 30,49; B 20 4 22,70; B 4 34 30,89; B 4 18 30,249; B 20 4 22,270; B 8 4 28,312; B 4 38 30,291; B 8 6 44,37; B 4 206 42,143; B 4 2 50,255; B 14 4 55,258; B 12 12 56,266; B 8 4 58,216; B 8 4 60,16; B 18 4 63,220; B 18 4 65,20; B 8 6 66,37; B 8 6 66,197; B 8 6 66,237; B 4 18 68,49; B 4 122 68,133; B 4 18 68,249; B 8 6 82,37; B 4 18 84,49; B 4 38 84,91; B 22 4 101,86; B 8 8 108,4; B 4 76 110,46; B 6 8 11,50; B 6 8 11,98; B 6 8 11,250; B 6 8 11,294; B 12 4 46,248; B 6 4 49,252; B 8 2 56,55; B 20 4 60,70; B 12 12 56,62; B 8 2 96,55; B 20 4 92,70; B 12 12 96,62; B 6 10 103,93; L ND; B 8 8 18,78; B 8 8 18,278; B 4 20 18,64; B 4 12 18,88; B 4 20 18,264; B 4 8 18,286; B 8 8 26,204; B 8 8 28,32; B 8 8 28,232; B 4 20 28,218; B 16 4 36,14; B 4 12 30,22; B 8 8 34,120; B 8 8 34,320; B 4 14 36,109; B 4 40 36,74; B 14 8 31,302; B 4 10 36,311; B 4 36 36,272; B 4 12 42,22; B 8 8 44,32; B 8 8 56,278; B 4 20 56,264; B 8 8 66,32; B 8 8 66,202; B 8 8 66,232; B 4 22 66,217; B 16 4 74,14; B 4 12 68,22; B 4 8 68,8; B 8 8 72,320; B 8 18 76,115; B 8 36 76,72; B 4 62 74,285; B 4 12 80,22; B 8 8 82,32; B 18 8 81,102; B 4 8 96,86; B 26 8 25,50; B 26 8 25,98; B 8 8 94,78; B 30 8 87,94; B 26 8 25,250; B 26 8 63,250; B 8 8 56,78; B 4 18 56,65; B 8 2 56,55; B 4 18 96,65; B 8 2 96,55; B 52 8 76,50; B 26 8 25,294; L NM; B 114 8 57,4; B 114 8 57,18; B 114 8 57,120; B 114 8 57,218; B 114 8 57,278; B 114 8 57,320; B 20 8 20,204; B 8 12 28,34; B 8 12 28,234; B 8 12 44,34; B 8 12 66,34; B 8 12 66,200; B 8 12 66,234; B 8 12 82,34; B 12 8 14,50; B 12 8 14,98; B 12 8 100,94; B 12 8 14,250; B 8 12 56,54; B 114 8 57,78; B 8 12 96,54; B 12 8 14,294; B 12 8 52,250; L NI; B 12 24 18,64; B 12 16 18,88; B 12 24 18,264; B 12 12 18,286; B 12 24 56,264; B 12 12 96,86; B 12 26 56,63; B 12 26 96,63; L NC; B 4 4 14,204; B 4 4 18,78; B 4 4 18,278; B 4 4 20,218; B 4 4 22,18; B 4 4 26,204; B 4 8 28,34; B 4 8 28,234; B 4 4 34,120; B 4 4 34,320; B 4 8 44,34; B 4 4 56,278; B 4 4 58,218; B 4 4 60,18; B 4 8 66,34; B 4 8 66,200; B 4 8 66,234; B 4 4 72,320; B 4 4 76,120; B 4 8 82,34; B 4 4 108,4; B 8 4 14,50; B 8 4 14,98; B 4 4 94,78; B 8 4 100,94; B 8 4 14,250; B 8 4 14,294; B 8 4 52,250; B 4 4 56,78; B 4 8 56,54; B 4 8 96,54; DF;

S_AddSubNeg

DS 110 125/1; L NP; B 8 4 134,6; B 8 4 96,6; B 8 4 36,6; B 20 4 90,10; B 44 4 30,10; B 32 4 136,10; B 4 18 122,21; B 6 8 109,24; B 6 8 135,24; B 8 6 28,27; B 8 4 86,26; B 8 2 14,25; B 8 6 44,27; B 8 2 148,25; B 16 4 10,28; B 4 198 4,129; B 4 20 110,38; B 10 4 149,28; B 4 16 152,38; B 14 2 89,29; B 10 2 91,31; B 4 88 136,72; B 4 18 42,39; B 4 18 30,39; B 4 16 94,40; B 4 2 76,45; B 4 2 60,45; B 8 4 150,48; B 14 4 123,48; B 12 8 122,54; B 14 4 81,48; B 12 8 82,54; B 14 4 55,48; B 12 8 54,54; B 4 14 148,57; B 20 4 118,60; B 20 4 86,60; B 20 4 22,60; B 30 4 55,60; B 4 60 68,92; B 12 4 122,76; B 12 4 82,76; B 12 4 18,76; B 12 4 54,76; B 14 4 123,80; B 14 4 81,80; B 14 4 17,80; B 14 4 55,80; B 4 82 78,135; B 4 82 126,135; B 4 32 20,116; B 4 154 144,197; B 14 4 35,118; B 12 4 140,118; B 4 26 40,133; B 8 8 68,126; B 14 4 25,134; B 8 6 30,147; B 22 4 37,160; B 4 8 28,154; B 12 4 106,176; B 4 10 102,183; B 48 4 42,176; B 4 2 64,179; B 34 4 79,182; B 72 2 68,191; B 44 2 82,189; B 32 2 48,193; B 26 2 73,201; B 50 2 61,203; B 28 2 50,205; B 4 28 38,220; B 8 8 98,208; B 40 4 94,214; B 24 4 56,216; B 4 56 46,246; B 6 8 55,226; B 8 6 78,227; B 8 4 108,226; B 4 44 54,252; B 20 4 12,230; B 18 2 103,229; B 14 2 101,231; B 4 16 96,240; B 4 18 80,239; B 4 16 20,240; B 4 2 132,245; B 14 4 127,248; B 12 12 126,256; B 14 4 33,248; B 12 8 32,254; B 20 4 104,260; B 4 34 96,279; B 4 34 80,279; B 20 4 28,260; B 4 34 20,279; B 44 4 124,276; B 22 4 37,276; B 22 4 63,276; B 14 4 109,280; B 4 2 114,283; B 14 4 67,280; B 4 2 62,283; B 14 4 33,280; B 4 2 38,283; B 8 2 18,45; B 12 12 18,52; B 6 8 61,40; B 6 8 59,90; B 4 54 30,89; B 4 4 60,84; B 4 4 12,84; B 12 4 16,98; B 4 36 42,80; B 6 8 127,90; B 6 8 77,90; B 4 118 94,121; B 4 112 110,118; B 4 4 76,84; B 4 4 128,84; B 6 8 75,40; B 6 8 127,38; B 4 4 128,44; B 8 2 108,245; B 12 12 108,252; B 6 8 133,240; B 6 8 115,288; B 8 4 60,290; B

4 26 58,305; B 6 4 61,286; B 8 2 68,245; B 20 4 72,260; B 12 12 68,252; B 6 8 37,238; B 4 4 38,244; B 6 8 39,288; B 6 10 13,91; L ND; B 4 18 146,9; B 4 18 86,9; B 4 18 46,9; B 4 18 28,9; B 4 18 14,9; B 8 8 86,22; B 8 8 14,22; B 8 8 28,22; B 8 8 44,22; B 8 8 148,22; B 8 2 114,27; B 8 2 114,21; B 24 4 122,24; B 8 2 130,27; B 8 2 130,21; B 8 8 144,40; B 8 20 102,104; B 4 12 36,100; B 4 8 142,48; B 4 20 82,54; B 4 20 54,54; B 16 8 148,56; B 4 46 154,83; B 8 8 122,68; B 8 8 82,68; B 8 8 18,68; B 8 8 54,68; B 4 42 52,115; B 8 8 152,110; B 8 74 12,143; B 8 8 38,110; B 8 8 134,176; B 8 34 134,147; B 8 8 28,126; B 24 8 126,126; B 24 8 94,126; B 4 8 26,134; B 26 8 77,140; B 8 2 54,143; B 8 2 54,137; B 24 8 110,140; B 34 4 41,140; B 8 4 30,144; B 8 8 70,154; B 40 8 102,154; B 32 4 38,154; B 4 8 52,160; B 22 10 33,161; B 8 6 40,169; B 20 8 80,168; B 4 6 72,161; B 8 8 26,192; B 8 14 26,173; B 24 8 126,168; B 32 4 122,182; B 8 16 40,188; B 22 8 19,184; B 8 32 54,180; B 4 10 108,189; B 8 8 120,194; B 32 4 94,196; B 4 8 120,202; B 36 4 40,198; B 4 8 80,214; B 4 8 80,202; B 16 4 114,208; B 4 8 108,214; B 16 2 74,207; B 22 2 71,209; B 10 2 65,211; B 4 10 62,217; B 8 8 108,222; B 8 8 78,222; B 8 8 60,226; B 8 22 88,303; B 8 40 88,264; B 8 22 12,303; B 8 40 12,264; B 4 20 126,254; B 4 20 32,254; B 8 8 126,268; B 8 8 108,268; B 8 8 68,268; B 8 8 32,268; B 4 12 108,278; B 4 12 68,278; B 4 12 32,278; B 4 18 18,55; B 8 2 18,45; B 48 8 36,40; B 44 8 36,90; B 4 14 54,79; B 4 14 18,79; B 4 42 36,65; B 48 8 102,90; B 4 14 82,79; B 4 14 122,79; B 8 42 102,65; B 50 6 101,39; B 10 2 121,35; B 40 2 96,43; B 4 22 122,53; B 4 18 108,255; B 8 2 108,245; B 4 26 120,223; B 14 4 125,238; B 8 4 128,242; B 52 8 88,288; B 4 18 68,255; B 8 2 68,245; B 28 6 22,239; B 8 2 32,235; B 26 2 21,243; B 30 8 23,288; B 50 8 87,240; L NM; B 156 8 78,8; B 8 12 28,24; B 8 12 86,24; B 8 12 14,24; B 8 12 44,24; B 8 12 148,24; B 12 8 112,24; B 12 8 132,24; B 156 8 78,68; B 156 8 78,110; B 66 8 57,126; B 24 8 110,126; B 8 14 68,137; B 8 14 54,143; B 24 8 94,140; B 24 8 126,140; B 8 12 30,144; B 40 8 70,154; B 24 8 126,154; B 8 26 54,177; B 8 24 40,176; B 40 8 102,168; B 74 8 87,194; B 156 8 78,208; B 8 12 78,224; B 8 12 108,224; B 12 8 58,226; B 156 8 78,268; B 156 8 78,310; B 8 12 18,44; B 12 8 58,40; B 12 8 16,90; B 12 8 56,90; B 12 8 124,90; B 12 8 80,90; B 12 8 78,40; B 30 6 133,39; B 16 2 140,43; B 20 2 128,35; B 8 12 108,244; B 12 8 130,240; B 12 8 112,288; B 12 8 64,288; B 8 12 68,244; B 12 8 34,238; B 12 8 36,288; L NI; B 12 24 122,54; B 12 24 82,54; B 12 24 54,54; B 12 16 122,78; B 12 16 82,78; B 12 16 18,78; B 12 16 54,78; B 12 24 32,254; B 12 16 108,278; B 12 16 68,278; B 12 16 32,278; B 12 26 18,53; B 12 26 108,253; B 12 24 126,254; B 12 26 68,253; L NC; B 4 4 134,8; B 4 4 96,8; B 4 4 36,8; B 4 8 28,24; B 4 8 86,24; B 4 8 14,24; B 4 8 44,24; B 4 8 148,24; B 8 4 112,24; B 8 4 132,24; B 4 4 144,40; B 4 4 122,68; B 4 4 82,68; B 4 4 18,68; B 4 4 54,68; B 4 4 152,110; B 4 4 102,110; B 4 4 38,110; B 4 4 12,110; B 4 4 68,126; B 4 4 28,126; B 4 4 102,126; B 4 4 118,126; B 4 4 86,126; B 4 4 68,140; B 4 4 54,140; B 4 4 86,140; B 4 4 102,140; B 4 4 118,140; B 4 4 134,140; B 4 8 30,144; B 4 4 70,154; B 4 4 118,154; B 4 4 86,154; B 4 4 134,154; B 4 4 54,168; B 4 4 40,168; B 4 4 118,168; B 4 4 86,168; B 4 4 40,184; B 4 4 120,194; B 4 4 98,208; B 4 8 78,224; B 4 8 108,224; B 8 4 58,226; B 4 4 126,268; B 4 4 108,268; B 4 4 68,268; B 4 4 32,268; B 4 4 88,310; B 4 4 12,310; B 4 8 18,44; B 8 4 58,40; B 8 4 16,90; B 8 4 56,90; B 8 4 124,90; B 8 4 80,90; B 8 4 78,40; B 8 4 124,38; B 4 8 108,244; B 8 4 130,240; B 8 4 112,288; B 8 4 64,288; B 4 8 68,244; B 8 4 34,238; B 8 4 36,288; DF;

LastDelay

DS 111 125/1; L NP; B 4 72 26,36; B 4 16 12,8; B 8 8 12,20; B 8 8 6,48; B 4 104 4,104; B 8 8 28,76; B 8 4 96,88; B 8 8 72,90; B 8 8 34,90; B 18 4 101,92; B 4 14 72,101; B 4 14 34,101; B 8 6 104,109; B 8 6 82,109; B 8 6 62,109; B 8 6 44,109; B 8 6 24,109; B 4 244 80,234; B 4 206 42,215; B 8 4 66,114; B 8 4 28,114; B 4 14 68,123; B 4 18 106,121; B 4 14 30,123; B 4 38 106,163; B 4 34 68,161; B 20 4 22,142; B 4 38 30,163; B 12 4 56,158; B 22 4 13,158; B 12 4 94,158; B 14 4 55,162; B 14 4 93,162; B 4 6 52,179; B 4 30 90,191; B 8 4 104,184; B 4 80 102,226; B 8 4 28,184; B 4 80 26,226; B 16 4 58,184; B 8 4 66,200; B 4 64 68,234; B 4 12 64,192; B 4 38 14,217; B 8 8 90,210; B 8 8 14,240; B 4 40 56,266; B 8 6 104,269; B 8 6 66,269; B 8 6 28,269; B 8 4 96,288; B 8 4 58,288; B 8 4 20,288; B 18 4 101,292; B 18 4 63,292; B 18 4 25,292; B 8 6 104,309; B 8 6 66,309; B 8 6 28,309; B 4 18 68,321; B 4 18 106,321; B 4 18 30,321; B 4 38 68,363; B 20 4 22,342; B 4 38 30,363; B 4 38 106,363; B 22 4 89,358; B 14 4 17,358; B 4 28 12,386; B 14 4 55,358; B 4 28 50,386; B 8 4 28,384; B 4 14 26,393; B 8 2 18,127; B 12 12 18,134; B 8 2 56,127; B 20 4 60,142; B 12 12 56,134; B 8 2 94,127; B 20 4 98,142; B 12 12 94,134; B 6 10 11,165; B 6 8 51,172; B 4 4 50,166; B 6 8 89,172; B 4 4 88,166; B 8 2 18,327; B 12 12 18,334; B 8 4 44,320; B 8 2 56,327; B 20 4 60,342; B 12 8 56,336; B 18 4 53,330; B 4 6 46,325; B 8 2 94,327; B 20 4 98,342; B 12 12 94,334; B 6 8 13,368; B 4 4 12,362; B 6 8 51,368; B 4 4 50,362; B 6 8 89,368; B 4 4 88,362; L ND; B 8 8 82,34; B 4 62 82,69; B 8 8 102,76; B 4 20 104,90; B 8 8 104,104; B 8 4 82,106; B 28 4 72,102; B 8 4 62,106; B 8 4 44,106; B 28 4 34,102; B 8 4 24,106; B 4 14 74,181; B 4 40 74,146; B 4 14 112,181; B 4 40 112,146; B 4 62 36,157; B 8 8 56,150; B 8 8 94,150; B 8 8 18,150; B 4 8 18,158; B 4 28 20,188; B 8 8 72,192; B 8 8 110,192; B 8 8 34,192; B 8 8 8,192; B 4 10 74,201; B 16 4 14,204; B 4 6 8,199; B 12 28 14,220; B 20 8 66,210; B 4 36 62,232; B 8 8 50,240; B 16 4 56,252; B 4 6 50,247; B 8 8 104,274; B 8 8 66,274; B 8 8 28,274; B 4 22 104,289; B 4 22 66,289; B 4 22 28,289; B 8 8 104,304; B 8 8 66,304; B 8 8 28,304; B 4 10 74,383; B 4 36 74,344; B 4 10 112,383; B 4 36 112,344; B 4 10 36,383; B 4 36 36,344; B 8 8 56,350; B 8 8 18,350; B 8 8 94,350; B 14 6 107,375; B 14 6 31,375; B 14 6 69,375; B 8 8 72,392; B 8 8 110,392; B 8 8 34,392; B 4 18 18,137; B 8 2 18,127; B 26 8 25,122; B 4 18 56,137; B 8 2 56,127; B 26 8 63,122; B 4 18 94,137; B 8 2 94,127; B 26 8 101,122; B 8 8 16,166; B 6 4 19,172; B 24 6 64,171; B 8 2 56,175; B 22 2 65,167; B 4 12 56,160; B 24 6 102,171; B 8 2 94,175; B 22 2 103,167; B 4 12 94,160; B 4 18 18,337; B 26 8 25,322; B 8 2 18,327; B 4 18 56,337; B 26 8 63,322; B 8 2 56,327; B 4 18 94,337; B 26 8 101,322; B 8 2 94,327; B 22 2 27,363; B 4 8 18,358; B 24 8 26,368; B 22 2 65,363; B 4 8 56,358; B 24 8 64,368; B 4 8 94,358; B 24 8 102,368; B 22 2 103,363; L NM; B 114 8 57,20; B 114 8 57,34; B 114 8 57,48; B 82 8 65,76; B 114 8 57,90; B 8 12 104,106; B 8 12 82,106; B 8 12 62,106; B 8 12 44,106; B 8 12 24,106; B 114 8 57,150; B 8 18 14,179; B 114 8 57,192; B 38 8 75,210; B 44 8 32,240; B 8 12 104,272; B 8 12

66,272; B 8 12 28,272; B 114 8 57,290; B 8 12 104,306; B 8 12 66,306; B 8 12 28,306; B 114 8 57,350; B 114 8 57,392; B 8 12 18,126; B 8 12 56,126; B 8 12 94,126; B 12 8 14,166; B 12 8 54,172; B 12 8 92,172; B 8 12 18,326; B 8 12 56,326; B 8 12 94,326; B 12 8 16,368; B 12 8 54,368; B 12 8 92,368; L NI; B 12 16 56,160; B 12 12 18,158; B 12 16 94,160; B 12 12 94,358; B 12 12 18,358; B 12 12 56,358; B 12 26 18,135; B 12 26 56,135; B 12 26 94,135; B 12 26 18,335; B 12 26 56,335; B 12 26 94,335; L NC; B 4 4 12,20; B 4 4 82,34; B 4 4 6,48; B 4 4 28,76; B 4 4 102,76; B 4 4 96,90; B 4 4 72,90; B 4 4 34,90; B 4 8 104,106; B 4 8 82,106; B 4 8 62,106; B 4 8 44,106; B 4 8 24,106; B 4 4 56,150; B 4 4 94,150; B 4 4 18,150; B 4 4 72,192; B 4 4 110,192; B 4 4 34,192; B 4 4 8,192; B 4 4 90,210; B 4 4 60,210; B 4 4 14,240; B 4 4 50,240; B 4 8 104,272; B 4 8 66,272; B 4 8 28,272; B 4 4 96,290; B 4 4 58,290; B 4 4 20,290; B 4 8 104,306; B 4 8 66,306; B 4 8 28,306; B 4 4 56,350; B 4 4 18,350; B 4 4 94,350; B 4 4 72,392; B 4 4 110,392; B 4 4 34,392; B 4 8 18,126; B 4 8 56,126; B 4 8 94,126; B 8 4 14,166; B 8 4 54,172; B 8 4 92,172; B 4 8 18,326; B 4 8 56,326; B 4 8 94,326; B 8 4 16,368; B 8 4 54,368; B 8 4 92,368; DF;

S_Add

DS 129 125/1; L NP; B 48 4 26,176; B 72 4 52,192; B 20 4 26,10; B 8 4 20,6; B 8 6 28,27; B 4 18 26,39; B 8 4 28,96; B 4 64 30,130; B 4 32 26,78; B 12 8 38,54; B 12 4 38,76; B 4 56 42,122; B 14 4 39,48; B 4 2 44,45; B 14 4 39,80; B 4 2 46,245; B 4 2 46,283; B 14 4 51,248; B 12 8 52,254; B 14 4 51,280; B 12 4 52,276; B 12 4 46,152; B 6 4 49,156; B 4 2 48,179; B 8 8 52,126; B 30 4 39,60; B 4 60 52,92; B 40 4 78,214; B 8 6 62,227; B 4 2 60,45; B 14 4 65,48; B 12 8 66,54; B 14 4 65,80; B 12 4 66,76; B 4 82 62,135; B 4 18 64,239; B 20 4 56,260; B 4 34 64,279; B 8 4 70,26; B 14 2 73,29; B 10 2 75,31; B 20 4 74,10; B 8 4 80,6; B 4 16 78,40; B 34 4 63,182; B 20 4 70,60; B 4 118 78,121; B 14 2 85,231; B 8 8 82,208; B 4 16 80,240; B 4 34 80,279; B 4 12 86,184; B 20 4 88,260; B 18 2 87,229; B 8 4 92,226; B 6 8 93,24; B 4 20 94,38; B 12 4 90,176; B 4 112 94,118; B 14 4 93,280; B 4 2 98,283; B 20 4 102,60; B 12 8 106,54; B 12 4 106,76; B 12 12 110,256; B 32 4 120,10; B 4 18 106,21; B 4 82 110,135; B 14 4 107,48; B 14 4 107,80; B 8 4 118,6; B 14 4 111,248; B 4 2 116,245; B 6 8 119,24; B 4 88 120,72; B 44 4 108,276; B 12 4 124,118; B 4 154 128,197; B 8 2 132,25; B 8 4 134,48; B 4 14 132,57; B 10 4 133,28; B 4 16 136,38; B 6 8 45,40; B 6 8 59,40; B 6 8 111,38; B 4 4 112,44; B 6 8 111,90; B 4 4 112,84; B 6 8 61,90; B 4 4 60,84; B 6 8 43,90; B 4 4 44,84; B 6 8 45,240; B 6 8 117,240; B 8 2 92,245; B 12 12 92,252; B 8 4 44,290; B 4 26 42,305; B 6 4 45,286; B 6 8 99,288; L ND; B 36 4 24,198; B 8 24 10,184; B 8 8 22,110; B 4 14 20,99; B 4 40 20,64; B 22 4 17,170; B 8 4 24,166; B 8 16 24,188; B 4 6 24,161; B 8 8 28,22; B 4 18 30,9; B 8 8 38,68; B 8 32 38,180; B 16 8 30,154; B 4 6 36,161; B 4 20 38,54; B 26 8 61,140; B 8 8 52,268; B 8 8 54,154; B 4 20 52,254; B 4 12 52,278; B 20 8 64,168; B 4 6 56,161; B 8 8 62,222; B 8 8 66,68; B 4 18 64,209; B 4 20 66,54; B 40 8 86,154; B 24 8 78,126; B 8 8 70,22; B 8 22 72,303; B 8 40 72,264; B 4 18 70,9; B 24 8 94,140; B 8 22 86,103; B 8 40 86,64; B 8 8 92,222; B 8 8 92,268; B 32 4 78,198; B 4 12 92,190; B 4 8 92,214; B 4 12 92,278; B 8 2 98,27; B 8 2 98,21; B 24 4 106,24; B 8 8 104,194; B 8 8 106,68; B 16 4 98,208; B 4 8 104,202; B 4 20 106,54; B 8 8 110,268; B 4 20 110,254; B 8 2 114,27; B 8 2 114,21; B 32 4 106,182; B 24 8 110,126; B 24 8 110,168; B 8 8 118,176; B 8 34 118,147; B 8 8 128,40; B 16 8 132,56; B 4 8 126,48; B 8 8 132,22; B 4 18 130,9; B 8 8 136,110; B 4 46 138,83; B 26 8 31,40; B 50 6 85,39; B 8 2 106,35; B 48 2 84,43; B 8 2 106,93; B 4 12 106,78; B 48 6 86,89; B 8 2 66,93; B 4 12 66,78; B 24 6 30,89; B 8 2 38,93; B 22 2 29,85; B 4 12 38,78; B 44 2 86,85; B 4 26 104,223; B 14 4 109,238; B 8 4 112,242; B 4 18 92,255; B 50 8 71,240; B 8 2 92,245; B 52 8 72,288; L NM; B 140 8 70,8; B 140 8 70,68; B 140 8 70,110; B 140 8 70,208; B 140 8 70,268; B 140 8 70,310; B 8 74 10,151; B 8 24 24,176; B 8 12 28,24; B 74 8 71,194; B 8 26 38,177; B 28 8 60,154; B 26 8 61,126; B 8 14 52,137; B 8 12 62,224; B 40 8 86,168; B 24 8 78,140; B 8 12 70,24; B 24 8 94,126; B 8 12 92,224; B 12 8 96,24; B 24 8 110,140; B 24 8 110,154; B 12 8 116,24; B 8 12 132,24; B 12 8 42,40; B 12 8 62,40; B 30 6 117,39; B 16 2 124,43; B 20 2 112,35; B 12 8 108,90; B 12 8 64,90; B 12 8 40,90; B 12 8 48,240; B 12 8 114,240; B 8 12 92,244; B 54 8 27,288; B 12 8 96,288; L NI; B 12 24 38,54; B 12 16 38,78; B 12 24 52,254; B 12 16 52,278; B 12 24 66,54; B 12 16 66,78; B 12 16 92,278; B 12 24 106,54; B 12 16 106,78; B 12 24 110,254; B 12 26 92,253; L NC; B 4 4 10,184; B 4 4 20,8; B 4 4 22,110; B 4 4 24,168; B 4 4 24,184; B 4 8 28,24; B 4 4 38,68; B 4 4 38,168; B 8 4 52,154; B 4 4 52,126; B 4 4 52,140; B 4 4 52,268; B 4 8 62,224; B 4 4 66,68; B 4 8 70,24; B 4 4 70,126; B 4 4 70,140; B 4 4 70,154; B 4 4 70,168; B 4 4 72,310; B 4 4 80,8; B 4 4 82,208; B 4 4 86,110; B 4 4 86,126; B 4 4 86,140; B 4 8 92,224; B 4 4 92,268; B 8 4 96,24; B 4 4 102,126; B 4 4 102,140; B 4 4 102,154; B 4 4 102,168; B 4 4 104,194; B 4 4 106,68; B 4 4 110,268; B 8 4 116,24; B 4 4 118,8; B 4 4 118,140; B 4 4 118,154; B 4 4 128,40; B 4 8 132,24; B 4 4 136,110; B 8 4 42,40; B 8 4 62,40; B 8 4 108,38; B 8 4 108,90; B 8 4 64,90; B 8 4 40,90; B 8 4 48,240; B 8 4 114,240; B 4 8 92,244; B 8 4 48,288; B 8 4 96,288; DF;

S_AddSub

DS 141 125/1; L NP; B 8 4 134,6; B 8 4 68,6; B 8 4 36,6; B 50 4 47,10; B 42 4 117,10; B 4 18 124,21; B 8 6 100,27; B 8 2 114,25; B 8 2 134,25; B 8 6 28,27; B 8 2 58,25; B 8 6 44,27; B 30 4 69,28; B 10 4 113,28; B 10 4 135,28; B 4 86 138,73; B 4 18 42,39; B 4 18 30,39; B 4 22 102,41; B 4 2 12,45; B 18 4 77,48; B 12 8 80,54; B 14 4 17,48; B 12 8 18,54; B 4 22 110,41; B 22 4 119,60; B 4 112 110,118; B 18 4 83,60; B 20 4 22,60; B 4 54 30,89; B 4 34 42,79; B 30 4 55,60; B 4 60 68,92; B 8 4 92,68; B 4 110 94,125; B 4 4 90,64; B 12 4 122,76; B 12 4 82,76; B 12 4 18,76; B 12 4 54,76; B 14 4 123,80; B 14 4 81,80; B 14 4 17,80; B 4 2 12,83; B 14 4 55,80; B 4 2 12,93; B 4 82 78,135; B 4 82 126,135; B 12 4 16,96; B 4 34 20,115; B 4 154 144,197; B 14 4 35,118; B 10 4 141,118; B 4 26 40,133; B 8 8 68,126; B 14 4 25,134; B 8 6 30,147; B 22 4 37,160; B 4 8 28,154; B 12 4 106,176; B 4 12 102,184; B 48 4 42,176; B 4 2 64,179; B 34 4 79,182; B 72 4 68,192; B 8 8 98,208; B 4 30 52,219; B 20 4 104,214; B 8 6 78,227; B 18 2 103,229; B 8 4 108,226; B 14 2 101,231; B 4 16 96,240; B 4 18 80,239; B 14 4 47,248; B 12 12 46,256; B 14 4 127,248; B 12 12 126,256; B 4 34 96,279; B 20 4 72,260; B 4 34

80,279; B 44 4 124,276; B 12 4 68,276; B 14 4 109,280; B 4 2 114,283; B 14 4 67,280; B 4 2 62,283;
B 6 8 11,40; B 8 2 54,45; B 12 12 54,52; B 6 10 71,41; B 8 2 124,45; B 12 12 124,52; B 6 8 11,88; B 6
8 59,90; B 4 4 60,84; B 6 8 77,90; B 4 4 76,84; B 6 8 127,90; B 4 4 128,84; B 6 8 51,238; B 4 4 52,244;
B 8 2 68,245; B 12 12 68,252; B 8 2 108,245; B 20 4 104,260; B 12 12 108,252; B 6 8 131,238; B 4 4
132,244; B 8 4 60,290; B 4 26 58,305; B 6 4 61,286; B 6 8 115,288; L ND; B 4 18 102,9; B 4 18 58,9; B
4 18 46,9; B 4 18 28,9; B 8 8 100,22; B 8 8 58,22; B 8 8 28,22; B 8 8 44,22; B 8 4 114,24; B 28 4
124,20; B 8 4 134,24; B 18 4 81,22; B 4 8 88,28; B 22 4 105,46; B 4 10 96,53; B 30 4 101,34; B 4 14
36,99; B 4 40 36,64; B 10 2 77,43; B 6 2 79,45; B 4 18 80,55; B 4 20 18,54; B 12 4 100,60; B 8 22
102,103; B 8 22 102,73; B 8 8 122,68; B 8 8 82,68; B 8 8 18,68; B 8 8 54,68; B 4 12 122,78; B 4 12
18,78; B 4 12 54,78; B 4 42 52,115; B 8 74 12,143; B 8 8 38,110; B 8 8 134,176; B 8 34 134,147; B 8
8 28,126; B 24 8 126,126; B 24 8 94,126; B 4 8 26,134; B 26 8 77,140; B 8 2 54,143; B 8 2 54,137; B
24 8 110,140; B 34 4 41,140; B 8 4 30,144; B 8 8 70,154; B 40 8 102,154; B 32 4 38,154; B 4 8
52,160; B 22 10 33,161; B 8 6 40,169; B 20 8 80,168; B 4 6 72,161; B 8 8 26,192; B 8 14 26,173; B 24
8 126,168; B 32 4 122,182; B 8 16 40,188; B 22 8 19,184; B 8 32 54,180; B 4 12 108,190; B 8 8
120,194; B 4 8 120,202; B 32 4 94,198; B 36 4 40,198; B 4 18 80,209; B 16 4 114,208; B 4 8 108,214;
B 8 8 108,222; B 8 8 78,222; B 8 22 88,303; B 8 40 88,264; B 8 8 46,268; B 8 8 126,268; B 8 8
108,268; B 8 8 68,268; B 4 12 108,278; B 4 12 68,278; B 4 18 54,55; B 48 8 36,40; B 8 2 54,45; B 4
18 124,55; B 36 8 112,40; B 8 2 124,45; B 8 2 54,93; B 46 6 35,89; B 44 2 34,85; B 8 2 82,93; B 4 12
82,78; B 48 6 102,89; B 8 2 122,93; B 8 8 46,238; B 4 22 46,253; B 4 18 68,255; B 8 2 68,245; B 4 18
108,255; B 8 2 108,245; B 52 8 88,240; B 4 22 126,253; B 52 8 88,288; B 44 2 102,85; B 8 4 126,240;
B 4 16 120,218; B 10 4 125,236; B 6 4 121,228; B 4 4 122,232; B 8 18 76,33; L NM; B 156 8 78,8; B 8
12 100,24; B 8 12 28,24; B 8 12 58,24; B 8 12 44,24; B 8 12 114,24; B 8 12 134,24; B 12 8 74,40; B
156 8 78,68; B 156 8 78,110; B 66 8 57,126; B 24 8 110,126; B 8 14 68,137; B 8 14 54,143; B 24 8
94,140; B 24 8 126,140; B 8 12 30,144; B 40 8 70,154; B 24 8 126,154; B 8 26 54,177; B 8 24 40,176;
B 40 8 102,168; B 74 8 87,194; B 156 8 78,208; B 8 12 78,224; B 8 12 108,224; B 156 8 78,268; B
156 8 78,310; B 12 8 14,40; B 8 12 54,44; B 8 12 124,44; B 12 8 14,88; B 12 8 56,90; B 12 8 80,90; B
12 8 124,90; B 12 8 48,238; B 8 12 68,244; B 8 12 108,244; B 12 8 128,238; B 12 8 64,288; B 12 8
112,288; L NI; B 12 24 80,54; B 12 24 18,54; B 12 16 122,78; B 12 16 82,78; B 12 16 18,78; B 12 16
54,78; B 12 24 126,254; B 12 24 46,254; B 12 16 108,278; B 12 16 68,278; B 12 26 54,53; B 12 26
124,53; B 12 26 68,253; B 12 26 108,253; L NC; B 4 4 134,8; B 4 4 68,8; B 4 4 36,8; B 4 8 100,24; B 4
8 28,24; B 4 8 58,24; B 4 8 44,24; B 4 8 114,24; B 4 8 134,24; B 8 4 74,40; B 4 4 122,68; B 4 4 82,68;
B 4 4 18,68; B 4 4 54,68; B 4 4 102,110; B 4 4 38,110; B 4 4 12,110; B 4 4 68,126; B 4 4 28,126; B 4 4
102,126; B 4 4 118,126; B 4 4 86,126; B 4 4 68,140; B 4 4 54,140; B 4 4 86,140; B 4 4 102,140; B 4 4
118,140; B 4 4 134,140; B 4 8 30,144; B 4 4 70,154; B 4 4 118,154; B 4 4 86,154; B 4 4 134,154; B 4
4 54,168; B 4 4 40,168; B 4 4 118,168; B 4 4 86,168; B 4 4 40,184; B 4 4 120,194; B 4 4 98,208; B 4 8
78,224; B 4 8 108,224; B 4 4 46,268; B 4 4 126,268; B 4 4 108,268; B 4 4 68,268; B 4 4 88,310; B 8 4
14,40; B 4 8 54,44; B 4 8 124,44; B 8 4 14,88; B 8 4 56,90; B 8 4 80,90; B 8 4 124,90; B 8 4 48,238; B
4 8 68,244; B 4 8 108,244; B 8 4 128,238; B 8 4 64,288; B 8 4 112,288; DF;

PadBlank

DS 202 125/1; L NM; B 200 16 100,8; B 200 32 100,196; B 108 108 100,106; L NG; B 92 92 100,106;
DF;

PadOut4

DS 204 125/1; L ND; B 16 54 8,239; B 200 56 100,28; B 50 156 25,134; B 20 8 22,262; B 16 24
22,230; B 12 8 24,276; B 4 30 30,275; B 24 4 42,232; B 24 4 42,288; B 100 56 100,184; B 48 32
76,246; B 32 14 68,283; B 12 8 58,262; B 12 10 72,225; B 52 8 100,274; B 48 32 124,246; B 32 14
132,283; B 12 10 128,225; B 12 8 142,262; B 24 4 158,288; B 50 156 175,134; B 16 8 164,234; B 20
8 178,262; B 4 30 170,275; B 16 24 178,230; B 12 8 176,276; B 16 54 192,239; L NM; B 8 154 4,93; B
200 16 100,8; B 200 32 100,196; B 8 12 20,56; B 8 12 20,106; B 8 12 20,156; B 8 12 22,274; B 26 8
33,156; B 26 8 33,106; B 26 8 33,56; B 12 8 38,224; B 28 8 50,262; B 8 40 40,240; B 12 8 50,28; B 8
22 50,41; B 108 108 100,106; B 26 8 65,224; B 52 8 100,246; B 52 8 100,274; B 16 68 100,242; B 12
8 100,28; B 8 22 100,41; B 26 8 135,224; B 28 8 150,262; B 12 8 150,28; B 8 22 150,41; B 26 8
167,56; B 26 8 167,106; B 26 8 167,156; B 12 18 162,229; B 8 30 160,245; B 8 12 178,274; B 8 12
180,56; B 8 12 180,106; B 8 12 180,156; B 8 154 196,93; L NC; B 4 8 4,164; B 4 8 4,130; B 4 8 4,80; B
4 8 4,32; B 8 4 10,8; B 4 8 20,56; B 4 8 20,106; B 4 8 20,156; B 4 8 22,274; B 8 4 30,8; B 8 4 38,224; B
8 4 50,28; B 8 4 50,8; B 8 4 58,262; B 8 4 58,224; B 8 4 70,8; B 8 4 72,224; B 8 4 76,208; B 8 4
76,200; B 8 4 80,246; B 8 4 80,274; B 8 4 90,8; B 8 4 92,208; B 8 4 92,200; B 8 4 100,246; B 8 4
100,274; B 8 4 100,28; B 8 4 108,208; B 8 4 108,200; B 8 4 110,8; B 8 4 120,246; B 8 4 120,274; B 8
4 124,208; B 8 4 124,200; B 8 4 128,224; B 8 4 130,8; B 8 4 142,262; B 8 4 142,224; B 8 4 150,28; B
8 4 150,8; B 8 4 162,224; B 8 4 162,234; B 8 4 170,8; B 4 8 178,274; B 4 8 180,56; B 4 8 180,106; B 4
8 180,156; B 8 4 190,8; B 4 8 196,32; B 4 8 196,80; B 4 8 196,130; B 4 8 196,164; L NP; B 180 4
100,20; B 34 4 27,202; B 4 184 12,112; B 8 6 22,271; B 4 56 22,242; B 132 4 88,254; B 148 4 100,36;
B 38 4 45,190; B 4 156 28,114; B 12 8 38,224; B 4 18 42,217; B 4 12 42,206; B 106 4 101,238; B 52 4
74,282; B 12 8 58,224; B 4 18 62,217; B 4 24 62,200; B 4 10 100,285; B 42 4 133,282; B 12 8
142,224; B 4 18 138,217; B 38 4 155,190; B 4 24 138,200; B 28 4 164,246; B 4 28 152,268; B 4 8
152,242; B 12 8 162,224; B 4 18 158,217; B 34 4 173,202; B 4 12 158,206; B 4 156 172,114; B 8 6
178,271; B 4 56 178,242; B 4 184 188,112; L NI; B 12 32 22,230; B 12 16 22,262; B 12 32 178,230; B
12 16 178,262; L NG; B 92 92 100,106; DF;

PadIn4

DS 205 125/1; L NM; B 200 16 100,8; B 200 32 100,196; B 108 108 100,106; B 8 34 150,41; L ND; B 4 192 14,116; B 134 4 81,22; B 16 26 22,185; B 154 36 105,186; B 8 16 150,28; L NP; B 150 4 93,178; B 150 4 93,192; B 8 8 22,204; B 4 10 20,173; B 4 16 20,200; B 4 14 166,185; L NC; B 4 4 22,204; B 8 4 36,200; B 8 4 56,200; B 8 4 76,200; B 8 4 96,200; B 8 4 116,200; B 8 4 136,200; B 4 8 150,30; B 8 4 156,200; B 8 4 176,200; L NG; B 92 92 100,106; DF;

PadGnd

DS 206 125/1; L NM; B 200 16 100,8; B 200 32 100,196; B 164 40 100,176; B 16 36 54,178; B 108 108 100,106; B 16 36 146,178; L NG; B 92 92 100,106; DF;

PadVdd

DS 207 125/1; L NM; B 16 200 8,100; B 40 108 32,100; B 108 108 106,100; B 36 16 34,54; B 36 16 34,146; L NG; B 92 92 106,100; DF;

PadTriState4

DS 208 125/1; L ND; B 16 56 8,240; B 200 56 100,28; B 50 156 25,134; B 16 8 22,264; B 16 20 22,242; B 12 8 24,278; B 18 8 37,248; B 4 22 30,271; B 6 14 43,253; B 12 8 48,256; B 12 14 52,219; B 100 56 100,184; B 112 8 108,336; B 32 34 68,323; B 14 66 59,275; B 68 72 100,272; B 32 34 132,323; B 14 56 141,280; B 50 156 175,134; B 16 8 164,248; B 12 40 170,316; B 4 22 168,271; B 12 8 174,278; B 16 8 178,264; B 16 24 178,240; B 8 8 178,334; B 24 10 188,319; B 18 56 191,240; B 12 54 194,287; L NM; B 8 154 4,93; B 200 16 100,8; B 200 32 100,196; B 42 6 27,223; B 6 120 9,280; B 8 12 20,56; B 8 12 20,106; B 8 12 20,156; B 8 54 22,297; B 26 8 33,156; B 26 8 33,106; B 26 8 33,56; B 38 8 45,320; B 12 8 38,236; B 12 8 38,236; B 28 8 50,288; B 8 54 40,265; B 12 8 50,28; B 12 8 52,222; B 8 22 50,41; B 108 108 100,106; B 96 8 100,336; B 12 28 58,246; B 52 8 100,240; B 52 8 100,272; B 52 8 100,304; B 16 124 100,274; B 12 8 100,28; B 8 22 100,41; B 36 8 154,320; B 22 8 147,288; B 12 28 142,246; B 12 8 150,28; B 8 22 150,41; B 26 8 167,56; B 26 8 167,106; B 26 8 167,156; B 12 20 162,242; B 8 52 160,266; B 8 54 176,297; B 12 8 180,334; B 8 12 180,56; B 8 12 180,106; B 8 12 180,156; B 8 154 196,93; L NC; B 4 8 4,164; B 4 8 4,130; B 4 8 4,80; B 4 8 4,32; B 8 4 10,8; B 4 8 20,56; B 4 8 20,106; B 4 8 20,156; B 4 8 22,276; B 8 4 30,8; B 8 4 38,236; B 8 4 50,28; B 8 4 50,8; B 8 4 52,222; B 8 4 58,256; B 8 4 58,288; B 8 4 58,320; B 8 4 58,336; B 8 4 58,236; B 8 4 70,8; B 8 4 76,208; B 8 4 76,200; B 8 4 78,336; B 8 4 80,272; B 8 4 80,240; B 8 4 80,304; B 8 4 90,8; B 8 4 92,208; B 8 4 92,200; B 8 4 100,336; B 8 4 100,272; B 8 4 100,240; B 8 4 100,304; B 8 4 100,28; B 8 4 108,208; B 8 4 108,200; B 8 4 110,8; B 8 4 120,272; B 8 4 120,240; B 8 4 120,304; B 8 4 122,336; B 8 4 124,208; B 8 4 124,200; B 8 4 130,8; B 8 4 142,256; B 8 4 142,288; B 8 4 142,320; B 8 4 142,336; B 8 4 142,236; B 8 4 150,28; B 8 4 150,8; B 8 4 162,236; B 8 4 162,248; B 8 4 170,8; B 4 8 176,276; B 8 4 180,334; B 4 8 180,56; B 4 8 180,106; B 4 8 180,156; B 8 4 190,8; B 4 8 196,32; B 4 8 196,80; B 4 8 196,130; B 4 8 196,164; L NP; B 180 4 100,20; B 34 4 27,202; B 4 184 12,112; B 8 6 22,273; B 4 50 22,249; B 16 4 30,256; B 148 4 100,36; B 38 4 45,190; B 4 156 28,114; B 12 8 38,236; B 4 44 36,276; B 118 4 95,296; B 48 4 64,312; B 4 22 42,223; B 4 30 42,325; B 4 12 42,206; B 114 4 103,280; B 4 18 48,271; B 114 4 105,328; B 104 4 100,248; B 104 4 100,264; B 12 8 58,236; B 4 22 62,223; B 4 24 62,200; B 42 4 133,312; B 12 8 142,236; B 4 22 138,223; B 38 4 155,190; B 4 24 138,200; B 4 16 152,304; B 4 18 152,271; B 4 12 152,252; B 26 4 165,256; B 12 8 162,236; B 4 22 158,223; B 34 4 173,202; B 4 12 158,206; B 4 50 160,303; B 26 4 173,286; B 4 48 170,316; B 4 156 172,114; B 8 6 176,273; B 4 50 178,249; B 6 8 183,334; B 4 42 184,309; B 4 184 188,112; L NI; B 12 16 22,264; B 12 32 22,240; B 12 16 178,264; B 12 32 178,240; B 12 18 184,319; L NG; B 92 92 100,106; DF;

PadClock

DS 209 125/1; L NM; B 200 16 100,8; B 16 152 8,92; B 16 108 8,274; B 200 32 100,196; B 14 12 17,322; B 102 8 75,354; B 80 12 64,304; B 80 12 64,322; B 88 8 70,224; B 12 8 40,282; B 8 12 48,242; B 8 12 48,232; B 108 108 100,106; B 8 28 50,38; B 102 8 125,340; B 102 8 125,340; B 8 12 80,266; B 8 12 80,240; B 10 6 89,269; B 8 20 90,282; B 78 12 135,304; B 80 12 136,322; B 10 6 111,269; B 8 20 110,282; B 8 12 120,266; B 8 12 120,240; B 8 14 144,351; B 12 8 160,282; B 8 98 178,261; B 12 8 182,164; L ND; B 16 80 8,192; B 8 8 30,224; B 26 4 43,214; B 16 22 38,299; B 4 10 32,219; B 8 8 42,282; B 8 6 42,287; B 4 20 44,270; B 14 4 51,262; B 8 8 48,240; B 136 4 114,22; B 8 12 50,30; B 14 4 57,238; B 8 12 56,322; B 4 34 54,305; B 18 32 63,200; B 4 30 58,277; B 20 4 68,262; B 8 12 66,304; B 4 24 64,250; B 32 16 80,208; B 14 4 71,238; B 12 32 74,308; B 8 8 78,340; B 8 8 80,264; B 8 8 80,238; B 4 8 80,332; B 4 10 80,323; B 18 8 89,320; B 8 4 84,262; B 4 36 82,222; B 20 16 92,192; B 12 16 92,256; B 8 12 100,322; B 8 114 100,253; B 38 16 117,204; B 18 8 111,320; B 12 16 108,256; B 8 4 116,262; B 8 8 120,264; B 8 8 120,238; B 4 36 118,222; B 8 8 122,354; B 4 22 120,339; B 4 10 120,323; B 12 32 126,308; B 20 4 132,262; B 20 4 132,238; B 8 12 134,304; B 8 12 144,322; B 4 56 142,264; B 14 4 149,262; B 4 34 146,305; B 16 22 162,299; B 8 8 158,282; B 4 20 156,270; B 8 6 158,287; B 16 82 182,241; B 8 8 184,164; B 4 140 184,90; B 4 8 188,164; B 4 122 192,221; L NC; B 4 8 4,226; B 4 8 4,160; B 4 8 12,226; B 4 8 12,160; B 4 4 28,354; B 4 4 30,224; B 4 4 38,282; B 4 8 38,304; B 4 4 42,282; B 4 4 48,240; B 4 4 48,244; B 4 8 50,30; B 4 8 56,322; B 4 8 66,304; B 4 8 68,204; B 4 4 78,340; B 4 4 80,264; B 4 4 80,268; B 4 4 80,238; B 4 4 80,242; B 4 4 90,288; B 4 8 96,204; B 4 8 100,304; B 4 8 100,322; B 4 8 104,204; B 4 4 110,224; B 4 4 110,288; B 4 4 120,264; B 4 4 120,268; B 4 4 120,238; B 4 4 120,242; B 4 4 122,354; B 4 8 132,204; B 4 8 134,304; B 4 8 144,322; B 4 4 158,282; B 4 4 162,282; B 4 8 162,304; B 4 4 172,340; B 4 4 178,194; B 4 6 178,207; B 4 8 178,276; B 4 8 178,244; B 8 4 182,164; L NP; B 8 8 28,354; B 4 22 26,339; B 4 36 26,310; B 24 4 38,294; B 6 8 37,282; B 42 4 57,226; B 4 10 38,275; B 4 46 38,249; B 12 4 44,272; B 8

6 48,245; B 10 4 55,244; B 4 12 58,238; B 32 4 74,180; B 4 38 60,201; B 10 4 73,268; B 10 4 73,244; B 4 12 70,262; B 4 12 70,238; B 18 4 81,278; B 4 40 74,308; B 4 8 74,284; B 4 30 76,211; B 8 6 80,269; B 8 6 80,243; B 12 4 88,244; B 8 8 90,288; B 4 38 88,201; B 88 4 132,180; B 4 22 92,257; B 4 12 92,274; B 4 36 92,310; B 4 4 92,180; B 12 4 112,244; B 8 8 110,224; B 8 8 110,288; B 4 22 108,257; B 4 12 108,274; B 4 36 108,310; B 18 4 119,278; B 4 32 112,208; B 8 6 120,269; B 8 6 120,243; B 42 4 143,216; B 10 4 127,268; B 10 4 127,244; B 4 24 124,204; B 4 40 126,308; B 4 8 126,284; B 4 12 130,262; B 4 12 130,238; B 24 4 162,294; B 12 4 156,272; B 6 8 163,282; B 4 10 162,275; B 4 58 162,243; B 8 8 172,340; B 4 8 174,332; B 4 36 174,310; B 8 8 178,194; B 6 8 179,164; B 4 14 178,175; B 6 4 185,196; B 4 88 186,242; L NI; B 12 12 44,272; B 12 12 58,238; B 16 12 72,238; B 12 12 70,262; B 12 16 92,320; B 12 16 108,320; B 16 12 128,238; B 12 12 130,262; B 12 12 156,272; L NG; B 92 92 100,106; DF;

PadClockOut

DS 210 125/1; L ND; B 16 46 8,235; B 200 56 100,28; B 50 156 25,134; B 4 12 8,284; B 22 4 20,280; B 22 4 20,284; B 22 4 23,256; B 16 24 22,230; B 14 8 25,268; B 14 8 31,268; B 12 4 33,286; B 24 4 42,232; B 4 18 36,263; B 16 4 46,274; B 4 4 40,270; B 8 8 42,286; B 100 56 100,184; B 48 32 76,246; B 42 4 73,286; B 32 14 68,279; B 12 8 58,262; B 12 10 72,225; B 16 16 100,282; B 48 32 124,246; B 42 4 127,276; B 32 14 132,283; B 12 10 128,225; B 12 8 142,262; B 24 4 158,288; B 50 156 175,134; B 16 8 164,234; B 20 8 178,262; B 4 30 170,275; B 16 24 178,230; B 12 8 176,276; B 16 54 192,239; L NM; B 8 154 4,93; B 200 16 100,8; B 200 32 100,196; B 8 12 20,56; B 8 12 20,106; B 8 12 20,156; B 8 12 22,266; B 26 8 33,156; B 26 8 33,106; B 26 8 33,56; B 12 8 38,224; B 28 8 50,262; B 8 40 40,240; B 8 12 42,284; B 12 8 50,28; B 8 22 50,41; B 108 108 100,106; B 26 8 65,224; B 52 8 100,246; B 16 82 100,249; B 12 8 100,28; B 8 22 100,41; B 26 8 135,224; B 28 8 150,262; B 12 8 150,28; B 8 22 150,41; B 26 8 167,56; B 26 8 167,106; B 26 8 167,156; B 12 18 162,229; B 8 30 160,245; B 8 12 178,274; B 8 12 180,56; B 8 12 180,106; B 8 12 180,156; B 8 154 196,93; L NC; B 4 8 4,164; B 4 8 4,130; B 4 8 4,80; B 4 8 4,32; B 8 4 10,8; B 4 8 20,56; B 4 8 20,106; B 4 8 20,156; B 4 8 22,266; B 8 4 30,8; B 8 4 38,224; B 4 8 42,284; B 8 4 50,28; B 8 4 50,8; B 8 4 58,262; B 8 4 58,224; B 8 4 70,8; B 8 4 72,224; B 8 4 76,208; B 8 4 76,200; B 8 4 80,246; B 8 4 90,8; B 8 4 92,208; B 8 4 92,200; B 8 4 100,246; B 8 4 100,278; B 8 4 100,286; B 8 4 100,28; B 8 4 108,208; B 8 4 108,200; B 8 4 110,8; B 8 4 120,246; B 8 4 124,208; B 8 4 124,200; B 8 4 128,224; B 8 4 130,8; B 8 4 142,262; B 8 4 142,224; B 8 4 150,28; B 8 4 150,8; B 8 4 162,224; B 8 4 162,234; B 8 4 170,8; B 4 8 178,274; B 4 8 180,56; B 4 8 180,106; B 4 8 180,156; B 8 4 190,8; B 4 8 196,32; B 4 8 196,80; B 4 8 196,130; B 4 8 196,164; L NP; B 180 4 100,20; B 34 4 27,202; B 4 184 12,112; B 8 6 22,263; B 4 48 22,238; B 4 16 22,282; B 28 4 36,246; B 148 4 100,36; B 38 4 45,190; B 4 156 28,114; B 12 8 38,224; B 50 4 63,280; B 8 4 42,282; B 4 18 42,217; B 4 12 42,206; B 4 8 48,252; B 106 4 101,254; B 106 4 101,238; B 12 8 58,224; B 4 18 62,217; B 4 24 62,200; B 42 4 133,282; B 12 8 142,224; B 4 18 138,217; B 38 4 155,190; B 4 24 138,200; B 28 4 164,246; B 4 28 152,268; B 4 8 152,242; B 12 8 162,224; B 4 18 158,217; B 34 4 173,202; B 4 12 158,206; B 4 156 172,114; B 8 6 178,271; B 4 56 178,242; B 4 184 188,112; L NI; B 12 32 22,230; B 12 12 22,256; B 12 32 178,230; B 12 16 178,262; L NG; B 92 92 100,106; DF;

PadOut8

DS 250 125/1; L ND; B 16 54 8,239; B 200 56 100,28; B 50 156 25,134; B 20 6 22,263; B 16 24 22,230; B 12 8 24,276; B 4 30 30,275; B 24 4 42,232; B 24 4 42,288; B 48 14 64,283; B 100 56 100,184; B 48 32 76,246; B 12 8 58,262; B 12 10 72,225; B 52 8 100,274; B 48 32 124,246; B 32 14 132,283; B 12 10 128,225; B 12 8 142,262; B 24 4 158,288; B 50 156 175,134; B 16 8 164,234; B 20 8 178,262; B 4 30 170,275; B 16 24 178,230; B 12 8 176,276; B 16 54 192,239; L NM; B 8 154 4,93; B 200 16 100,8; B 200 32 100,196; B 8 12 20,56; B 8 12 20,106; B 8 12 20,156; B 8 12 22,274; B 26 8 33,156; B 26 8 33,106; B 26 8 33,56; B 12 8 38,224; B 28 8 50,262; B 8 40 40,240; B 12 8 50,28; B 8 22 50,41; B 108 108 100,106; B 26 8 65,224; B 52 8 100,246; B 52 8 100,274; B 16 68 100,242; B 12 8 100,28; B 8 22 100,41; B 26 8 135,224; B 28 8 150,262; B 12 8 150,28; B 8 22 150,41; B 26 8 167,56; B 26 8 167,106; B 26 8 167,156; B 12 18 162,229; B 8 30 160,245; B 8 12 178,274; B 8 12 180,56; B 8 12 180,106; B 8 12 180,156; B 8 154 196,93; L NC; B 4 8 4,164; B 4 8 4,130; B 4 8 4,80; B 4 8 4,32; B 8 4 10,8; B 4 8 20,56; B 4 8 20,106; B 4 8 20,156; B 4 8 22,274; B 8 4 30,8; B 8 4 38,224; B 8 4 50,28; B 8 4 50,8; B 8 4 58,262; B 8 4 58,224; B 8 4 70,8; B 8 4 72,224; B 8 4 76,208; B 8 4 76,200; B 8 4 80,246; B 8 4 80,274; B 8 4 90,8; B 8 4 92,208; B 8 4 92,200; B 8 4 100,246; B 8 4 100,274; B 8 4 100,28; B 8 4 108,208; B 8 4 108,200; B 8 4 110,8; B 8 4 120,246; B 8 4 120,274; B 8 4 124,208; B 8 4 124,200; B 8 4 128,224; B 8 4 130,8; B 8 4 142,262; B 8 4 142,224; B 8 4 150,28; B 8 4 150,8; B 8 4 162,224; B 8 4 162,234; B 8 4 170,8; B 4 8 178,274; B 4 8 180,56; B 4 8 180,106; B 4 8 180,156; B 8 4 190,8; B 4 8 196,32; B 4 8 196,80; B 4 8 196,130; B 4 8 196,164; L NP; B 180 4 100,20; B 34 4 27,202; B 4 184 12,112; B 8 6 22,271; B 4 56 22,242; B 132 4 88,254; B 148 4 100,36; B 38 4 45,190; B 4 156 28,114; B 12 8 38,224; B 64 4 68,282; B 4 18 42,217; B 4 12 42,206; B 106 4 101,238; B 12 8 58,224; B 4 18 62,217; B 4 24 62,200; B 4 10 100,285; B 42 4 133,282; B 12 8 142,224; B 4 18 138,217; B 38 4 155,190; B 4 24 138,200; B 28 4 164,246; B 4 28 152,268; B 4 8 152,242; B 12 8 162,224; B 4 18 158,217; B 34 4 173,202; B 4 12 158,206; B 4 156 172,114; B 8 6 178,271; B 4 56 178,242; B 4 184 188,112; L NI; B 12 32 22,230; B 12 16 22,262; B 12 32 178,230; B 12 16 178,262; L NG; B 92 92 100,106; DF;

PadIn8

DS 280 125/1; L NP; B 6 4 21,224; B 150 4 93,178; B 150 4 93,192; B 8 8 22,204; B 4 8 20,172; B 4 6 20,197; B 94 4 65,228; B 14 2 113,237; B 10 2 115,235; B 28 2 130,227; B 4 4 118,232; B 4 10

128,241; B 62 2 147,229; B 38 2 159,231; B 4 10 166,185; B 18 6 105,249; B 8 8 110,242; B 52 2 152,247; B 8 2 136,251; B 36 2 160,245; B 20 2 136,249; L ND; B 8 8 16,226; B 4 24 14,210; B 4 148 14,98; B 170 26 97,185; B 154 6 105,201; B 154 4 105,170; B 16 10 100,229; B 8 4 120,246; B 16 10 130,229; B 16 12 154,230; B 70 4 127,222; B 16 2 154,219; B 142 4 83,22; B 8 12 150,30; B 32 12 184,220; B 12 8 174,254; B 4 10 172,245; B 36 4 176,238; B 8 4 190,242; B 12 52 194,188; B 8 4 100,246; B 4 6 102,237; B 44 4 118,242; B 8 4 136,246; B 4 6 136,237; L NM; B 200 16 100,8; B 200 32 100,196; B 12 8 18,226; B 108 108 100,106; B 12 14 152,219; B 8 28 150,38; B 8 28 190,250; B 12 154 194,93; B 8 12 100,246; B 12 40 174,238; B 8 40 120,238; B 64 6 148,261; B 8 12 136,246; L NI; B 16 12 110,242; B 12 12 128,242; B 12 12 172,246; L NC; B 8 4 18,226; B 4 4 22,204; B 8 4 36,200; B 8 4 56,200; B 8 4 76,200; B 8 4 96,200; B 8 4 116,200; B 4 4 120,244; B 8 4 136,200; B 8 4 152,222; B 4 8 150,30; B 8 4 156,200; B 8 4 174,222; B 8 4 174,254; B 8 4 176,200; B 4 4 190,240; B 8 4 194,166; B 4 8 100,246; B 4 8 136,246; L NG; B 92 92 100,106; DF;

PadTriState8

DS 281 125/1; L ND; B 16 56 8,240; B 200 56 100,28; B 50 156 25,134; B 16 8 22,264; B 16 20 22,242; B 12 8 24,278; B 32 22 36,313; B 18 8 37,248; B 4 22 30,271; B 6 14 43,253; B 12 8 48,256; B 12 14 52,219; B 100 56 100,184; B 112 8 108,336; B 32 34 68,323; B 14 66 59,275; B 68 72 100,272; B 32 34 132,323; B 14 56 141,280; B 50 156 175,134; B 16 8 164,248; B 12 40 170,316; B 4 22 168,271; B 12 8 174,278; B 16 8 178,264; B 16 24 178,240; B 8 8 178,334; B 24 10 188,319; B 18 56 191,240; B 12 54 194,287; L NM; B 8 154 4,93; B 200 16 100,8; B 200 32 100,196; B 42 6 27,223; B 6 120 9,280; B 8 12 20,56; B 8 12 20,106; B 8 12 20,156; B 8 54 22,297; B 26 8 33,156; B 26 8 33,106; B 26 8 33,56; B 38 8 45,320; B 12 8 38,236; B 12 8 38,236; B 28 8 50,288; B 8 54 40,265; B 12 8 50,28; B 12 8 52,222; B 8 22 50,41; B 108 108 100,106; B 96 8 100,336; B 12 28 58,246; B 52 8 100,240; B 52 8 100,272; B 52 8 100,304; B 16 124 100,274; B 12 8 100,28; B 8 22 100,41; B 36 8 154,320; B 22 8 147,288; B 12 28 142,246; B 12 8 150,28; B 8 22 150,41; B 26 8 167,56; B 26 8 167,106; B 26 8 167,156; B 12 20 162,242; B 8 52 160,266; B 8 54 176,297; B 12 8 180,334; B 8 12 180,56; B 8 12 180,106; B 8 12 180,156; B 8 154 196,93; L NC; B 4 8 4,164; B 4 8 4,130; B 4 8 4,80; B 4 8 4,32; B 8 4 10,8; B 4 8 20,56; B 4 8 20,106; B 4 8 20,156; B 4 8 22,276; B 8 4 30,8; B 8 4 38,236; B 8 4 50,28; B 8 4 50,8; B 8 4 52,222; B 8 4 58,256; B 8 4 58,288; B 8 4 58,320; B 8 4 58,336; B 8 4 58,236; B 8 4 70,8; B 8 4 76,208; B 8 4 76,200; B 8 4 78,336; B 8 4 80,272; B 8 4 80,240; B 8 4 80,304; B 8 4 90,8; B 8 4 92,208; B 8 4 92,200; B 8 4 100,336; B 8 4 100,272; B 8 4 100,240; B 8 4 100,304; B 8 4 100,28; B 8 4 108,208; B 8 4 108,200; B 8 4 110,8; B 8 4 120,272; B 8 4 120,240; B 8 4 120,304; B 8 4 122,336; B 8 4 124,208; B 8 4 124,200; B 8 4 130,8; B 8 4 142,256; B 8 4 142,288; B 8 4 142,320; B 8 4 142,336; B 8 4 142,236; B 8 4 150,28; B 8 4 150,8; B 8 4 162,236; B 8 4 162,248; B 8 4 170,8; B 4 8 176,276; B 8 4 180,334; B 4 8 180,56; B 4 8 180,106; B 4 8 180,156; B 8 4 190,8; B 4 8 196,32; B 4 8 196,80; B 4 8 196,130; B 4 8 196,164; L NP; B 180 4 100,20; B 34 4 27,202; B 4 184 12,112; B 32 4 28,312; B 4 18 14,323; B 32 4 28,330; B 8 6 22,273; B 4 50 22,249; B 16 4 30,256; B 148 4 100,36; B 38 4 45,190; B 4 156 28,114; B 12 8 38,236; B 4 44 36,276; B 118 4 95,296; B 48 4 64,312; B 4 22 42,223; B 4 12 42,334; B 4 12 42,206; B 114 4 103,280; B 4 18 48,271; B 114 4 105,328; B 104 4 100,248; B 104 4 100,264; B 12 8 58,236; B 4 22 62,223; B 4 24 62,200; B 42 4 133,312; B 12 8 142,236; B 4 22 138,223; B 38 4 155,190; B 4 24 138,200; B 4 16 152,304; B 4 18 152,271; B 4 12 152,252; B 26 4 165,256; B 12 8 162,236; B 4 22 158,223; B 34 4 173,202; B 4 12 158,206; B 4 50 160,303; B 26 4 173,286; B 4 48 170,316; B 4 156 172,114; B 8 6 176,273; B 4 50 178,249; B 6 8 183,334; B 4 42 184,309; B 4 184 188,112; L NI; B 12 16 22,264; B 12 32 22,240; B 12 16 178,264; B 12 32 178,240; B 12 18 184,319; L NG; B 92 92 100,106; DF;

PadMux

DS 282 125/1; L ND; B 16 56 8,240; B 200 56 100,28; B 50 156 25,134; B 16 56 8,240; B 200 56 100,28; B 50 156 25,134; B 8 8 8,344; B 8 8 8,344; B 20 4 22,342; B 20 4 22,342; B 16 8 22,264; B 16 20 22,242; B 16 8 22,264; B 16 20 22,242; B 12 8 24,278; B 12 8 24,278; B 32 22 36,313; B 32 22 36,313; B 18 8 37,248; B 4 22 30,271; B 18 8 37,248; B 4 22 30,271; B 8 8 36,344; B 8 8 36,344; B 6 14 43,253; B 6 14 43,253; B 12 8 48,256; B 12 8 48,256; B 12 14 52,219; B 12 14 52,219; B 100 56 100,184; B 100 56 100,184; B 112 8 108,336; B 32 34 68,323; B 14 66 59,275; B 112 8 108,336; B 32 34 68,323; B 14 66 59,275; B 68 72 100,272; B 68 72 100,272; B 32 34 132,323; B 32 34 132,323; B 14 56 141,280; B 14 56 141,280; B 50 156 175,134; B 50 156 175,134; B 16 8 164,248; B 16 8 164,248; B 12 40 170,316; B 12 40 170,316; B 4 22 168,271; B 4 22 168,271; B 12 8 174,278; B 12 8 174,278; B 16 8 178,264; B 16 24 178,240; B 16 8 178,264; B 16 24 178,240; B 8 8 178,334; B 8 8 178,334; B 24 10 188,319; B 24 10 188,319; B 18 56 191,240; B 18 56 191,240; B 12 54 194,287; B 12 54 194,287; L NM; B 8 154 4,93; B 200 16 100,8; B 200 32 100,196; B 8 154 4,93; B 200 16 100,8; B 200 32 100,196; B 8 8 8,344; B 8 8 8,344; B 42 6 27,223; B 6 120 9,280; B 42 6 27,223; B 6 120 9,280; B 8 12 20,56; B 8 12 20,106; B 8 12 20,156; B 8 12 20,56; B 8 12 20,106; B 8 12 20,156; B 8 8 22,354; B 162 6 99,357; B 8 54 22,297; B 8 8 22,354; B 162 6 99,357; B 8 54 22,297; B 26 8 33,156; B 26 8 33,106; B 26 8 33,56; B 26 8 33,156; B 26 8 33,106; B 26 8 33,56; B 38 8 45,320; B 38 8 45,320; B 12 8 38,344; B 12 8 38,236; B 12 8 38,236; B 12 8 38,344; B 12 8 38,236; B 12 8 38,236; B 28 8 50,288; B 8 54 40,265; B 28 8 50,288; B 8 54 40,265; B 12 8 50,28; B 12 8 50,28; B 12 8 52,222; B 8 22 50,41; B 108 108 100,106; B 12 8 52,222; B 8 22 50,41; B 108 108 100,106; B 96 8 100,336; B 12 28 58,246; B 96 8 100,336; B 12 28 58,246; B 52 8 100,240; B 52 8 100,272; B 52 8 100,304; B 52 8 100,240; B 52 8 100,272; B 52 8 100,304; B 16 124 100,274; B 16 124 100,274; B 12 8 100,28; B 12 8 100,28; B 8 22 100,41; B 8 22 100,41; B 36 8 154,320; B 22 8 147,288; B 12 28 142,246; B 36 8 154,320; B 22 8 147,288; B 12 28 142,246; B 12 8 150,28; B 12 8 150,28; B 8 22 150,41; B 8 22 150,41; B 26 8 167,56; B 26 8 167,106; B 26 8 167,156; B 26 8 167,56; B 26 8 167,106; B 26 8 167,156; B 12 20 162,242; B 8 52 160,266; B 12 20 162,242; B 8 52 160,266; B 8 54 176,297; B 8 54

176,297; B 12 8 180,334; B 12 8 180,334; B 8 12 180,56; B 8 12 180,106; B 8 12 180,156; B 8 12 180,56; B 8 12 180,106; B 8 12 180,156; B 6 20 180,347; B 6 6 180,339; B 6 6 180,357; B 6 20 180,347; B 6 6 180,339; B 6 6 180,357; B 8 154 196,93; B 8 154 196,93; L NC; B 4 8 4,164; B 4 8 4,130; B 4 8 4,80; B 4 8 4,32; B 4 8 4,164; B 4 8 4,130; B 4 8 4,80; B 4 8 4,32; B 4 4 8,344; B 8 4 10,8; B 4 4 8,344; B 8 4 10,8; B 4 8 20,56; B 4 8 20,106; B 4 8 20,156; B 4 8 20,56; B 4 8 20,106; B 4 8 20,156; B 4 4 22,354; B 4 8 22,276; B 4 4 22,354; B 4 8 22,276; B 8 4 30,8; B 8 4 30,8; B 8 4 38,344; B 8 4 38,236; B 8 4 38,344; B 8 4 38,236; B 8 4 50,28; B 8 4 50,8; B 8 4 50,28; B 8 4 50,8; B 8 4 52,222; B 8 4 52,222; B 8 4 58,256; B 8 4 58,288; B 8 4 58,320; B 8 4 58,336; B 8 4 58,236; B 8 4 58,256; B 8 4 58,288; B 8 4 58,320; B 8 4 58,336; B 8 4 58,236; B 8 4 70,8; B 8 4 70,8; B 8 4 76,208; B 8 4 76,200; B 8 4 76,208; B 8 4 76,200; B 8 4 78,336; B 8 4 78,336; B 8 4 80,272; B 8 4 80,240; B 8 4 80,304; B 8 4 80,272; B 8 4 80,240; B 8 4 80,304; B 8 4 90,8; B 8 4 90,8; B 8 4 92,208; B 8 4 92,200; B 8 4 92,208; B 8 4 92,200; B 8 4 100,336; B 8 4 100,272; B 8 4 100,240; B 8 4 100,304; B 8 4 100,28; B 8 4 100,336; B 8 4 100,272; B 8 4 100,240; B 8 4 100,304; B 8 4 100,28; B 8 4 108,208; B 8 4 108,200; B 8 4 108,208; B 8 4 108,200; B 8 4 110,8; B 8 4 110,8; B 8 4 120,272; B 8 4 120,240; B 8 4 120,304; B 8 4 120,272; B 8 4 120,240; B 8 4 120,304; B 8 4 122,336; B 8 4 122,336; B 8 4 124,208; B 8 4 124,200; B 8 4 124,208; B 8 4 124,200; B 8 4 130,8; B 8 4 130,8; B 8 4 142,256; B 8 4 142,288; B 8 4 142,320; B 8 4 142,336; B 8 4 142,236; B 8 4 142,256; B 8 4 142,288; B 8 4 142,320; B 8 4 142,336; B 8 4 142,236; B 8 4 150,28; B 8 4 150,8; B 8 4 150,28; B 8 4 150,8; B 8 4 162,236; B 8 4 162,248; B 8 4 162,236; B 8 4 162,248; B 8 4 170,8; B 8 4 170,8; B 4 8 176,276; B 4 8 176,276; B 8 4 180,334; B 8 4 180,334; B 4 8 180,56; B 4 8 180,106; B 4 8 180,156; B 4 8 180,56; B 4 8 180,106; B 4 8 180,156; B 8 4 190,8; B 8 4 190,8; B 4 8 196,32; B 4 8 196,80; B 4 8 196,130; B 4 8 196,164; B 4 8 196,32; B 4 8 196,80; B 4 8 196,130; B 4 8 196,164; L NP; B 180 4 100,20; B 34 4 27,202; B 4 184 12,112; B 180 4 100,20; B 34 4 27,202; B 4 184 12,112; B 32 4 28,312; B 4 18 14,323; B 32 4 28,330; B 32 4 28,312; B 4 18 14,323; B 32 4 28,330; B 8 8 22,354; B 8 6 22,273; B 8 8 22,354; B 8 6 22,273; B 4 14 22,343; B 4 50 22,249; B 4 14 22,343; B 4 50 22,249; B 16 4 30,256; B 16 4 30,256; B 148 4 100,36; B 38 4 45,190; B 4 156 28,114; B 148 4 100,36; B 38 4 45,190; B 4 156 28,114; B 12 8 38,236; B 12 8 38,236; B 4 44 36,276; B 4 44 36,276; B 118 4 95,296; B 118 4 95,296; B 6 8 41,344; B 6 8 41,344; B 4 6 42,343; B 4 4 42,346; B 48 4 64,312; B 4 22 42,223; B 4 12 42,334; B 4 12 42,206; B 4 6 42,343; B 4 4 42,346; B 48 4 64,312; B 4 22 42,223; B 4 12 42,334; B 4 12 42,206; B 58 4 71,346; B 58 4 71,346; B 114 4 103,280; B 4 18 48,271; B 114 4 103,280; B 4 18 48,271; B 114 4 105,328; B 104 4 100,248; B 104 4 100,264; B 114 4 105,328; B 104 4 100,248; B 104 4 100,264; B 12 8 58,236; B 12 8 58,236; B 4 22 62,223; B 4 24 62,200; B 4 22 62,223; B 4 24 62,200; B 4 14 100,353; B 4 4 100,346; B 4 14 100,353; B 4 4 100,346; B 42 4 133,312; B 42 4 133,312; B 12 8 142,236; B 4 22 138,223; B 38 4 155,190; B 4 24 138,200; B 12 8 142,236; B 4 22 138,223; B 38 4 155,190; B 4 24 138,200; B 4 16 152,304; B 4 18 152,271; B 4 12 152,252; B 4 16 152,304; B 4 18 152,271; B 4 12 152,252; B 26 4 165,256; B 26 4 165,256; B 12 8 162,236; B 4 22 158,223; B 34 4 173,202; B 4 12 158,206; B 12 8 162,236; B 4 22 158,223; B 34 4 173,202; B 4 12 158,206; B 4 50 160,303; B 4 50 160,303; B 26 4 173,286; B 26 4 173,286; B 4 20 170,350; B 4 48 170,316; B 4 20 170,350; B 4 48 170,316; B 4 156 172,114; B 4 156 172,114; B 8 6 176,273; B 8 6 176,273; B 4 50 178,249; B 4 50 178,249; B 6 8 183,334; B 6 8 183,334; B 4 42 184,309; B 4 42 184,309; B 4 184 188,112; B 4 184 188,112; L NI; B 12 16 22,264; B 12 32 22,240; B 12 16 22,264; B 12 32 22,240; B 12 16 178,264; B 12 32 178,240; B 12 16 178,264; B 12 32 178,240; B 12 18 184,319; B 12 18 184,319; L NG; B 92 92 100,106; B 92 92 100,106; DF;

ScanClkIn

DS 510 125/1; L NP; B 4 8 30,6; B 4 4 2,34; B 8 8 4,22; B 4 12 2,184; B 4 58 2,65; B 6 8 3,36; B 4 4 2,92; B 4 4 2,34; B 4 4 2,188; B 4 18 2,199; B 4 4 2,176; B 2 4 3,176; B 12 4 8,176; B 12 4 8,158; B 6 4 7,92; B 4 12 6,14; B 4 2 6,19; B 8 6 8,189; B 4 2 8,93; B 4 4 8,138; B 8 8 10,140; B 4 4 8,138; B 4 4 8,92; B 4 40 8,112; B 8 6 12,71; B 16 4 16,84; B 12 16 14,64; B 6 4 13,138; B 4 4 14,158; B 12 4 18,6; B 4 4 14,160; B 4 2 14,161; B 2 4 15,158; B 12 4 22,138; B 4 4 18,188; B 8 4 20,160; B 4 18 18,199; B 8 6 24,109; B 4 24 22,176; B 4 4 22,160; B 4 2 22,163; B 4 4 22,188; B 4 4 22,188; B 4 2 22,189; B 2 4 23,6; B 4 4 24,84; B 4 4 24,160; B 6 8 25,154; B 8 8 26,6; B 4 4 24,160; B 4 20 24,96; B 2 4 25,160; B 2 4 25,84; L ND; B 4 4 2,126; B 4 4 2,14; B 4 26 2,141; B 4 4 2,152; B 8 8 6,168; B 2 4 3,126; B 4 32 4,112; B 6 4 7,152; B 4 4 6,14; B 8 8 8,36; B 8 8 8,184; B 4 4 8,172; B 4 4 8,180; B 4 4 8,154; B 4 4 8,162; B 4 2 8,183; B 4 4 8,182; B 8 8 10,202; B 4 4 8,182; B 4 2 8,153; B 4 4 8,152; B 4 2 8,171; B 8 2 10,207; B 4 4 8,158; B 4 4 8,176; B 4 32 8,112; B 4 4 10,14; B 8 8 12,50; B 8 8 12,76; B 4 20 12,26; B 6 4 13,182; B 4 4 12,34; B 6 4 13,152; B 4 4 12,14; B 4 32 12,112; B 4 4 14,54; B 4 4 14,74; B 4 2 14,79; B 4 2 14,89; B 4 4 14,98; B 2 4 13,34; B 2 4 13,14; B 8 8 16,92; B 8 8 16,92; B 4 2 14,53; B 8 4 16,80; B 8 4 16,88; B 4 4 14,78; B 4 4 14,98; B 4 16 14,64; B 8 4 16,84; B 6 4 17,14; B 2 4 15,98; B 4 16 18,176; B 4 4 18,2; B 4 4 18,10; B 4 4 18,144; B 4 4 18,148; B 4 4 18,14; B 4 4 18,14; B 14 4 23,78; B 4 4 18,6; B 8 8 20,154; B 4 4 22,144; B 4 2 22,145; B 4 14 22,125; B 4 4 22,144; B 4 4 22,134; B 4 4 22,142; B 4 16 22,176; B 4 4 22,138; B 8 8 24,114; B 8 8 26,202; B 4 16 26,176; B 4 64 28,48; B 4 4 28,180; B 4 4 28,14; B 4 4 28,78; B 4 16 28,190; B 2 4 29,180; B 2 4 31,14; L NM; B 12 8 6,36; B 8 8 4,50; B 8 8 4,22; B 22 8 11,6; B 6 8 3,202; B 2 8 1,168; B 12 8 6,92; B 6 8 3,140; B 8 8 6,168; B 8 12 8,186; B 8 8 10,140; B 8 8 10,202; B 8 8 12,50; B 8 12 12,74; B 24 8 20,22; B 22 8 21,168; B 8 8 16,92; B 8 6 16,91; B 8 8 16,92; B 18 8 23,140; B 8 8 18,202; B 16 8 24,50; B 12 8 22,154; B 12 8 26,92; B 8 12 24,112; B 8 8 26,6; B 8 8 26,202; B 2 8 31,6; B 2 8 31,202; L NI; B 12 12 8,176; B 12 12 8,158; B 12 10 12,73; B 12 24 14,64; L NC; B 4 4 4,22; B 8 4 6,36; B 4 4 6,168; B 4 8 8,186; B 4 4 10,140; B 4 4 10,202; B 4 8 12,74; B 4 4 12,50; B 4 4 16,92; B 4 4 16,92; B 8 4 22,154; B 4 8 24,112; B 4 4 26,202; B 4 4 26,6; DF;

ScanIn

DS 511 125/1; L NP; B 4 18 2,199; B 4 12 2,184; B 4 4 2,176; B 4 4 2,188; B 12 4 8,176; B 12 4 8,158; B 2 4 3,176; B 8 6 8,189; B 4 4 10,42; B 6 8 11,36; B 4 4 10,42; B 4 40 14,90; B 4 4 14,158; B 4 4 14,42; B 4 2 14,161; B 4 26 14,57; B 4 4 14,42; B 4 4 14,160; B 2 4 15,158; B 8 4 20,160; B 4 4 18,188; B 4 18 18,199; B 4 2 22,163; B 4 2 22,189; B 4 4 22,188; B 4 4 22,188; B 4 24 22,176; B 4 4 22,160; B 6 8 25,154; B 4 4 26,160; B 4 4 26,160; B 4 4 26,160; L ND; B 6 4 3,14; B 8 8 6,168; B 8 8 6,64; B 4 44 4,38; B 4 4 4,14; B 8 8 8,184; B 4 2 8,183; B 4 4 8,172; B 4 4 8,180; B 4 4 8,154; B 4 4 8,162; B 8 8 10,202; B 8 2 10,207; B 4 2 8,153; B 4 4 8,76; B 4 4 8,152; B 4 76 8,116; B 4 2 8,171; B 4 4 8,182; B 4 4 8,182; B 4 4 8,158; B 4 4 8,176; B 4 32 10,90; B 2 4 9,76; B 6 4 13,152; B 6 4 13,182; B 4 32 14,90; B 8 8 16,36; B 4 36 18,18; B 4 32 18,90; B 4 4 18,34; B 4 16 18,176; B 8 8 20,154; B 4 4 20,98; B 4 10 20,55; B 4 4 20,98; B 4 4 20,48; B 8 8 22,64; B 8 8 22,92; B 2 4 21,98; B 4 16 22,176; B 8 4 26,48; B 8 8 26,202; B 4 16 26,176; B 4 4 28,180; B 4 34 28,33; B 4 4 28,48; B 4 4 28,14; B 4 16 28,190; B 2 4 29,180; B 2 4 31,14; L NM; B 18 8 9,92; B 32 8 16,140; B 2 8 1,168; B 6 8 3,202; B 32 8 16,22; B 32 8 16,50; B 32 8 16,6; B 8 8 6,168; B 8 8 6,64; B 8 12 8,186; B 8 8 10,202; B 12 8 14,36; B 8 6 14,63; B 22 8 21,168; B 8 8 18,202; B 12 8 22,154; B 8 8 22,64; B 8 8 22,92; B 8 8 26,202; B 6 8 29,92; B 2 8 31,202; L NI; B 12 12 8,158; B 12 12 8,176; L NC; B 4 4 6,168; B 4 4 6,64; B 4 8 8,186; B 4 4 10,202; B 8 4 14,36; B 8 4 22,154; B 4 4 22,64; B 4 4 22,92; B 4 4 26,202; DF;

ScanUnClkIn

DS 512 125/1; L NP; B 4 4 2,34; B 4 12 2,184; B 4 58 2,65; B 6 8 3,36; B 4 4 2,188; B 4 4 2,92; B 4 4 2,34; B 4 18 2,199; B 4 4 2,176; B 8 8 4,22; B 2 4 3,176; B 12 4 8,176; B 12 4 8,158; B 6 4 7,92; B 4 12 6,14; B 4 2 6,19; B 8 6 8,189; B 4 2 8,93; B 4 4 8,138; B 8 8 10,140; B 4 4 8,138; B 4 4 8,92; B 4 40 8,112; B 16 4 16,84; B 8 6 12,71; B 12 16 14,64; B 6 4 13,138; B 4 4 14,158; B 4 4 14,160; B 4 2 14,161; B 2 4 15,158; B 4 4 18,188; B 12 4 22,138; B 8 4 20,160; B 4 18 18,199; B 4 2 22,189; B 8 6 24,109; B 4 24 22,176; B 4 4 22,160; B 4 2 22,163; B 4 4 22,188; B 4 4 22,188; B 4 4 24,84; B 4 4 24,160; B 6 8 25,154; B 4 4 24,160; B 4 20 24,96; B 2 4 25,160; B 2 4 25,84; L ND; B 4 4 2,126; B 4 26 2,141; B 4 4 2,14; B 4 4 2,152; B 8 8 6,168; B 2 4 3,126; B 4 32 4,112; B 6 4 7,152; B 4 4 6,14; B 8 8 8,184; B 8 8 8,36; B 4 4 8,172; B 4 4 8,180; B 4 4 8,154; B 4 4 8,162; B 4 2 8,183; B 4 4 8,182; B 8 8 10,202; B 4 4 8,182; B 4 2 8,153; B 4 4 8,152; B 4 2 8,171; B 8 2 10,207; B 4 4 8,176; B 4 32 8,112; B 4 4 8,158; B 8 8 12,50; B 4 4 10,14; B 8 8 12,76; B 4 20 12,26; B 6 4 13,182; B 4 4 12,34; B 6 4 13,152; B 4 4 12,14; B 4 32 12,112; B 4 4 14,54; B 4 2 14,79; B 4 2 14,89; B 4 4 14,74; B 4 4 14,98; B 2 4 13,14; B 8 8 16,92; B 8 8 16,92; B 2 4 13,34; B 4 2 14,53; B 8 4 16,80; B 8 4 16,88; B 4 4 14,78; B 4 4 14,98; B 8 4 16,84; B 4 16 14,64; B 6 4 17,14; B 2 4 15,98; B 4 16 18,176; B 4 4 18,144; B 4 4 18,14; B 4 4 18,148; B 14 4 23,78; B 4 16 18,8; B 8 8 20,154; B 4 4 22,144; B 4 2 22,145; B 4 14 22,125; B 4 4 22,144; B 4 4 22,134; B 4 4 22,142; B 4 4 22,138; B 4 16 22,176; B 8 8 24,114; B 8 8 26,202; B 4 16 26,176; B 4 64 28,48; B 4 4 28,180; B 4 4 28,14; B 4 4 28,78; B 4 16 28,190; B 2 4 29,180; B 2 4 31,14; L NM; B 6 8 3,140; B 2 8 1,168; B 12 8 6,36; B 8 8 4,50; B 8 8 4,22; B 6 8 3,202; B 32 8 16,6; B 12 8 6,92; B 8 8 6,168; B 8 12 8,186; B 8 8 10,140; B 8 8 10,202; B 8 8 12,50; B 24 8 20,22; B 8 12 12,74; B 22 8 21,168; B 8 8 16,92; B 8 8 16,92; B 8 6 16,91; B 8 8 18,202; B 18 8 23,140; B 12 8 22,154; B 16 8 24,50; B 12 8 26,92; B 8 12 24,112; B 8 8 26,202; B 2 8 31,202; L NI; B 12 12 8,176; B 12 12 8,158; B 12 10 12,73; B 12 24 14,64; L NC; B 4 4 4,22; B 8 4 6,36; B 4 4 6,168; B 4 8 8,186; B 4 4 10,140; B 4 4 10,202; B 4 8 12,74; B 4 4 12,50; B 4 4 16,92; B 4 4 16,92; B 8 4 22,154; B 4 8 24,112; B 4 4 26,202; DF;

ScanJoin

DS 513 125/1; L NP; B 4 16 2,216; B 4 8 2,198; B 6 8 3,52; B 8 2 4,25; B 8 2 4,19; B 14 4 7,192; B 12 6 6,205; B 10 4 5,58; B 4 88 8,104; B 18 4 9,36; B 8 2 14,39; B 8 2 14,33; B 24 4 12,22; B 4 6 16,165; B 4 18 18,215; B 8 4 20,204; B 4 30 22,187; B 36 8 36,52; B 8 44 46,138; B 8 20 50,38; B 24 4 12,170; B 8 6 18,159; L ND; B 8 4 4,30; B 4 6 2,147; B 8 8 6,184; B 4 26 6,167; B 4 10 6,37; B 8 4 8,202; B 8 10 10,219; B 4 8 8,192; B 14 4 7,142; B 14 4 9,114; B 12 24 8,128; B 8 8 16,108; B 20 4 14,44; B 22 4 11,152; B 8 4 18,156; B 4 10 18,21; B 4 12 22,36; B 8 4 20,28; B 8 8 26,214; B 26 4 17,198; B 4 10 28,205; B 14 12 23,190; B 8 8 40,198; B 20 4 46,192; B 8 10 8,51; L NM; B 8 8 4,156; B 12 8 6,52; B 8 8 4,22; B 8 6 4,157; B 40 8 20,66; B 6 30 5,139; B 8 6 8,205; B 8 2 14,33; B 36 8 18,184; B 20 8 10,108; B 8 14 16,97; B 30 4 15,220; B 8 12 18,156; B 24 8 12,38; B 6 6 21,45; B 8 8 22,52; B 44 8 34,86; B 8 34 32,163; B 56 4 28,216; B 34 4 39,212; B 8 54 36,35; B 40 6 24,199; B 8 2 40,195; B 28 8 42,142; B 24 8 44,4; B 48 6 26,121; B 8 2 46,117; B 8 8 46,156; B 6 22 45,171; B 8 2 50,29; B 8 6 52,221; B 14 6 49,185; B 10 6 51,33; L NI; B 12 12 6,170; B 12 12 8,192; L NC; B 8 4 6,52; B 4 4 4,22; B 4 4 6,184; B 4 8 8,202; B 4 4 10,218; B 4 4 14,36; B 4 4 16,108; B 4 8 18,156; B 4 4 22,52; B 4 4 26,214; B 4 4 40,198; B 4 4 46,120; B 4 4 46,156; B 4 4 50,32; DF;

StScanClkIn

DS 514 125/1; L NP; B 4 16 2,214; B 4 136 2,124; B 6 8 3,52; B 16 4 8,194; B 8 8 10,20; B 4 10 8,11; B 4 44 10,86; B 6 2 11,63; B 18 4 9,34; B 8 4 14,38; B 8 4 12,110; B 8 4 16,160; B 4 38 14,131; B 12 2 16,59; B 14 2 15,61; B 8 2 20,91; B 4 14 18,215; B 12 16 22,82; B 8 6 22,55; B 14 2 15,175; B 14 4 23,206; B 22 4 23,156; B 8 6 30,123; B 4 24 28,192; B 14 2 25,179; B 24 2 20,177; B 18 4 27,6; B 8 4 32,10; B 6 8 31,172; B 18 4 25,102; B 4 16 32,112; B 6 10 3,201; L ND; B 4 6 2,17; B 4 12 6,36; B 8 4 8,204; B 8 10 10,217; B 4 22 8,157; B 8 8 12,186; B 20 4 14,28; B 4 8 10,194; B 4 10 14,177; B 8 8 20,68; B 14 4 13,144; B 12 24 14,130; B 8 4 20,92; B 14 4 15,116; B 8 6 22,111; B 8 8 22,50; B 4 16 22,38; B 4 12 22,20; B 10 2 23,107; B 4 18 22,81; B 8 8 24,102; B 26 4 13,12; B 4 4 24,166; B 4 10 24,5; B 24 4 18,170; B 8 4 26,174; B 18 4 25,96; B 8 4 26,162; B 4 28 28,146; B 8 8 30,128; B 8 8 32,216; B 6 4 33,62; B 4 30 32,79; B 32 4 20,200; B 4 10 34,207; B 4 40 34,40; B 14 12 29,192; B 6 4

35,18; B 8 14 8,49; L NM; B 12 8 6,202; B 12 8 6,52; B 8 2 10,17; B 38 6 19,21; B 38 8 19,186; B 8 12 20,92; B 38 8 19,36; B 8 12 22,52; B 38 8 19,158; B 12 8 28,172; B 38 8 19,68; B 38 8 19,110; B 8 12 30,126; B 8 2 32,11; B 38 6 19,7; B 38 8 19,216; L NI; B 12 12 10,194; B 12 12 14,176; B 14 8 21,90; B 12 2 20,95; B 12 16 22,78; L NC; B 8 4 6,202; B 8 4 6,52; B 4 4 10,20; B 4 4 10,216; B 4 4 12,186; B 4 4 14,36; B 4 4 16,158; B 4 8 20,92; B 4 4 20,68; B 4 8 22,52; B 4 4 22,110; B 8 4 28,172; B 4 8 30,126; B 4 4 32,8; B 4 4 32,216; DF;

StScanIn

DS 515 125/1; L NP; B 4 16 2,214; B 16 4 8,194; B 14 2 9,175; B 4 38 14,111; B 10 4 17,90; B 18 2 21,179; B 28 2 16,177; B 4 14 18,215; B 4 30 20,73; B 8 6 22,55; B 14 4 23,206; B 6 8 25,172; B 4 24 28,192; B 6 10 3,201; L ND; B 8 4 4,18; B 4 58 6,49; B 8 8 8,82; B 8 4 8,204; B 8 8 10,186; B 8 10 10,217; B 4 10 8,177; B 4 42 8,147; B 4 8 10,194; B 14 4 13,124; B 12 8 14,118; B 18 4 15,170; B 8 4 20,174; B 18 8 17,110; B 14 12 15,100; B 8 8 22,50; B 4 46 24,23; B 8 8 28,82; B 4 14 26,71; B 8 8 32,216; B 12 4 30,62; B 4 40 34,40; B 32 4 20,200; B 4 10 34,207; B 14 12 29,192; B 6 4 35,18; L NM; B 38 6 19,7; B 38 6 19,159; B 38 8 19,68; B 38 6 19,21; B 12 8 6,202; B 38 6 19,37; B 8 2 8,85; B 38 8 19,186; B 12 8 22,172; B 8 12 22,52; B 28 6 18,81; B 8 2 28,85; B 38 8 19,110; B 38 8 19,216; L NI; B 12 12 8,176; B 12 12 10,194; L NC; B 8 4 6,202; B 4 4 8,82; B 4 4 10,216; B 4 4 10,186; B 8 4 22,172; B 4 4 22,110; B 4 8 22,52; B 4 4 28,82; B 4 4 32,216; DF;

StScanUnClkIn

DS 516 125/1; L NP; B 4 16 2,214; B 4 136 2,124; B 6 8 3,52; B 16 4 8,194; B 8 8 10,20; B 4 10 8,11; B 4 44 10,86; B 18 4 9,34; B 8 4 14,38; B 6 2 11,63; B 8 4 12,110; B 4 38 14,131; B 8 4 16,160; B 14 2 15,61; B 12 2 16,59; B 8 2 20,91; B 4 14 18,215; B 12 16 22,82; B 14 2 15,175; B 8 6 22,55; B 14 4 23,206; B 22 4 23,156; B 8 6 30,123; B 4 24 28,192; B 14 2 25,179; B 24 2 20,177; B 6 8 31,172; B 4 16 32,112; B 18 4 25,102; B 6 10 3,201; L ND; B 4 6 2,17; B 4 12 6,36; B 8 4 8,204; B 4 22 8,157; B 8 10 10,217; B 20 4 14,28; B 8 8 12,186; B 4 8 10,194; B 4 10 14,177; B 8 8 20,68; B 14 4 13,144; B 12 24 14,130; B 8 4 20,92; B 8 6 22,111; B 14 4 15,116; B 8 8 22,50; B 4 12 22,20; B 4 16 22,38; B 10 2 23,107; B 4 18 22,81; B 8 8 24,102; B 26 4 13,12; B 4 4 24,166; B 4 10 24,5; B 24 4 18,170; B 8 4 26,174; B 20 4 26,96; B 8 4 26,162; B 4 28 28,146; B 8 8 30,128; B 8 8 32,216; B 32 4 20,200; B 4 10 34,207; B 14 12 29,192; B 6 4 35,18; B 4 74 34,57; B 8 14 8,49; L NM; B 12 8 6,52; B 12 8 6,202; B 38 6 19,7; B 8 2 10,17; B 38 6 19,21; B 8 12 20,92; B 38 8 19,186; B 8 12 22,52; B 38 8 19,36; B 38 8 19,158; B 12 8 28,172; B 38 8 19,68; B 8 12 30,126; B 38 8 19,110; B 38 8 19,216; L NI; B 12 12 10,194; B 12 12 14,176; B 14 8 21,90; B 12 2 20,95; B 12 16 22,78; L NC; B 8 4 6,52; B 8 4 6,202; B 4 4 10,216; B 4 4 10,20; B 4 4 12,186; B 4 4 14,36; B 4 4 16,158; B 4 4 20,68; B 4 8 20,92; B 4 4 22,110; B 4 8 22,52; B 8 4 28,172; B 4 8 30,126; B 4 4 32,216; DF;

StScanJoin

DS 517 125/1; L NP; B 4 10 2,193; B 4 18 2,215; B 6 8 3,202; B 8 8 4,38; B 4 104 2,136; B 4 4 2,82; B 6 8 3,84; B 4 18 2,51; B 4 4 2,82; B 4 4 2,186; B 4 4 2,58; B 4 2 6,27; B 14 4 11,58; B 12 4 10,186; B 4 4 6,26; B 4 12 6,20; B 10 4 13,26; B 12 4 14,170; B 4 60 10,126; B 4 4 10,154; B 4 4 10,94; B 8 8 16,10; B 8 8 16,156; B 6 4 15,94; B 4 4 16,94; B 4 4 16,26; B 4 12 16,70; B 4 20 16,86; B 4 4 16,62; B 4 4 16,26; B 8 8 18,28; B 4 4 16,58; B 4 4 18,222; B 4 4 18,218; B 8 8 20,44; B 12 4 24,8; B 2 4 19,8; B 6 8 23,188; B 4 4 22,170; B 4 4 22,170; B 4 12 22,178; B 4 4 22,194; B 8 4 24,218; B 4 2 22,103; B 4 40 22,122; B 4 4 22,102; B 4 4 22,194; B 4 24 24,92; B 4 4 24,102; B 4 4 24,78; B 4 2 26,219; B 4 4 26,218; B 2 4 25,102; B 14 8 31,44; B 4 2 26,195; B 4 4 26,194; B 4 24 26,206; B 4 4 26,194; B 4 4 28,78; B 6 8 29,72; B 4 4 28,78; B 4 4 28,78; B 8 8 32,154; B 8 10 34,43; B 8 8 34,44; B 8 8 34,34; B 4 4 36,152; B 4 4 36,198; B 4 46 36,177; B 2 4 37,152; B 4 4 40,198; B 6 8 41,34; B 8 8 46,200; B 8 8 48,34; B 8 8 48,124; B 4 56 48,152; B 4 4 48,122; B 8 8 50,184; B 2 4 51,122; L ND; B 4 4 2,20; B 4 4 6,20; B 8 8 8,84; B 8 8 8,202; B 8 2 10,223; B 8 8 10,178; B 8 8 10,218; B 4 4 8,176; B 4 4 8,176; B 4 8 10,76; B 4 2 10,181; B 4 4 10,20; B 4 4 10,182; B 4 4 10,190; B 4 4 10,70; B 4 10 10,197; B 4 4 10,200; B 4 4 10,200; B 4 4 10,200; B 4 4 10,186; B 2 4 11,70; B 4 4 12,52; B 4 16 12,46; B 4 4 12,70; B 4 4 12,36; B 6 4 13,176; B 4 2 14,165; B 14 4 19,20; B 4 4 14,164; B 4 4 14,166; B 4 4 14,174; B 8 4 16,200; B 4 2 14,177; B 4 4 14,176; B 4 4 14,170; B 4 38 16,129; B 14 4 21,36; B 10 4 19,52; B 4 4 16,146; B 4 4 16,108; B 4 4 16,70; B 2 4 17,108; B 4 32 18,122; B 16 4 24,164; B 10 4 21,164; B 8 4 22,146; B 4 4 20,70; B 4 4 22,52; B 4 4 22,70; B 4 16 22,206; B 4 4 22,70; B 4 18 22,63; B 4 32 22,122; B 8 8 24,72; B 4 18 24,157; B 4 4 24,20; B 2 4 23,70; B 4 4 24,146; B 4 4 24,164; B 4 4 24,4; B 4 4 24,12; B 4 2 24,1; B 4 8 24,18; B 4 4 24,8; B 4 4 26,36; B 4 4 26,20; B 4 16 26,30; B 4 32 26,122; B 4 16 26,206; B 8 8 28,188; B 2 4 27,20; B 4 6 30,107; B 4 16 30,206; B 4 4 30,108; B 8 8 32,86; B 4 4 30,108; B 8 8 32,100; B 4 4 30,186; B 4 4 30,164; B 4 22 30,177; B 8 8 34,214; B 4 4 32,208; B 4 4 32,204; B 2 4 33,204; B 10 4 41,84; B 4 20 40,124; B 4 4 40,132; B 4 4 40,112; B 4 4 42,192; B 4 4 42,132; B 4 60 42,164; B 4 28 44,100; B 4 4 44,112; B 4 4 44,112; B 4 4 44,84; B 2 4 43,132; B 12 4 50,192; L NM; B 6 6 3,161; B 6 6 3,85; B 8 8 4,58; B 8 8 4,70; B 8 8 4,100; B 8 8 4,112; B 6 18 3,155; B 12 8 6,202; B 12 8 6,84; B 10 8 5,188; B 20 6 10,23; B 6 6 3,161; B 6 8 3,218; B 8 12 4,68; B 8 8 4,70; B 8 8 4,38; B 8 8 4,112; B 12 6 6,9; B 8 12 4,110; B 6 6 3,143; B 8 8 6,188; B 8 2 6,179; B 8 6 6,177; B 8 8 6,174; B 8 12 6,186; B 6 6 9,143; B 8 8 10,218; B 6 6 9,143; B 6 18 9,137; B 8 8 10,178; B 28 6 20,85; B 6 6 9,125; B 4 6 8,177; B 30 8 23,58; B 20 8 18,100; B 24 8 22,174; B 32 6 28,125; B 8 8 16,10; B 8 8 16,156; B 8 8 18,28; B 6 6 17,23; B 6 6 17,27; B 6 6 17,27; B 6 4 17,28; B 24 8 26,218; B 6 8 19,36; B 8 8 20,44; B 12 8 26,188; B 12 8 26,72; B 8 6 24,155; B 8 8 30,174; B 8 6 30,175; B 8 8 30,168; B 8 8 32,100; B 6 6 31,85; B 8 8 32,86; B 6 2 31,89; B 8 8 32,154; B 8 8 34,214; B 8 8 34,58; B 8 8 34,218; B 8 48 34,38; B 8 4 34,220; B 8 8 34,10; B 16 8 42,168; B 14 8 43,100; B 18 8 47,214; B 10 8 43,10; B 8 8 44,10; B 8 8 44,4; B 8 6 44,11; B 8 8 46,100; B 8 8 46,168; B 8 8 46,142; B 8 8

46,86; B 8 14 46,97; B 8 8 46,200; B 8 26 46,159; B 8 8 48,124; B 8 8 48,34; B 8 8 50,184; B 8 6 52,221; B 8 8 52,4; B 8 8 52,214; B 6 8 53,142; B 6 8 53,86; B 6 6 53,199; B 4 6 54,33; B 2 6 55,185; L NC; B 4 4 4,38; B 8 4 6,202; B 8 4 6,84; B 4 4 10,178; B 4 4 10,218; B 4 4 16,10; B 4 4 16,156; B 4 4 18,28; B 4 4 20,44; B 8 4 26,72; B 8 4 26,188; B 4 4 32,154; B 4 4 32,100; B 4 4 32,86; B 4 4 34,214; B 4 4 46,200; B 4 4 48,34; B 4 4 48,124; B 4 4 50,184; L NI; B 12 12 10,186; B 12 12 14,170; DF;

ScanClkOut
DS 518 125/1; L NP; B 8 8 4,184; B 6 8 3,170; B 4 86 2,123; B 4 10 6,193; B 10 4 5,78; B 4 22 8,65; B 6 4 9,20; B 6 8 11,156; B 4 10 10,5; B 8 2 12,119; B 16 4 14,16; B 14 4 15,12; B 14 4 15,150; B 4 14 14,103; B 14 2 13,205; B 18 4 15,32; B 8 4 20,36; B 18 2 15,203; B 8 6 20,199; B 6 8 19,48; B 10 4 17,94; B 4 40 20,72; B 4 12 20,142; B 14 16 15,128; L ND; B 4 30 2,43; B 8 4 10,88; B 4 8 8,94; B 8 8 12,142; B 8 8 12,228; B 8 4 12,118; B 14 4 7,192; B 10 8 9,170; B 4 16 12,182; B 10 4 11,84; B 4 30 12,209; B 4 16 12,36; B 8 8 14,48; B 16 4 8,60; B 14 4 15,114; B 14 12 9,68; B 4 8 14,78; B 4 18 14,129; B 8 2 16,153; B 18 4 9,26; B 8 8 18,4; B 4 16 16,16; B 16 4 14,100; B 14 4 15,104; B 4 6 20,109; B 12 6 18,157; B 6 4 23,192; B 4 30 22,175; B 8 10 4,19; L NM; B 12 8 6,170; B 8 2 4,181; B 12 8 6,18; B 26 8 13,214; B 8 12 12,118; B 12 8 14,156; B 8 8 12,228; B 26 6 13,185; B 12 8 16,48; B 26 8 13,86; B 26 8 13,142; B 8 2 20,203; B 8 2 20,37; B 26 8 13,4; B 26 6 13,33; B 26 6 13,199; L NI; B 14 8 13,120; B 12 2 12,115; B 12 16 14,132; B 12 16 16,14; L NC; B 4 4 4,184; B 8 4 6,170; B 8 4 6,18; B 4 4 10,86; B 4 8 12,118; B 4 4 12,142; B 8 4 14,156; B 4 4 12,228; B 8 4 16,48; B 4 4 18,4; B 4 4 20,34; B 4 4 20,200; DF;

ScanOut
DS 519 125/1; L NP; B 18 4 9,164; B 10 4 5,74; B 4 86 2,119; B 4 42 8,51; B 6 4 9,20; B 6 8 17,228; B 4 10 10,5; B 16 4 14,16; B 14 4 15,12; B 4 58 16,195; L ND; B 8 8 12,228; B 8 4 4,192; B 4 6 2,31; B 4 14 6,183; B 8 8 8,172; B 18 4 9,26; B 8 4 10,158; B 16 4 8,36; B 8 8 16,86; B 14 12 9,44; B 4 32 14,66; B 8 8 18,4; B 18 4 15,154; B 4 16 16,16; B 6 4 23,192; B 4 34 22,173; B 8 10 4,19; L NM; B 26 8 13,214; B 12 8 6,18; B 26 6 13,185; B 26 6 13,199; B 26 8 13,142; B 26 6 13,33; B 8 8 8,172; B 8 8 10,156; B 6 8 9,164; B 12 8 14,228; B 26 8 13,86; B 26 8 13,4; L NI; B 12 16 16,14; L NC; B 8 4 6,18; B 4 4 8,172; B 4 4 10,156; B 8 4 14,228; B 4 4 16,86; B 4 4 18,4; DF;

PlaCell
DS 12 125/1; L NM; B 32 8 16,26; B 32 8 16,10; L NP; B 4 32 4,16; B 4 32 20,16; L ND; B 8 32 12,16; DF;

PlaGround
DS 13 125/1; L NM; B 32 8 16,10; L NP; B 4 16 4,8; B 4 16 20,8; L ND; B 8 16 12,8; L NC; B 4 4 12,10; DF;

PlaPullup
DS 14 125/1; L NM; B 8 32 4,16; B 12 8 24,26; B 12 8 24,10; L ND; B 8 8 4,26; B 8 8 4,10; B 18 4 15,26; B 18 4 15,10; B 8 8 26,10; B 8 8 26,26; L NC; B 4 4 4,10; B 4 4 4,26; B 8 4 24,10; B 8 4 24,26; L NI; B 18 12 19,26; B 18 12 19,10; L NP; B 10 12 19,26; B 10 12 19,10; DF;

PlaConnect
DS 15 125/1; L NM; B 10 8 5,26; B 10 8 5,10; B 8 32 20,16; L NP; B 8 8 6,26; B 8 8 6,10; B 18 4 17,28; B 18 4 17,12; L NC; B 4 4 6,26; B 4 4 6,10; B 4 4 20,20; L ND; B 10 8 21,20; DF;

PlaConnectSpace
DS 16 125/1; L NM; B 18 8 9,10; B 8 16 20,8; L NP; B 8 8 6,10; B 18 4 17,12; L NC; B 4 4 6,10; DF;

PlaOrSpace
DS 17 125/1; L NP; B 32 4 16,12; L NM; B 8 16 10,8; B 8 16 26,8; DF;

PlaGroundSpace
DS 18 125/1; L NP; B 16 4 8,12; B 8 8 10,12; L NM; B 8 16 10,8; L NC; B 4 4 10,12; DF;

PlaIn
DS 19 125/1; L NM; B 32 8 16,76; B 32 8 16,38; B 12 8 8,18; B 12 8 8,58; L NP; B 8 8 4,18; B 8 8 4,58; B 4 12 2,26; B 4 12 2,50; B 18 4 11,30; B 12 4 8,46; B 4 22 4,71; B 4 54 20,55; B 4 24 20,12; L NI; B 12 12 8,30; B 12 12 8,46; L NC; B 8 4 8,18; B 8 4 8,58; B 4 4 8,38; B 4 4 12,76; B 4 4 28,76; L ND; B 8 8 8,38; B 8 12 10,20; B 8 12 10,56; B 4 12 8,30; B 4 2 8,25; B 4 12 8,44; B 4 2 8,51; B 8 10 12,77; B 12 16 20,58; B 12 16 20,12; B 8 8 28,76; B 6 62 27,43; DF;

PlaClockIn
DS 20 125/1; L NP; B 32 4 16,6; B 8 8 4,54; B 8 8 4,94; B 4 12 2,62; B 4 12 2,86; B 18 4 11,66; B 12 4 8,82; B 4 22 4,107; B 8 6 20,19; B 4 20 20,30; B 4 54 20,91; B 4 24 20,48; L NM; B 32 8 16,112; B 32 8 16,74; B 12 8 8,54; B 12 8 8,94; B 8 12 20,16; L NI; B 12 12 8,66; B 12 12 8,82; L NC; B 8 4 8,54; B 8 4 8,94; B 4 4 8,74; B 4 4 12,112; B 4 8 20,16; B 4 4 28,112; L ND; B 8 8 8,74; B 8 12 10,56; B 8 12 10,92; B 4 12 8,66; B 4 2 8,61; B 4 12 8,80; B 4 2 8,87; B 8 10 12,113; B 12 16 20,32; B 12 16 20,94; B 12 16

20,48; B 8 8 20,14; B 4 12 20,6; B 8 8 28,112; B 6 62 27,79; DF;

PlaProgTop
DS 21 125/1; L ND; B 8 18 4,9; L NC; B 4 4 4,14; DF;

PlaProgBottom
DS 22 125/1; L ND; B 8 18 4,9; L NC; B 4 4 4,4; DF;

PlaProgLeft
DS 23 125/1; L ND; B 18 8 9,4; L NC; B 4 4 4,4; DF;

PlaProgRight
DS 24 125/1; L ND; B 18 8 9,4; L NC; B 4 4 14,4; DF;

PlaPullupSpace
DS 25 125/1; L NM; B 8 16 4,8; B 8 8 26,10; DF;

PlaOut
DS 26 125/1; L NP; B 32 4 16,102; B 12 2 6,33; B 8 2 8,35; B 8 4 4,12; B 4 14 2,21; B 4 10 6,5; B 22 4 11,30; B 8 8 10,90; B 4 30 12,71; B 6 8 21,6; B 4 22 22,55; B 4 2 22,1; B 4 14 22,17; B 8 8 26,90; B 10 4 25,68; B 4 16 28,78; L ND; B 4 18 6,51; B 12 8 12,68; B 8 2 14,25; B 4 8 16,30; B 18 4 15,74; B 8 4 20,78; B 24 4 16,62; B 12 8 22,56; B 8 8 26,6; B 24 4 18,18; B 14 4 23,50; B 4 28 28,34; B 4 6 28,13; B 12 4 12,22; B 14 8 11,38; L NM; B 32 8 16,76; B 32 8 16,20; B 8 12 8,36; B 8 20 10,96; B 12 8 24,6; B 8 20 26,96; L NI; B 12 12 16,30; B 12 12 22,18; L NC; B 4 8 8,36; B 4 4 10,90; B 4 4 10,20; B 4 4 20,76; B 8 4 24,6; B 4 4 26,90; DF;

PlaClockOut
DS 27 125/1; L NP; B 32 4 16,102; B 8 4 4,12; B 4 16 2,22; B 4 10 6,5; B 8 6 10,89; B 4 30 12,71; B 4 22 22,55; B 8 6 26,89; B 10 4 25,68; B 4 16 28,78; B 6 8 21,4; B 8 12 20,14; B 24 4 12,32; B 12 2 18,29; B 20 2 14,35; L ND; B 4 18 6,51; B 12 8 12,68; B 8 8 10,110; B 8 8 10,94; B 8 6 10,19; B 4 8 10,102; B 18 4 15,74; B 8 4 20,78; B 24 4 16,62; B 12 8 22,56; B 8 8 26,110; B 8 8 26,94; B 4 8 26,102; B 20 4 20,14; B 14 4 23,50; B 4 32 28,32; B 8 8 26,4; B 4 4 28,10; B 4 12 18,32; B 16 4 12,40; B 8 4 8,36; B 14 2 13,23; B 10 2 15,25; L NM; B 32 8 16,76; B 8 8 10,110; B 8 12 10,92; B 8 8 26,110; B 8 12 26,92; B 12 8 24,4; B 8 12 8,36; B 32 8 16,20; L NI; B 16 12 20,14; B 12 16 18,32; L NC; B 4 8 8,36; B 4 4 10,110; B 4 8 10,92; B 4 4 10,20; B 4 4 20,76; B 4 4 26,110; B 4 8 26,92; B 8 4 24,4; DF;

PlaHoleWires
DS 28 125/1; L NM; B 10 8 5,46; B 18 8 9,84; B 8 42 6,21; B 16 8 18,4; B 8 28 20,76; B 10 8 21,60; DF;

PlaOutSpace
DS 29 125/1; L NP; B 16 4 8,86; L NM; B 16 8 8,60; B 16 8 8,4; B 8 28 10,76; DF;

PlaNorOut
DS 30 125/1; L NM; B 32 8 16,60; B 32 8 16,4; B 8 12 8,20; B 8 20 10,80; B 8 8 26,74; B 8 20 26,80; L NP; B 32 4 16,86; B 8 6 8,17; B 4 18 6,9; B 8 8 10,74; B 4 32 12,56; B 16 4 18,18; B 4 26 22,41; B 6 4 25,52; B 8 8 26,74; B 4 22 28,61; L ND; B 4 24 6,36; B 8 8 8,22; B 12 16 12,52; B 20 4 18,24; B 8 12 16,6; B 8 8 20,60; B 12 16 22,40; B 4 18 20,17; B 4 14 28,29; L NC; B 4 8 8,20; B 4 4 10,74; B 4 4 16,4; B 4 4 20,60; B 4 4 26,74; L NI; B 12 12 20,18; DF;

PlaClockNorOut
DS 31 125/1; L NP; B 32 4 16,86; B 4 14 6,7; B 8 6 10,73; B 4 30 12,55; B 4 22 22,39; B 8 6 26,73; B 10 4 25,52; B 4 16 28,62; B 20 6 14,17; B 12 2 18,13; L ND; B 4 18 6,35; B 8 8 10,94; B 8 8 10,78; B 12 8 12,52; B 4 8 10,86; B 8 8 16,4; B 24 4 16,46; B 12 8 22,40; B 18 4 15,58; B 8 4 20,62; B 8 8 26,94; B 8 8 26,78; B 4 8 26,86; B 14 4 23,34; B 4 6 28,29; B 4 14 18,15; B 26 4 17,24; B 8 4 8,20; L NM; B 32 8 16,60; B 32 8 16,4; B 8 8 10,94; B 8 12 10,76; B 8 8 26,94; B 8 12 26,76; B 8 12 8,20; L NI; B 12 16 18,16; L NC; B 4 8 8,20; B 4 4 10,94; B 4 8 10,76; B 4 4 16,4; B 4 4 20,60; B 4 4 26,94; B 4 8 26,76; DF;

PlaStaticIn
DS 80 125/1; L NP; B 4 6 2,111; B 4 20 4,132; B 8 8 4,118; B 14 4 7,106; B 4 34 20,125; B 6 4 21,106; B 4 12 22,98; B 16 4 16,90; B 32 4 16,6; B 4 4 4,48; B 6 4 3,52; B 8 6 6,43; B 8 8 8,16; B 4 12 10,26; B 6 8 11,80; B 8 6 22,33; B 4 38 22,55; B 4 50 2,79; B 4 4 10,86; L ND; B 4 8 8,106; B 8 4 10,100; B 8 10 12,137; B 12 4 12,96; B 8 8 28,136; B 24 12 18,116; B 16 4 22,124; B 4 6 28,129; B 4 6 4,31; B 8 8 6,38; B 8 8 16,80; B 4 6 16,73; B 14 2 21,65; B 12 26 22,51; B 24 4 14,26; B 8 4 22,30; B 16 4 22,68; B 4 10 16,89; B 4 40 28,90; B 4 24 22,12; L NM; B 32 8 16,98; B 32 8 16,136; B 12 8 8,118; B 32 6 16,15; B 8 12 6,40; B 8 2 8,19; B 12 8 14,80; B 8 12 22,30; L NI; B 12 12 8,106; B 12 12 16,90; L NC; B 8 4 8,118; B 4 4 10,98; B 4 4 12,136; B 4 4 28,136; B 4 8 6,40; B 4 4 8,16; B 8 4 14,80; B 4 8 22,30; DF;

PlaShiftIn

DS 81 125/1; L NM; B 32 8 16,100; B 32 8 16,216; B 32 6 16,189; B 32 8 16,152; B 32 8 16,62; B 32 6 16,29; B 32 6 16,15; B 8 8 4,16; B 12 8 8,202; B 8 12 10,46; B 12 8 14,166; B 8 8 14,188; B 12 8 16,80; B 8 8 20,28; B 12 8 24,120; B 8 12 26,46; L ND; B 18 14 9,71; B 32 4 16,36; B 8 8 4,62; B 8 8 10,202; B 8 8 10,44; B 8 4 10,40; B 8 8 12,100; B 8 10 12,217; B 12 4 14,180; B 4 8 10,174; B 8 8 12,166; B 8 8 12,152; B 4 10 10,145; B 14 18 15,131; B 4 12 12,90; B 8 8 14,80; B 4 42 12,21; B 10 4 19,200; B 8 8 20,100; B 4 14 20,111; B 8 8 22,120; B 4 24 22,190; B 8 8 26,44; B 8 4 26,40; B 8 10 28,217; L NP; B 4 22 2,31; B 8 8 4,16; B 32 4 16,6; B 4 132 4,156; B 20 4 12,92; B 4 8 6,202; B 4 14 6,75; B 10 4 9,70; B 12 4 12,88; B 8 6 10,49; B 8 8 14,188; B 4 20 12,62; B 4 16 14,126; B 12 8 20,110; B 4 14 16,181; B 6 8 17,166; B 8 4 18,144; B 4 12 16,136; B 6 10 19,81; B 4 10 18,37; B 8 8 20,28; B 4 24 20,154; B 12 4 24,206; B 4 14 20,215; B 4 6 20,87; B 8 6 26,49; B 6 10 27,119; B 4 86 28,165; B 4 12 28,112; B 4 54 28,79; L NC; B 4 4 4,16; B 4 4 4,62; B 8 4 8,202; B 4 8 10,46; B 4 4 12,216; B 8 4 14,166; B 4 4 12,152; B 4 4 12,100; B 4 4 14,188; B 8 4 16,80; B 4 4 20,100; B 4 4 20,28; B 8 4 24,120; B 4 8 26,46; B 4 4 28,216; L NI; B 12 16 12,90; B 12 16 20,110; DF;

Afterburner

DS 82 125/1; L NP; B 4 68 2,34; B 16 4 8,70; B 8 8 4,96; B 4 30 2,115; B 10 4 9,38; B 10 4 15,54; B 4 12 14,78; B 4 26 18,13; B 4 24 18,118; B 6 8 21,110; B 4 34 22,39; L NM; B 32 8 16,46; B 8 8 4,96; B 32 8 16,124; B 8 6 6,63; B 6 34 5,83; B 8 8 10,30; B 14 6 15,17; B 6 6 11,23; B 8 8 14,62; B 12 8 24,110; B 8 8 26,16; L NI; B 12 12 8,38; B 12 12 16,54; L NC; B 4 4 4,96; B 4 4 10,46; B 4 4 10,30; B 4 4 10,124; B 4 4 14,62; B 8 4 24,110; B 4 4 26,16; B 4 4 26,124; L ND; B 8 8 10,30; B 4 8 8,38; B 8 8 10,46; B 14 16 13,72; B 4 14 8,81; B 8 4 10,88; B 8 10 10,125; B 4 4 12,2; B 16 16 18,12; B 8 8 14,62; B 4 30 12,105; B 4 16 16,50; B 8 8 26,16; B 8 8 26,110; B 8 8 26,124; B 4 86 28,63; DF;

PlaShiftOut

DS 83 125/1; L NP; B 32 4 16,118; B 32 4 16,100; B 32 4 16,30; B 4 10 2,107; B 8 8 10,90; B 4 40 10,66; B 4 22 22,75; B 10 4 25,62; B 6 8 27,40; B 4 16 28,52; B 16 4 22,12; B 4 10 28,5; B 6 10 27,19; L ND; B 14 4 7,106; B 10 8 7,40; B 4 6 4,47; B 14 8 11,62; B 8 8 10,126; B 8 6 10,111; B 4 8 12,118; B 24 12 16,72; B 20 4 12,52; B 18 4 13,56; B 8 8 22,40; B 4 12 20,30; B 8 8 22,20; B 8 6 26,111; B 8 8 26,92; B 10 4 27,106; B 26 4 17,80; B 4 6 28,85; B 4 8 20,12; B 8 8 18,4; L NM; B 32 8 16,54; B 32 8 16,4; B 8 8 8,40; B 6 12 9,30; B 8 8 10,126; B 8 28 10,100; B 12 8 24,40; B 24 6 18,21; B 12 2 24,17; B 8 42 26,109; L NI; B 12 12 20,12; L NC; B 4 4 8,40; B 4 4 10,126; B 4 4 10,90; B 4 4 10,110; B 4 4 18,4; B 4 4 18,54; B 8 4 24,40; B 8 4 24,20; B 4 4 26,92; B 4 4 26,110; DF;

PlaShiftJoin

DS 84 125/1; L NM; B 32 8 16,98; B 32 8 16,214; B 38 6 19,187; B 32 8 16,150; B 32 8 16,60; B 32 6 16,27; B 42 6 21,13; B 8 8 4,14; B 12 8 8,200; B 8 12 10,44; B 12 8 14,164; B 8 8 14,186; B 12 8 16,78; B 8 8 20,26; B 12 8 24,118; B 8 12 26,44; B 8 8 30,200; B 26 8 45,150; B 26 8 45,98; B 24 8 44,214; B 26 6 45,185; B 8 8 36,26; B 14 6 41,199; B 8 8 46,12; B 8 8 52,198; B 8 10 52,215; B 8 8 54,144; B 8 8 54,94; B 8 8 54,164; B 6 22 55,173; L ND; B 18 14 9,69; B 32 4 16,34; B 8 8 4,60; B 8 8 10,200; B 8 8 10,42; B 8 4 10,38; B 8 8 12,98; B 8 10 12,215; B 12 4 14,178; B 4 8 10,172; B 8 8 12,164; B 8 8 12,150; B 4 10 10,143; B 14 18 15,129; B 4 12 12,88; B 8 8 14,78; B 4 40 12,20; B 10 4 19,198; B 8 8 20,98; B 4 14 20,109; B 8 8 22,118; B 4 24 22,188; B 8 8 26,42; B 8 4 26,38; B 8 10 28,215; B 8 8 52,198; B 10 4 53,196; L NP; B 4 22 2,29; B 8 8 4,14; B 32 4 16,4; B 4 132 4,154; B 20 4 12,90; B 4 8 6,200; B 4 14 6,73; B 10 4 9,68; B 12 4 12,86; B 8 6 10,47; B 8 8 14,186; B 4 20 12,60; B 4 16 14,124; B 12 8 20,108; B 4 14 16,179; B 6 8 17,164; B 8 4 18,142; B 4 12 16,134; B 6 10 19,79; B 4 10 18,35; B 8 8 20,26; B 4 24 20,152; B 12 4 24,204; B 4 14 20,213; B 4 6 20,85; B 8 6 26,47; B 6 10 27,117; B 4 86 28,163; B 4 12 28,110; B 4 54 28,77; B 8 8 30,200; B 32 4 42,2; B 8 8 36,26; B 4 162 36,111; B 20 4 48,208; B 4 22 40,199; B 6 4 45,180; B 4 148 44,108; B 8 8 46,12; B 4 30 46,25; B 4 14 48,185; B 10 4 53,190; B 8 8 54,164; B 6 4 55,120; B 4 46 54,145; L NC; B 4 4 4,14; B 4 4 4,60; B 8 4 8,200; B 4 8 10,44; B 4 4 12,214; B 8 4 14,164; B 4 4 12,150; B 4 4 12,98; B 4 4 14,186; B 8 4 16,78; B 4 4 20,98; B 4 4 20,26; B 8 4 24,118; B 4 8 26,44; B 4 4 28,214; B 4 4 30,200; B 4 4 36,26; B 4 4 46,12; B 4 4 52,198; B 4 4 54,164; L NI; B 12 16 12,88; B 12 16 20,108; DF;